Evolutionary Biogeography of the Andean Region

CRC BIOGEOGRAPHY SERIES

Series Editor

Malte C. Ebach

School of Biological, Earth and Environmental Sciences, Australia, University of New South Wales

Biogeography and Evolution in New Zealand
by Michael Heads

Handbook of Australasian Biogeography
edited by Malte C. Ebach

Neotropical Biogeography: Regionalization and Evolution
by Juan J. Morrone

Evolutionary Biogeography of the Andean Region
by Juan J. Morrone

For more information about this series, please visit: https://www.crcpress.com/CRC-Biogeography-Series/book-series/CRCHANOFBIO

Evolutionary Biogeography of the Andean Region

Juan J. Morrone
Professor of Biogeography, Systematics and Comparative Biology
Museo de Zoología "Alfonso L. Herrera"
Facultad de Ciencias
UNAM

CRC Press
Taylor & Francis Group
Boca Raton London New York

CRC Press is an imprint of the
Taylor & Francis Group, an **informa** business

Cover photo caption: Páramo of Belmira, Colombia. Photography by courtesy of Mario Alberto Quijano.

CRC Press
Taylor & Francis Group
6000 Broken Sound Parkway NW, Suite 300
Boca Raton, FL 33487-2742

First issued in paperback 2020

ISBN-13: 978-1-138-59872-0 (hbk)
ISBN-13: 978-0-367-65716-1 (pbk)

Library of Congress Cataloging-in-Publication Data

Names: Morrone, Juan J., author.
Title: Evolutionary biogeography of the Andean region / Juan J. Morrone.
Description: Boca Raton : Taylor & Francis, 2018. | Series: CRC biogeography series | Includes bibliographical references.
Identifiers: LCCN 2018007915 | ISBN 9781138598720 (hardback : alk. paper)
Subjects: LCSH: Biogeography--Andes Region.
Classification: LCC QH111 .M68 2018 | DDC 577.2/2--dc23
LC record available at https://lccn.loc.gov/2018007915

Visit the Taylor & Francis Web site at
http://www.taylorandfrancis.com

and the CRC Press Web site at
http://www.crcpress.com

The Andean region.

Contents

List of figures xv
Preface xxvii
About the author xxix

Chapter 1 Theoretical background 1
Evolutionary biogeography 1
Steps of evolutionary biogeography 2
Identification of biotas 2
Testing relationships among biotas 5
Regionalization 7
Identification of cenocrons 7
Construction of a geobiotic scenario 8
Biogeographical regionalization 8
Nomenclatural conventions 9
Format used in the book 9

Chapter 2 Historical background 11
Biogeographical regionalization of the world 11
Regionalization of the Andean region 14

Chapter 3 The Austral kingdom 29
Austral kingdom 29
Endemic and characteristic taxa 30
Biotic relationships 30
Regionalization 35
Geological scenario 35
Case study: Event-based biogeographical analysis of the Austral areas 37

Chapter 4 The Andean region **43**
Andean region 43
Endemic and characteristic taxa 45
Andean niche conservatism 46
Biotic relationships 47
Regionalization 52
Geological scenario 53
Case study: Event-based biogeographical analysis of the Andean region 58

Chapter 5 The Subantarctic subregion **63**
Subantarctic subregion 63
Endemic and characteristic taxa 65
Biotic relationships 66
Regionalization 68
Case study: Cladistic biogeographical analysis of the Subantarctic subregion 68
Maule province 72
Definition 73
Endemic and characteristic taxa 73
Vegetation 74
Biotic relationships 76
Regionalization 76
Angol district 76
Chillán Cordillera district 77
Pehuén district 77
Temuco district 79
Valdivian Forest province 79
Definition 80
Endemic and characteristic taxa 81
Vegetation 82
Biotic relationships 83
Regionalization 83
Aysén Cordillera district 84
Llanquihue district 84
South Chiloé district 85
Valdivian district 85
Valdivian Cordillera district 86
Cenocrons 86
Case study: Post-glacial recolonization of marsupial *Dromiciops gliroides* in the Valdivian Forest 86
Magellanic Forest province 88
Definition 90
Endemic and characteristic taxa 90

Vegetation ... 90
Biotic relationships ... 92
Case study: Integrative biogeographic analysis of the fly genus *Palpibracus* ... 92
Magellanic Moorland province ... 96
Definition ... 97
Endemic and characteristic taxa ... 98
Vegetation ... 98
Biotic relationships ... 100
Falkland Islands province ... 100
Definition ... 101
Endemic and characteristic taxa ... 102
Vegetation ... 103
Biotic relationships ... 104
Regionalization ... 104
Falkland Islands district ... 104
South Georgia Island district ... 105
Cenocrons ... 106
Case study: Integrative biogeographic analysis of the weevils of the Falkland Islands ... 106
Juan Fernández province ... 111
Definition ... 112
Endemic and characteristic taxa ... 112
Vegetation ... 114
Biotic relationships ... 114
Cenocrons ... 114
Case study: Modes of speciation of the plant genus *Robinsonia* in the Juan Fernández Islands ... 116

Chapter 6 The Central Chilean subregion ... 119
Central Chilean subregion ... 119
Endemic and characteristic taxa ... 120
Biotic relationships ... 121
Regionalization ... 121
Case study: Cladistic biogeographical analysis of Central Chile ... 122
Coquimban province ... 123
Definition ... 125
Endemic and characteristic taxa ... 125
Vegetation ... 126
Biotic relationships ... 127

Regionalization ... 128
Central Andean Cordillera district ... 128
Coquimban Desert district ... 129
Intermediate Desert district ... 130
Cenocrons ... 130
Santiagan province ... 130
Definition ... 131
Endemic and characteristic taxa ... 131
Vegetation ... 132
Biotic relationships ... 132
Regionalization ... 133
Central Coastal Cordillera district ... 133
Central Valley district ... 134
Southern Andean Cordillera district ... 135
Cenocrons ... 136

Chapter 7 The Patagonian subregion ... 137
Patagonian subregion ... 137
Endemic and characteristic taxa ... 138
Biotic relationships ... 138
Regionalization ... 138
Patagonian province ... 139
Definition ... 140
Endemic and characteristic taxa ... 141
Vegetation ... 142
Biotic relationships ... 143
Regionalization ... 143
Central Patagonian subprovince ... 147
Chubut district ... 148
San Jorge Gulf district ... 148
Santa Cruz district ... 149
Fuegian subprovince ... 150
Payunia subprovince ... 150
Northern Payunia district ... 151
Southern Payunia district ... 152
Subandean subprovince ... 152
Austral High Andean district ... 153
Meridional Subandean Patagonia district ... 154
Septentrional Subandean Patagonia district ... 154
Western Patagonian subprovince ... 154
Cenocrons ... 155
Case study: Areas of endemism in the Patagonian steppes based on insect taxa ... 157

Chapter 8 The South American transition zone 161
South American transition zone 161
Endemic and characteristic taxa 163
Biotic relationships 164
Regionalization 165
Cenocrons 165
Case study: Quaternary biogeography of the grass *Munroa argentina* 166
Páramo province 169
Definition 170
Endemic and characteristic taxa 170
Vegetation 170
Biotic relationships 173
Regionalization 174
Alto Cauca Highland district 174
Alto Patía district 174
Andalucía district 175
Awa district 175
Cañón Chicamocha district 175
Cañón del Cauca district 175
Catatumbo Mountains Forest district 175
Cauca and Valle Western Cordillera Andean Forest district 175
Cauca Pacific Slope Subandean Forest district 176
Eastern Andean district 176
Eastern Cordillera Páramos district 176
Farallones de Cali district 176
Frontino district 176
Paramillo del Sinú district 177
Páramos Huila-Tolima district 177
Perijá district 177
Quindío Páramo district 178
San Agustín district 178
San Juan Cloud Forest district 178
San Lucas Mountains district 178
Sierra Nevada district 178
Tachira district 179
Tolima district 179
Western Cordillera Northern Andean Forests district 179
Cenocrons 179
Case study: Evolutionary biogeography of the Páramo flora 180
Desert province 182
Definition 183
Endemic and characteristic taxa 183

Vegetation ... 184
Biotic relationships ... 184
Regionalization ... 184
Arequipa district ... 184
Callao district ... 184
Cardonales district ... 185
Coastal Desert district ... 185
Mollendo district ... 185
Porculla district ... 185
Surco district ... 185
Cenocrons ... 186
Case study: Diversification of the plant genus *Nolana* ... 186
Puna province ... 189
Definition ... 191
Endemic and characteristic taxa ... 191
Vegetation ... 192
Biotic relationships ... 192
Regionalization ... 192
Bolivian district ... 193
Cajamarca district ... 193
Central district ... 194
Cuyan district ... 194
Huancaspata district ... 194
Jujuyan district ... 194
Pasco district ... 194
Shimbe district ... 195
Atacama province ... 195
Definition ... 196
Endemic and characteristic taxa ... 196
Vegetation ... 197
Biotic relationships ... 199
Regionalization ... 199
Desventuradas Archipelago district ... 199
Interior Desert district ... 199
Northern Andean district ... 200
Northern Coast district ... 201
Northern Precordilleran district ... 201
Tamarugal district ... 202
Cenocrons ... 202
Cuyan High Andean province ... 202
Definition ... 203
Endemic and characteristic taxa ... 203
Vegetation ... 204

Biotic relationships....204
Cenocrons....204
Monte province....205
Definition....207
Endemic and characteristic taxa....207
Vegetation....208
Biotic relationships....209
Regionalization....210
Eremean district....210
Northern district....211
Prepuna district....211
Southern district....212
Cenocrons....213
Comechingones province....213
Definition....213
Endemic and characteristic taxa....213
Vegetation....214
Biotic relationships....215

References....217
Index....245

List of figures

Figure 1.1 Flowchart showing the steps of evolutionary biogeography 3

Figure 1.2 Steps of a parsimony analysis of endemicity used to identify generalized tracks. (a) Map with grid-cells, (b) data matrix, (c) cladogram obtained and (d) map with two generalized tracks and one node 4

Figure 1.3 Steps of a parsimony analysis of paralogy-free subtrees used to identify a general area cladogram. (a) Two taxon-area cladograms obtained from two different taxa, (b) four paralogy-free subtrees derived from the taxon-area cladograms, (c) data matrix and (d) general area cladogram obtained 6

Figure 2.1 Map of the Neotropical region and the Wallace (1876) subregions 12

Figure 2.2 Map of the Kuschel (1960) zoogeographical regionalization of Southern Chile 16

Figure 2.3 Map of the Ringuelet (1961) zoogeographical regionalization of Argentina 17

Figure 2.4 Map of the Peña (1966a,b) zoogeographical regionalization of Chile 18

Figure 2.5 Map of the Kuschel (1969) zoogeographical regionalization of South America 20

Figure 2.6 Map of the Cabrera (1971) phytogeographical regionalization of Argentina 21

Figure 2.7 Map of the Cabrera and Willink (1973) biogeographical regionalization of South America 22

Figure 2.8 Map of the Artigas (1975) zoogeographical regionalization of Chile 24

Figure 2.9 Map of the Rivas-Martínez and Tovar (1983) phytogeographical regionalization of South America 25

Figure 2.10 Map of the Roig (1998) phytogeographical regionalization of Argentinean Patagonia 26

Figure 3.1 The cladistic biogeographical analysis of the world by Amorim and Tozoni (1884). (a) Areas analyzed and (b) general area cladogram obtained 31

Figure 3.2 General area cladogram depicting the relationships of the biogeographical regions of the world........ 32

Figure 3.3 Distribution of the species of *Paralamyctes* (Chilopoda: Henicopidae), represented on a Triassic map of Gondwanaland. 1, *Paralamyctes quadridens*; 2, *P. tridens*; 3, *P. bipartitus*; 4, *P. newtoni*; 5, *P. prendinii*; 6, *P. asperulus*; 7, *P. weberi*; 8, *P. spenceri*; 9, *P. levigatus*; 10, *P. chilensis*; 11, *P. wellingtonensis*; 12, *P. trailli*; 13, *P. halli*; 14, *P. rahuensis*; 15, *P. validus*; 16, *P. harrisi*; 17, *P. subicolus*; 18, *P. mesibovi*; 19, *P. ginini*; 20, *P. grayi*; 21, *P. cassisi*; 22, *P. hornerae*; 23, *P. neverneverensis*; 24, *P. cammooensis*; 25, *P. monteithi* 33

Figure 3.4 Map of the Austral kingdom........ 35

Figure 3.5 Geological area cladogram of the Austral areas 36

Figure 3.6 Area cladograms representing different patterns of area relationships in the Southern Hemisphere. (a) Southern Gondwanaland pattern, (b) plant southern pattern, (c) inverted southern pattern, (d) northern Gondwanaland pattern, (e) tropical Gondwanaland pattern and (f) transamerican pattern........ 38

Figure 3.7 Area cladograms of the Austral areas. (a) Animals, (b) insects (exc. Eucnemidae), (c) non-insect animals and (d) plants........39

Figure 4.1 Maps with individual tracks in the Andean region. (a) *Epilobium denticulatum* (Onagraceae) and (b) *Gigantodax wrighti* (Diptera: Simuliidae) 46

Figure 4.2 Schematic representation of Austral individual tracks. (a) *Nothofagus* (Nothofagaceae), (b) *Aristotelia* (Elaeocarpaceae), (c) Cunoniaceae, (d) Donatiaceae, (e) Oxycorinini (Belidae) and (f) *Araucaria* (Araucariaceae) 48

Figure 4.3 Schematic representation of Austral individual tracks, generalized tracks and nodes. (a) *Azorella* (Apiaceae), (b) *Dunalia* (Solanaceae), (c) *Acicarpha* (Calyceraceae), (d) Empetraceae, (e) *Gutierrezia* (Asteraceae) and (f) representation of the most consistently found generalized tracks and the nodes delimited by them. A: Puna node, B: Subantarctic node and C: Patagonian node 50

Figure 4.4 *Trachodema tuberculosa,* a representative species of the Andean Listroderini (Coleoptera: Curculionidae) 51

Figure 4.5 Morrone's (1994) biogeographical analysis of the Andean region based on a track analysis and a parsimony analysis of endemicity of species of Curculionidae. (a) Original biota in the southern part of the region, (b) dispersal north and east, (c)–(e) successive isolation of the subregions and transition zone and (f) area cladogram showing the vicariant events and the progressive depauperation of the Andean biota. (CEN) Central Chilean subregion, (PAT) Patagonian subregion, (SATZ) South American transition zone and (SUB) Subantarctic subregion 52

Figure 4.6 Map of the biogeographical regionalization of the Andean region 54

Figure 4.7 Three perspectives of the Andean region. (a) Subantarctic subregion, (b) Andean region *sensu stricto* and (c) Andean region *sensu lato* (also including the South American transition zone) 55

Figure 4.8 Main South American areas affected by the Late Cretaceous–Early Paleocene marine transgressions 56

Figure 4.9 Main South American areas affected by the Middle Miocene–Late Miocene marine transgressions 57

Figure 4.10 Areas analyzed by the Donato et al. (2003) dispersal-vicariance analysis of the Andean region 59

Figure 4.11 Cladogram of the subtribe Listroderina used in the Donato et al. (2003) dispersal-vicariance analysis of the Andean region. D = Dispersal event, E = extinction events and V = vicariant event 60

Figure 5.1 Maps with individual tracks in the Subantarctic subregion. (a) *Cascellius* (Coleoptera: Carabidae) and (b) *Germainiellus* (Coleoptera: Curculionidae) 65

Figure 5.2 Geographical distribution of the species of the genus *Rhyephenes* (Coleoptera: Curculionidae) in the Subantarctic and Central Chilean subregions. (a) *Rhyephenes clathratus* (black circles) and *R. gayi* (open circles), (b) *R. goureaui,* (c) *R. humeralis* (black circles) and *R. lateralis* (open circles) and (d) *R. maillei* (black circles) and *R. squamiger* (open circles) 67

Figure 5.3 Maps showing the splitting of the Subantarctic generalized track based on species of Curculionidae. (a) Subantarctic generalized track, (b) initial vicariance of the Magellanic Moorland, (c) vicariance of the Valdivian Forest and (d) final vicariance of the Magellanic forest–Falkland Islands 69

Figure 5.4 Map of the provinces of the Subantarctic subregion 70

Figure 5.5 Areas analyzed in the Posadas and Morrone (2003) cladistic biogeographical analysis of the Subantarctic and Central Chilean subregions 70

Figure 5.6 *Falklandius antarcticus,* a representative species of the *Falklandius* generic group (Coleoptera: Curculionidae) distributed in the Subantarctic subregion 71

Figure 5.7 General area cladograms obtained in the Posadas and Morrone (2003) cladistic biogeographical analysis of the Subantarctic and Central Chilean subregions. (a) BPA and reconciled tree analysis and (b) DIVA analysis, showing the four most frequent dispersal events and on vicariant event 72

Figure 5.8 Maps with individual tracks in the Maule province. (a) *Chaetanthera serrata* (Asteraceae) and (b) *Triptilion achilleae* (Asteraceae) 73

Figure 5.9 *Nannomacer germaini,* a Nemonychidae (Coleoptera) endemic to the Maule province 74

Figure 5.10 Vegetation in the Maule province. (a) Forest with *Austrocedrus chilensis*, (b) forest with *Quillaja saponaria* and *Austrocedrus chilensis* and (c) forest with *Nothofagus dombeyi* 75

Figure 5.11 Forest of the Maule province, Reserva Nacional Los Ruiles, Chile 75

Figure 5.12 *Araucaria araucana*, endemic to the Pehuén district, Parque Nacional Los Paraguas, Chile 78

Figure 5.13 Maps with individual tracks in the Valdivian Forest province. (a) *Crinodendron hookerianum* (Elaeocarpaceae) and (b) *Misodendrum angulatum* (Misodendraceae) 81

Figure 5.14 Vegetation in the Valdivian Forest province. (a)–(c) Temperate forest with *Nothofagus obliqua* and *Laurelia sempervivens* and (d)–(f) forest-scrubland-steppe transition. (a) Sclerophyllous forest with *Nothofagus obliqua*, (b) more dense forest, (c) herbs, (d) forest with *Nothofagus pumilio* and *Berberis ilicifolia*, (e) shrubland with *Nothofagus antarctica* and *Berberis microphylla* and (f) steppe with *Festuca pallescens* and *Mulinum spinosum* 82

Figure 5.15 *Nothofagus* forest of the Valdivian Forest province, Lake Fagnano, Argentina 83

Figure 5.16 *Dromiciops gliroides*, the single living species of Microbiotheriidae (Microbiotheria), endemic to the Valdivian Forest province 87

Figure 5.17 Phylogenetic relationships of *Dromiciops gliroides* haplotypes over the geographical range of the species. (a) Map with 21 localities in Southern Chile and Argentina and (b) unrooted maximum-likelihood tree for the combined mtDNA data set 88

Figure 5.18 Maps with individual tracks in the Magellanic Forest province. (a) *Epilobium conjugens* (Onagraceae) and (b) *Germainiellus lugens* (Coleoptera: Curculionidae) 91

Figure 5.19 *Nothofagus* forest in the Magellanic Forest province, Southern Chile 91

Figure 5.20 Map with the provinces of the Subantarctic and Central Chilean subregions analyzed by Soares and de Carvalho (2005). (1) Coquimbo, (2) Santiago, (3) Maule, (4) Valdivian Forest, (5) Magellanic Forest, (6) Magellanic Moorland, (7) Falkland Islands and (8) Patagonia.......... 93

Figure 5.21 Taxon-area cladograms of taxa of the Subantarctic and Central Chilean subregions analyzed by Soares and de Carvalho (2005). (a) *Palpibracus*, (b) *Germainiellus*, (c) *Apsil* and (d) *Reynoldsia*. Coq = Coquimbo, Falk = Falkland Islands, MagF = Magellanic Forest, Mau = Maule, Pat = Patagonia, Sant = Santiago and ValF = Valdivian Forest.......... 94

Figure 5.22 Results of the Soares and de Carvalho (2005) analysis of the Subantarctic and Central Chilean subregions. (a) Generalized tracks and nodes and (b) consensus cladogram of the 10 cladograms obtained in the parsimony analysis of endemicity 95

Figure 5.23 Maps with individual tracks in the Magellanic Moorland province. (a) *Nothocascellius hyadesii* (Coleoptera: Carabidae) and (b) *Gigantodax brophyi* (Diptera: Simuliidae) 98

Figure 5.24 *Azorella filamentosa* (Apiaceae), cushion plant characteristic of the Magellanic Moorland province, Southern Chile.......... 99

Figure 5.25 Vegetation in the Magellanic Moorland province. (a) Coastal moorland with *Astelia pumila* and *Donatia fascicularis* and (b) interior moorland.......... 99

Figure 5.26 Map of the Falkland Islands.......... 101

Figure 5.27 Maps with individual tracks of weevil species (Coleoptera: Curculionidae) endemic to the Falkland Islands. (a) *Caneorhinus biangulatus*, (b) *Cylydrorhinus lemniscatus*, (c) *Malvinius compressiventris* and (d) *M. nordenskioeldi*.......... 102

Figure 5.28 *Lanteriella microphthalma*, species of Listroderini (Coleoptera: Curculionidae) endemic to the Falkland Islands.......... 103

Figure 5.29 Generalized tracks and nodes based on an analysis of plants, insects, crustaceans and mollusks showing that the Falkland Islands are a node, with generalized tracks connecting them with South Georgia, Tristan da Cunha-Gough, Crozet and Tierra del Fuego Islands. (CA) Campbell and other New Zealand Subantarctic Islands, (CR) Crozet, Marion and Prince Edward Islands, (FI) Falkland Islands, (JF) Juan Fernández Islands, (MA) Magellan area, (PA) Patagonia, (SG) South Georgia, (TC) Tristan da Cunha-Gough Islands and (TF) Tierra del Fuego 105

Figure 5.30 Generalized tracks and node based on species of Subantarctic Curculionidae. (a) Four generalized tracks and (b) two generalized tracks and one node. (1) Maule–Valdivian Forests generalized track, (2) Magellanic Forest generalized track, (3) Magellanic Moorland generalized track, (4) Falkland Islands generalized track, (5) Magellanic Forest–Magellanic Moorland generalized track, (6) node and (7) Magellanic Forest–Falkland Islands generalized track 107

Figure 5.31 *Megalometis spinifer,* weevil species endemic to the Maule and Valdivian Forests (Coleoptera: Curculionidae) 108

Figure 5.32 Geobiotic scenario explaining the biotic evolution of the Falkland biota. (a) Development of the original Subantarctic biota, (b) arrival of the Falkland Islands crustal block, (c) geodispersal of taxa from southern South America to the Falklands, (d) vicariance of the Magellanic Moorland, (e) vicariance of the Maule–Valdivian forests and (f) vicariance between the Magellanic forest and the Falkland Islands 110

Figure 5.33 Panoramic view of the Juan Fernández archipelago........ 112

Figure 5.34 Two endemic plant taxa of the Juan Fernández province. (a) *Dicksonia berteriana* (Dicksoniaceae) and (b) *Robinsonia* (Asteraceae) 113

Figure 5.35 Diagrammatic representation of plant lineages from the Juan Fernández Islands and continental Chile. (a) *Berberis* (Berberidaceae), (b) *Gunnera* (Gunneraceae), (c) *Myrceugenia* (Myrtaceae) and (d) *Sophora* (Fabaceae). (MF) Masafuera or Alejandro Selkirk Island and (MT) Masatierra or Robinson Crusoe Island 115

Figure 5.36 Map with the location of the Juan Fernández Islands and populations of *Robinsonia* (Asteraceae) sampled. (a) Juan Fernández archipelago, (b) Masafuera Island and (c) Masadentro Island. Open squares, *R. evenia;* open squares, *R. gayana;* closed circles, *R. gracilis;* closed pentagon, *R. saxatilis;* open pentagon, *R. thurifera* and closed squares, *R. masafuerae*......................117

Figure 5.37 Neighbor-joining of the 28 populations of *Robinsonia* subgen. *Robinsonia* based on distance of microsatellite data..118

Figure 6.1 Maps with individual tracks in the Central Chilean subregion. (a) *Triptilion spinosum* (Asteraceae), (b) *Platnickia elegans* (Araneae: Zodariidae) and (c) *Aegla papudo* (Decapoda: Aeglidae) 121

Figure 6.2 Map of the provinces of the Central Chilean subregion 122

Figure 6.3 Area cladograms in the Central Chilean subregion. Left, general area cladogram; right, area cladogram obtained from the parsimony analysis of endemicity 124

Figure 6.4 Maps with individual tracks in the Coquimban province. (a) *Triptilion globosum* (Asteraceae) and (b) *Listroderes robustus* (Coleoptera: Curculionidae).......... 125

Figure 6.5 Xeric vegetation typical of the Coquimban province, central Chile. (a) Shrubs and herbs and (b) cacti............... 126

Figure 6.6 Vegetation in the Coquimban province. (a)–(c) Desert scrubland, (d)–(f) xeric forest with *Acacia caven.* (a) Desert scrub, (b) sclerophyllous forest, (c) ravine forest with *Salix humboldtiana,* (d) sclerophyllous forest with *Cryptocarpa alba* and *Peumus boldus,* (e) shrubs with *Puya berteroniana* and *Colliguaja odorifera* and (f) ravine forest with *Beilschmiedia miersii* and *Crinodendron patagua*.. 127

Figure 6.7 Maps with individual tracks in the Santiagan province. (a) *Aegla laevis talcahuano* (Decapoda: Aeglidae) and (b) *Cyanoliseus patagonus byroni* (Psittaciformes: Psittacidae).. 132

Figure 6.8 Vegetation typical of the Santiagan province. (a) Shrubs and cacti and (b) shrubs 133

Figure 7.1 Map of the single province of the Patagonian subregion 139

Figure 7.2 Plains and plateaus characteristic of the Patagonian province, Santa Cruz river, Argentina 140

Figure 7.3 Maps with individual tracks in the Patagonian province. (a) *Azorella monantha* (Apiaceae), (b) *Acrostomus* (Coleoptera: Curculionidae), (c) *Epipedonota cristallisata* (Coleoptera: Tenebrionidae) and (d) *Mitragenius araneiformis* (Coleoptera: Tenebrionidae) 141

Figure 7.4 Shrub steppes of the Patagonian province, Lake Nahuel Huapi, Río Negro, Argentina 142

Figure 7.5 Generalized tracks and areas of endemism based on Coleoptera of the Patagonian province. (a) Generalized tracks and (b) districts, showing the localities of the species belonging to them. (1) Payunia generalized track and district, (2) Central Patagonian generalized track and district and (3) Fuegian generalized track and district 144

Figure 7.6 Generalized tracks based on plant taxa of the Patagonian province. (A) Northern Payunia district, (B) Southern Payunia district, (C) Western Patagonian subprovince, (D) Septentrional Subandean district, (E) Chubut district, (F) Santa Cruz district, (G) Meridional Subandean district and (H) Fuegian subprovince 146

Figure 7.7 *Anomophthalmus insolitus* (Coleoptera: Curculionidae), weevil species endemic to the Subandean subprovince of the Patagonian province 153

Figure 7.8 Stable areas and expansion routes detected in phylogeographic analyses in the Patagonian province. (a) Vertebrates and (b) plants. (1)–(11) Stable areas, (a)–(e) fragmentation and (f)–(i) areas of secondary contact 156

Figure 7.9 Areas of endemism in the Patagonian province found by Domínguez et al. (2006). (A) Western Patagonia, (B) northern Payunia, (E) southern Subandean, (F) Chubutian, (H) Austral Patagonia, (C) southern Payunia, (D) northern Subandean and (G) Santacrucian.......158

Figure 8.1 General area cladogram obtained by Urtubey et al. (2010) showing the relationships of the provinces of the South American transition zone and related areas. * denotes the provinces of the South American transition zone 164

Figure 8.2 Map of the provinces of the South American transition zone 165

Figure 8.3 Chronogram of *Munroa argentina* haplotypes and other Chloridoideae obtained by Amarilla et al. (2015) based on the consensus tree from the Bayesian dating analysis 167

Figure 8.4 Maps with individual tracks in the Páramo province. (a) Strengerianini (Decapoda: Pseudothelphusidae) and (b) *Gigantodax cervicornis* (Diptera: Simuliidae) 171

Figure 8.5 Paramo vegetation, Belmira, Colombia 171

Figure 8.6 Páramo vegetation. (A) Schematic representation of the vegetational zonation and (B) detail of the grass páramo. (a) *Chusquea tessellata*, (b) *Diplostephium schultzii*, (c) *Blechnum loxense*, (d) *Espeletia hartwegiana*, (e) *Neurolepis cf. aperta* and (f) *Puya* sp. 172

Figure 8.7 Distribution of the Colombian páramos analyzed by Londoño et al. (2014). (1) Perijá, (2) Jurisdicciones/Santurbán, (3) Tamá, (4) Almorzadero, (5) Yariguíes, (6) Cocuy, (7) Pisba, (8) Tota/Bijagual/Mamapacha, (9) Guantiva/Rusia, (10) Iguaque/Merchán, (11) Guerrero, (12) Rabanal and Bogotá river, (13) Chingaza, (14) Cruz Verde/Sumapaz, (15) Los Picachos, (16) Miraflores, (17) Belmira, (18) Nevados, (19) Chilí/Barragán, (20) Las Hermosas, (21) Nevado del Huila/Moras, (22) Guanacas/Puracé/Coconucos, (23) Sotará, (24) Doña Juana/Chimayoy, (25) La Cocha/Patascoy, (26) Chiles/Cumbal, (27) Paramillo, (28) Frontino/Urrao, (29) Citará, (30) Tatamá, (31) Duende, (32) Farallones de Cali, (33) Cerro Plateado, (34) Santa Marta, (A) Páramos de la Cordillera Oriental province, (B) Páramos del Macizo and Cordillera Central province, (C) Páramos de Antioquia province, (D) Páramos del Norte province and (E) Páramos de la Cordillera Occidental province 181

Figure 8.8 Map with the individual track of *Conepatus rex inca* (Carnivora: Mephitidae) in the desert province 183

Figure 8.9 Dispersal-vicariance analysis of *Nolana* by Dillon et al. (2009) 187

Figure 8.10 Map with the individual track of *Epilobium pedicelare* (Onagraceae) in the Puna province 192

Figure 8.11 Vegetation typical of the Puna province, western Peru. (a) Shrubs and herbs and (b) shrubs 193

Figure 8.12 Map with the individual track of *Listroderes robustior* (Coleoptera: Curculionidae) in the Atacama province..... 197

Figure 8.13 Vegetation of the Atacama province, 17 km east of Caldera, northern Chile 198

Figure 8.14 Vegetation in the scrub desert of the Atacama province. (a) Scrub, (b) xeric forest with *Acacia caven* and (c) ravine forest with *Salix humboldtiana* 198

Figure 8.15 Maps with individual tracks in the Cuyan High Andean province. (a) *Azorella cryptantha* (Apiaceae) and (b) *Cyanoliseus patagonus andinus* (Psittacidae)............ 204

Figure 8.16 Representative plant species of the Cuyan High Andean province, Mendoza, Argentina. (a) *Argylia uspallatensis* (Bignoniaceae) and (b) *Nastanthus ventosus* (Calyceraceae) 205

Figure 8.17 Maps with individual tracks in the Monte province. (a) *Enoplopactus lizeri* (Coleoptera: Curculionidae) and (b) *Bothrops ammodytoides* (Squamata: Viperidae) 208

Figure 8.18 Plant species characteristic of the Monte province, Argentina. (a) *Cereus forbesii* (Cactaceae), (b) *Aspidosperma* sp. (Apocynaceae) and (c) *Polylepis tarapacana* (Rosaceae) 209

Figure 8.19 Map with three districts of the Monte province (Prepuna district not represented) 210

Figure 8.20 Vegetation of the Comechingones province, Córdoba, Argentina. (a) High altitude grassland and (b) high altitude steppe 214

Preface

The Andean mountain chain represents the most striking geomorphological feature of South America, extending along the entire length of the continent, from Venezuela to southern Chile and Argentina. The topographic relief of the Andes is extremely complex, having provided the conditions for the evolution of an endemic, characteristic biota. Additionally, the general orientation of this cordillera favored the dispersal of temperate taxa from the south (Subantarctic subregion) to the north and from the northern hemisphere (Nearctic region) to the south, thus allowing the development of the South American transition zone. The Andean biota is particularly interesting, with some taxa endemic to the area, other taxa exhibiting close relationships with taxa living in other southern areas (Australia, Tasmania, New Zealand and South Africa), and other taxa being closely related to Neotropical taxa. Based on numerous distributional data of plant and animal taxa, authors have recognized an Andean region.

The Andean region comprises southern South America below 26°S, extending through the Andean highlands north of this latitude. It belongs to the Austral kingdom, which also includes the Australian, Cape and Antarctic regions. In this book, I address two central questions of evolutionary biogeography: which areas do researchers recognize within the Andean region and how did their biotas evolve? In the last decades molecular phylogenetics and parametric model-based biogeography have allowed the postulation of complex and idiosyncratic biogeographic scenarios for specific taxa; however, the search for biotic patterns has been somewhat neglected. In a previous book on Neotropical biogeography (Morrone, 2017), I argued that biogeographical regionalizations, based on the distributional patterns of plant and animal taxa, are still relevant in the twenty-first century, constituting the background knowledge of systematic, ecological, evolutionary and other kinds of studies. In this book, I continue with the analysis of another biogeographical region.

The Andean region consists of 3 subregions, 1 transition zone and 16 provinces. For each unit, I provide the valid name according to the International Code of Area Nomenclature (ICAN), followed by a list of citations and synonyms, a brief characterization, and some endemic and characteristic taxa. To deal with biotic evolution, I refer to the identification of biotas through areas of endemism and generalized tracks, their relationships based on track and cladistic biogeographical analyses and, when possible, the cenocrons or biotic subsets that have become integrated within them. This attempt of synthesis is based on a vast bibliography that I compiled during three decades. I feel grateful to many authors who provided insights on the regionalization and evolution of the Andean region. Particularly inspirational for my work were Jorge Artigas, Ángel L. Cabrera, Philip Darlington, René Jeannel, Guillermo Kuschel, Emilio Maury, Paul Müller, Rosendo Pascual, Eduardo Rapoport, Osvaldo Reig, Raúl Ringuelet, George G. Simpson, Armen Takhtajan and Abraham Willink.

I thank many friends and colleagues who during the last decades helped me with the bibliography, provided data and shared their ideas with me: Lone Aagesen, Dalton de Sousa Amorim, Marcelo Arana, María Marta Cigliano, Jorge Crisci, Juan M. Daza, Malte C. Ebach, Cecilia Ezcurra, Ignacio Ferro, Gustavo E. Flores, Michael Heads, Viviana Hechem, Carlos Jiménez-Rivillas, Liliana Katinas, Analía Lanteri, Estela Lopretto, Peter Löwenberg-Neto, Andrés Moreira-Muñoz, Paula Posadas, Mario Alberto Quijano, Sergio Roig-Juñent, Adriana Ruggiero, Claudia Szumik and Estrella Urtubey. I thank María Marta Cigliano, Mario Elgueta, Roberto Kiesling, Andrés Moreira-Muñoz, Mario Alberto Quijano, Mariano A. Rodríguez-Cabal, Sergio Roig-Juñent and Erick Yábar-Landa for providing color photographs and Sergio Roig-Juñent for providing the illustrations of the weevil species. Adrián Fortino kindly helped me with processing the illustrations and photographs. Finally, I am indebted to Mario Elgueta, Cecilia Ezcurra, Gustavo E. Flores and Sergio Roig-Juñent for providing useful comments to a preliminary version of the manuscript.

Juan J. Morrone
Facultad de Ciencias, UNAM, Mexico City

About the author

Juan J. Morrone is full professor of Biogeography, Systematics and Comparative Biology at the Facultad de Ciencias, Universidad Nacional Autónoma de México (UNAM), Mexico. He works on phylogenetic systematics of weevils (Coleoptera: Curculionidae) and evolutionary biogeography and regionalization of the Neotropical and Andean regions.

He joined the Museo de Zoología "Alfonso L. Herrera" of the Facultad de Ciencias, Universidad Nacional Autónoma de México (UNAM), Mexico in 1998, after working for some years at the Museo de La Plata, Universidad Nacional de La Plata (UNLP), Argentina, where he obtained his PhD degree. He is a member of the Academia Mexicana de Ciencias, Fellow of the Willi Hennig Society and research associate at the American Museum of Natural History. He has authored 270 scientific papers and authored or edited 30 books on evolutionary biogeography, phylogenetic systematics, biogeographical regionalization, biodiversity conservation and evolution.

chapter one

Theoretical background

Evolutionary biogeography integrates distributional, phylogenetic, molecular and paleontological data to discover biogeographical patterns exhibited by plant and animal taxa and to assess the historical changes that have shaped biotic assembly (Morrone, 2009). Biogeographical regionalizations are hierarchical classifications categorizing geographical areas in terms of their endemic taxa and their relationships. Regionalizations represent the syntheses of different kinds of biogeographical analyses and, at the same time, constitute the background knowledge of systematic, ecological and evolutionary studies, among others.

In this introductory chapter, I briefly characterize evolutionary biogeography and explain its steps. Parsimony analysis of endemicity allows the identification of biotas through generalized tracks or areas of endemism. A cladistic biogeographical approach elucidates the relationships among the biotas based on the phylogenetic hypotheses of the taxa that they contain. Based on the biotas recognized and their relationships, I obtained biogeographical regionalizations. The use of fossils and phylogeographical and molecular divergence dating analyses allows identification of cenocrons, the subsets of taxa that share a biogeography history. If geological data are available, I integrated the data to construct a geobiotic scenario that can explain how the episodes of dispersal and vicariance shaped biotic evolution.

Evolutionary biogeography

Evolutionary biogeography is the integrative study of distributional, phylogenetic, molecular and paleontological data aimed to discover biogeographical patterns and assess the historical changes that shaped them (Morrone, 2009, 2017). Its use follows a step-wise approach. The first step is the identification of areas of endemism or generalized tracks and the formulation of hypotheses about biotic identity based on the distributional congruence exhibited by different plant and animal taxa. Then, cladistic biogeographical analyses test these hypotheses, based on the available phylogenetic evidence on the taxa analyzed, and allow identification of vicariant events and their relative time sequence. Cladistic biogeographical hypotheses provide a biogeographical regionalization. As a fourth step, the molecular dating of divergences between lineages and fossil data allow identification of cenocrons, which represent subsets of taxa, identified by their common origin and evolutionary history that dispersed and integrated into the biota. Finally, after identification of the biotas and their cenocrons, one may construct a geobiotic scenario by accounting for biological and non-biological data to explain the episodes of vicariance/biotic divergence and dispersal/biotic convergence that shaped the evolution of the biotas analyzed.

The dispersal-vicariance model that I follow assumes that the relationship between Earth history and life is more complex than what studies assume in simpler models because the history of biotas is reticulate. The model treats vicariance as the default explanation for general biogeographical patterns, and dispersal as the process that shapes the distribution of cenocrons or individual species (Morrone, 2015a). The integration of vicariance and dispersal helps in understanding biotic assembly by incorporating the dating of the lineages and the identification of the cenocrons.

Steps of evolutionary biogeography

Figure 1.1 shows the five steps of evolutionary biogeography.

Identification of biotas

Biotas are sets of spatio-temporally integrated plant and animal taxa that coexist in given areas (Morrone, 2009). The identification of these sets of taxa constitutes the first step of an evolutionary biogeographical analysis. There are different ways to represent biotas graphically; generalized tracks and areas of endemism are two of the most common ways.

Generalized tracks indicate the pre-existence of ancestral biotas that became fragmented by geological or tectonic events (Craw et al., 1999). Generalized tracks result from the significant superposition of different individual tracks, which operationally correspond to line graphs connecting the different localities or distributional areas of species or supraspecific taxa according to their geographical proximity (Morrone, 2015a). Studies identify the areas where two or more generalized tracks superimpose as nodes. These studies usually interpret the nodes as tectonic and biotic convergence zones where different biotas contact. Areas of endemism are areas of non-random distributional congruence among different taxa. Studies identify these areas by plotting the distributional ranges of different species or supraspecific taxa on a map and finding the areas of congruence between them.

There are different methods used to identify generalized tracks or areas of endemism (reviewed by Morrone, 2009). The most commonly applied method for obtaining generalized tracks is parsimony analysis of endemicity (PAE; Echeverry and Morrone, 2010; Morrone, 2015a). Parsimony analysis of endemicity includes the following steps (Figure 1.2):

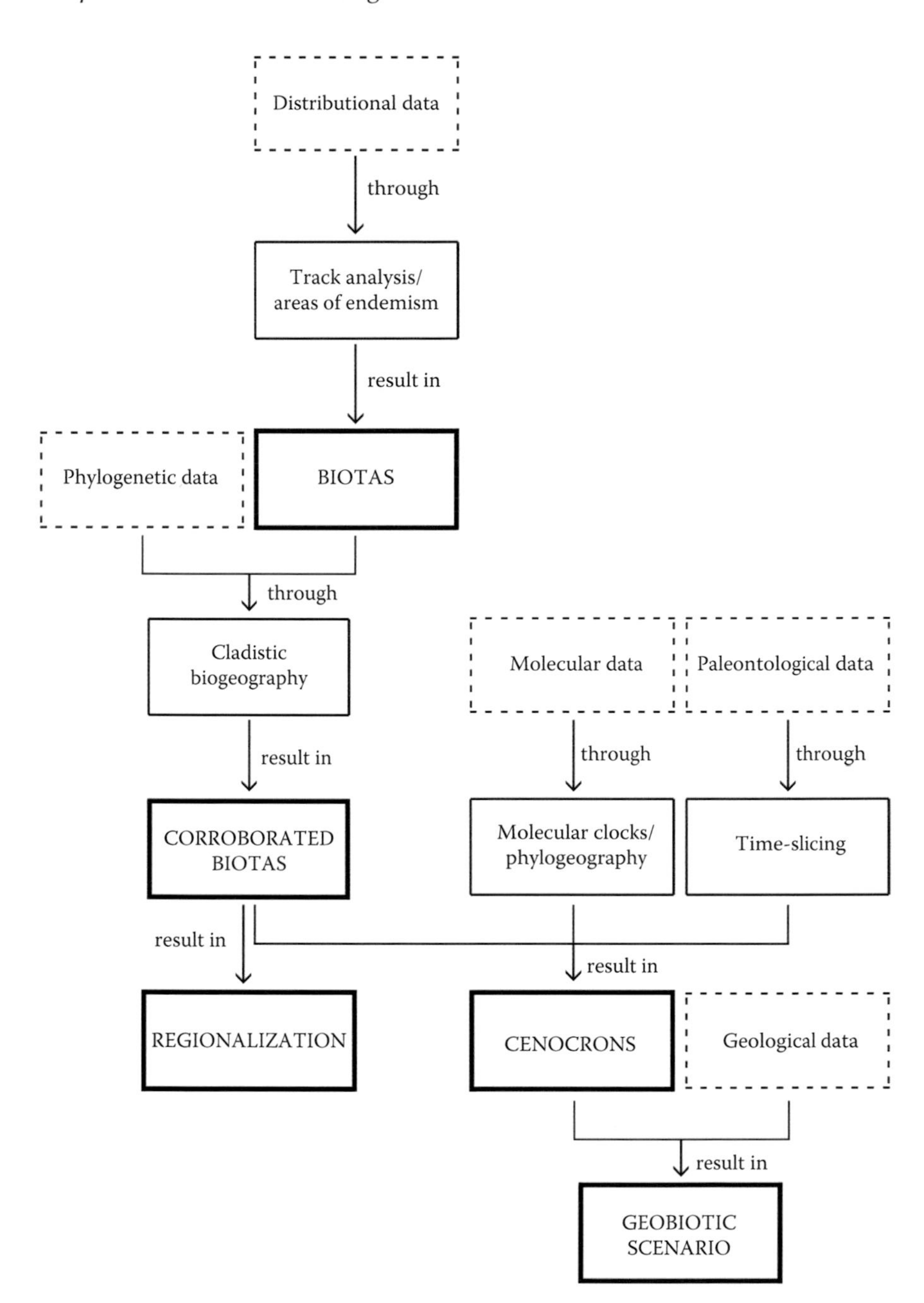

Figure 1.1 Flowchart showing the steps of evolutionary biogeography.

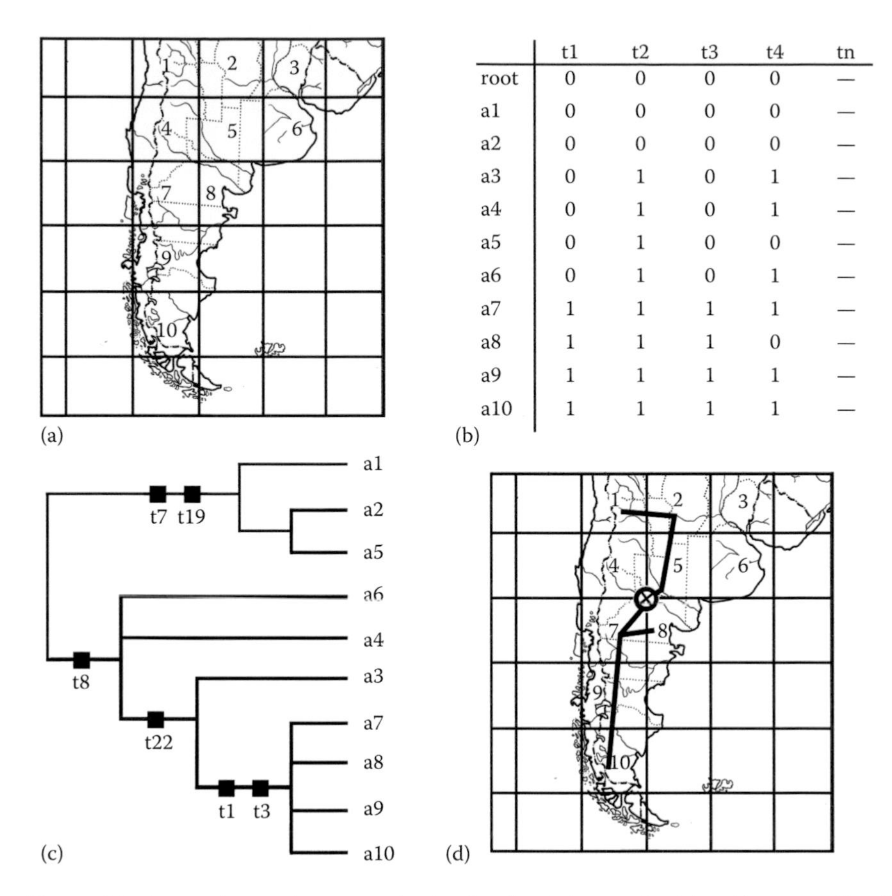

	t1	t2	t3	t4	tn
root	0	0	0	0	—
a1	0	0	0	0	—
a2	0	0	0	0	—
a3	0	1	0	1	—
a4	0	1	0	1	—
a5	0	1	0	0	—
a6	0	1	0	1	—
a7	1	1	1	1	—
a8	1	1	1	0	—
a9	1	1	1	1	—
a10	1	1	1	1	—

Figure 1.2 Steps of a parsimony analysis of endemicity used to identify generalized tracks. (a) Map with grid-cells, (b) data matrix, (c) cladogram obtained and (d) map with two generalized tracks and one node.

1. Choose a set of biogeographical units across the study area, for example localities, pre-defined areas of endemism, areas defined by physiographical criteria or grid-cells.
2. Construct individual tracks for species and/or supraspecific taxa, connecting the occurrence localities by a minimal spanning tree, either manually or using any of the programs available (e.g., Rojas-Parra, 2007; Liria, 2008; Echeverría-Londoño and Miranda-Esquivel, 2011).
3. Construct a data matrix, where rows represent the biogeographical units analyzed and columns represent the individual tracks. Code each entry as either "1" or "0," depending on whether each track is present or absent in the unit. A "?" may be included in case of doubtful occurrence in some geographical unit. The addition of a hypothetical unit coded as all zeros to the matrix roots the resulting cladogram(s).

4. Analyze the data matrix with a parsimony algorithm, applying a phylogenetic software (e.g., Swofford, 2003; Goloboff et al., 2008). For more than one cladogram found, calculate a strict consensus cladogram.
5. Identify generalized tracks in the resulting cladogram based on the monophyletic groups of units defined by at least two individual tracks.
6. Remove from the data matrix the synapomorphic individual tracks that support the clades previously obtained and repeat steps 4–6 until no more individual tracks support any clade.
7. If some areas result in the overlap of two or more generalized tracks, identify them as nodes.
8. Represent the identified generalized tracks and nodes on a map.

Testing relationships among biotas

Cladistic biogeography is an approach based on the correspondence between the phylogenetic relationships of different plant and animal taxa and the relationships between the areas that they inhabit (Parenti and Ebach, 2009). It uses information on the phylogenetic relationships between the taxa and their geographical distribution to postulate hypotheses on the relationships between the areas. When different taxa show the same pattern, such phylogenetic/geographical congruence is evidence of a common biotic history shaped mainly by vicariance. A cladistic biogeographical analysis begins by constructing taxon-area cladograms, from the taxonomic cladograms of two or more different taxa, and replacing their terminal taxa with the areas they inhabit. Then, one obtains resolved area cladograms from the taxon-area cladograms (when demanded by the method applied). Finally, one constructs a general area cladogram based on the information contained in the resolved area cladograms. General area cladograms based on the information from the different resolved area cladograms represent hypotheses on the biogeographical history of the taxa analyzed and the areas of endemism where they are distributed.

There are different methods for obtaining general area cladograms (reviewed by Morrone, 2009). One of them is parsimony analysis of paralogy-free subtrees (Morrone, 2009, 2014b). This analysis includes the following steps (Figure 1.3):

1. Replace the terminal species of the taxonomic cladograms analyzed to obtain a taxon-area cladogram for each taxon.
2. Identify the paralogy-free subtrees in the taxon-area cladograms by eliminating areas duplicated or redundant in the descendants of a node to eliminate or reduce significantly geographical paralogy so that data are associated only with informative nodes.

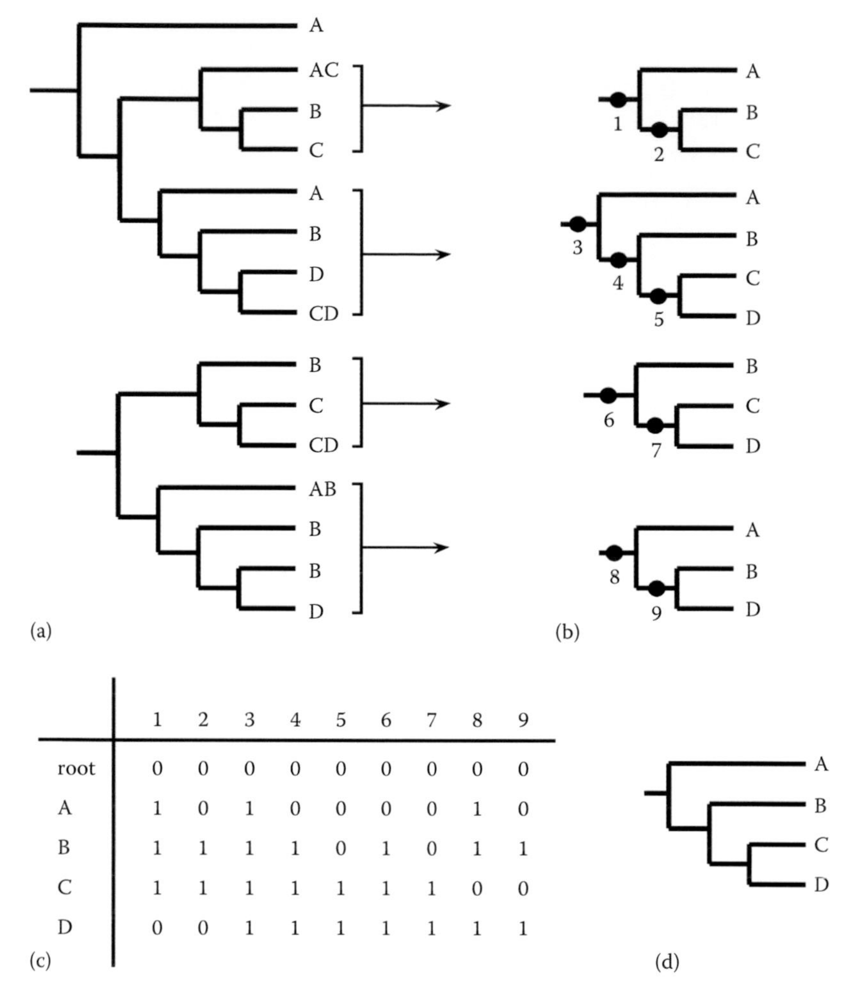

	1	2	3	4	5	6	7	8	9
root	0	0	0	0	0	0	0	0	0
A	1	0	1	0	0	0	0	1	0
B	1	1	1	1	0	1	0	1	1
C	1	1	1	1	1	1	1	0	0
D	0	0	1	1	1	1	1	1	1

Figure 1.3 Steps of a parsimony analysis of paralogy-free subtrees used to identify a general area cladogram. (a) Two taxon-area cladograms obtained from two different taxa, (b) four paralogy-free subtrees derived from the taxon-area cladograms, (c) data matrix and (d) general area cladogram obtained.

3. Identify the components (=nodes) on the paralogy-free subtrees obtained.
4. Compile a data matrix, scoring with "1" the presence of a component in an area and "0" its absence. The addition of a hypothetical unit coded as all zeros to the matrix then roots the resulting general area cladogram(s).

5. Analyze the data matrix with a parsimony algorithm, by applying a phylogenetic software (e.g., Swofford, 2003; Goloboff et al., 2008) to identify the general area cladogram(s).

Regionalization

As the geographical distributions of taxa have limits and these limits repeat for different taxa, they allow the recognition of biotas, which are graphically represented as areas of endemism. After one identifies them, one may order them hierarchically by a cladistic biogeographical analysis and use them to provide a biogeographical regionalization. It implies the recognition of successively nested areas, for which classically studies have used the following five categories: kingdom, region, dominion, province and district (Ebach et al., 2008).

In the cases where it is difficult to determine the exact boundaries of two regions, studies identified transition zones (Ferro and Morrone, 2014). They represent events of biotic *hybridization*, which historical and ecological changes promoted and allowed the mixture of different biotas. When one recognizes a transition zone between two regions, it means implicitly that the area belongs simultaneously to both regions.

Identification of cenocrons

Cenocrons are sets of taxa that share the same biogeographical history, constituting identifiable subsets within a biota by their common biotic origin and evolutionary history. After identifying the biotas, time-slicing, intraspecific phylogeography and molecular divergence dating help establish when the cenocrons dispersed and assembled in the identified biotas, incorporating a time perspective to the study of biotic evolution. Time-slicing consists of using fossils or molecular divergence dating to divide the complete set of taxa analyzed into slices according to the time when they occurred (Cecca et al., 2011). Intraspecific phylogeography, the analysis of the geographical distribution of genealogical lineages within and among closely related species based on molecular data (Avise, 2000), may offer insight into when and how relatively recent cenocrons incorporated into a biota. One may use phylogenetic molecular hypotheses to calculate molecular divergence dates, under the assumption that the rate of molecular evolution is approximately constant over time in all lineages (Zuckerland and Pauling, 1965), and that *ticks* correspond to mutations that do not occur at regular intervals, but rather at random points in time. One may measure time in arbitrary units and then calibrate in millions of years by reference to the fossil record or geological data (Magallón, 2004), thus giving minimum estimates of the age of a clade that one may use to elucidate the relative minimum age of the cenocron to which it belongs (Morrone, 2009).

Construction of a geobiotic scenario

After identification of the biotas and their cenocrons, one may construct a geobiotic scenario. By accounting biological and geological data, one can postulate a plausible scenario to help explain the episodes of vicariance/biotic divergence and dispersal/biotic convergence that have shaped biotic evolution. One may classify geographical features in terms of their impact on dispersal and vicariance. The most important are barriers (e.g., geographical or climatic features that hinder dispersal) and corridors (geographical or climatic features that facilitate dispersal). When dealing with long-term changes in broad biotic patterns, continental drift may be a relevant factor (Cox and Moore, 2005). The splitting and collision of land masses directly affect distributional patterns and also new mountains, oceans or land barriers change the climatic patterns upon the land masses.

Biogeographical regionalization

A biogeographical regionalization is a hierarchical system that categorizes geographical areas in terms of their endemic taxa (Escalante, 2009). During the past two centuries, studies proposed several biogeographical regionalizations for the world and for specific areas. Biogeographical regionalization is historically rooted in the nineteenth century, with classification and nomenclatural procedures similar to those of systematics. De Candolle (1820) proposed that phytogeographical kingdoms should harbor a minimum of endemic species and genera to be accepted. Sclater (1858) postulated that zoogeographical regions should be based on endemic families and Drude (1884) suggested the same for phytogeographical kingdoms. Wallace (1876) discussed some principles to apply for obtaining natural regionalizations.

In the last decades studies made different proposals for producing biogeographical regionalizations in a more objective way, although there is little agreement on the use of different methods available. Some studies considered that to produce natural biogeographical regionalizations, successively nested endemism should be the basic criterion used (Escalante, 2009), as has been the case traditionally. After recognizing natural areas based on endemic taxa (generalized tracks or areas of endemism), studies arranged the areas hierarchically (general area cladograms) and gave them names. To denote this biogeographical hierarchy, five basic levels may be used: kingdoms (also known as realms), regions, dominions, provinces and districts, and in some cases, subkingdoms, subregions or subprovinces (Ebach et al., 2008). In addition, studies recognized transition zones, representing areas of overlap, with a gradient of replacement and partial segregation between biotas belonging to different regions (Ferro and Morrone, 2014).

Nomenclatural conventions

In the biogeographical regionalization of the Andean region (Morrone, 2015b) followed in this book, I applied the nomenclatural conventions set out in the International Code of Area Nomenclature (ICAN; Ebach et al., 2008). ICAN provides a universal naming system to standardize area names used in biogeography and other disciplines in which the system groups names under more inclusive area names to represent a biogeographical hierarchy (kingdoms, regions, dominions, provinces and districts).

I followed the notion of priority for using the oldest available names instead of new names (ICAN, Art. 2.8). I used the dates from Sclater (1858) as the date of the starting point of biogeographical nomenclature, as it constitutes the first widely adopted world biogeographical regionalisation. In some cases, I kept widely used names instead of older synonyms, applying a criterion analogous to the *nomen conservandum* convention of taxonomical nomenclature to provide better stability (Morrone, 2014b, 2017).

Format used in the book

For each biogeographical unit recognized, I provide the valid name followed by a list of citations and synonyms arranged in chronological order, with a brief characterization of the type of analysis applied to them. I include maps that show the areas. For each province, I list some endemic and characteristic taxa and mention the main vegetational types. For the different areas recognized, I map individual tracks representing the distribution of selected taxa. Whenever I identified subprovinces and districts, I list and characterize them briefly. To deal with the biotic evolution of the Andean region, I refer to the identification of biotas (as either areas of endemism or generalized tracks), their relationships based on a track and cladistic biogeographical analyses and, when possible, the cenocrons or biotic subsets identified within the biotas. Finally, I provide some selected case studies to exemplify the diversity of evolutionary biogeographical analyses conducted in the Andean region.

chapter two

Historical background

The Andean region corresponds to southern South America below 26°S, extending through the Andean highlands north of this latitude (Morrone, 2015b). During the nineteenth and twentieth centuries, several studies recognized subregions, dominions and provinces in the Andean region or areas within it. Some of these studies considered it formally as a distinct region, whereas others treated this area as part of the Neotropical region, ranking it as a subregion or dominion.

The efforts of several studies, from Sclater (1858) to Moreira-Muñoz (2011), culminated in the recognition of the Andean region and its placement in the Austral kingdom. Within the Andean region, several studies on plant and animal distributional data recognized subregions, provinces and districts. This chapter discusses these biogeographical regionalizations, illustrating most of the proposed schemes.

Biogeographical regionalization of the world

Sclater (1858) divided the world into six zoogeographical regions based on bird taxa (Passeriformes). He named these regions Palearctic, Ethiopian, Indian, Australian, Nearctic and Neotropical. His Neotropical region included the West Indies, southern Mexico, Central America and the whole South America. Wallace (1876) followed the scheme of Sclater (1858) and applied it to other vertebrate taxa. According to the Sclater-Wallace system, the Neotropical region comprises South America, Central America and reaches as far north as central Mexico, where it reached limits within the Nearctic region. Within the Neotropical region, Wallace (1876) recognized the Chilean subregion (Figure 2.1), which corresponds in this book broadly to the Andean region. Many studies followed the Sclater-Wallace system (e.g., Murray, 1866; Huxley, 1868; Kirby, 1872; Allen, 1892; Sclater, 1894; Heilprin, 1887; Bartholomew et al., 1911; Mello-Leitão, 1937; Darlington, 1957; Morain, 1984; Fleming, 1987). Most studies considered it the *standard* system, especially for those analyzing the distribution of vertebrate taxa (Cox, 2001; Parenti and Ebach, 2009).

Figure 2.1 Map of the Neotropical region and the Wallace (1876) subregions. (From Wallace, A.R., *The Geographical Distribution of Animals. Vol. I & II*, Harper & Brothers, New York, 1876.)

Engler (1879, 1882) provided a phytogeographical regionalization of the world, recognizing the Holarctic, South American, Paleotropical and Old Oceanic kingdoms. Within them, Engler (1879, 1882) recognized 32 regions, which he further divided into provinces and, in some cases, districts. His South American phytogeographical kingdom is equivalent to the Neotropical zoogeographical region of Sclater (1858) and Wallace (1876), but he restricted the South American portion to the tropical areas. Engler (1879, 1882) assigned the southern temperate part of South America to the Old Oceanic kingdom, also comprising New Zealand's South Island, the Subantarctic islands, most of Australia and South Africa. He later changed the name of the Old Oceanic kingdom to the Austral kingdom (Engler, 1899). Phytogeographers widely adopted Engler's Austral, Antarctic or Holantarctic kingdom (Müller, 1986), although some studies preferred to exclude South Africa and Australia from it, leaving only southern South America and New Zealand (Good, 1947; Mattick, 1964; Cabrera and Willink, 1973; Takhtajan, 1986). Other researchers further separated southern South America and New Zealand into different kingdoms (Drude, 1884; Diels, 1908).

During the nineteenth and twentieth centuries, several studies based on South American plants and invertebrates suggested a more restrictive definition of the Neotropical region. As a result, several studies excluded the southern portion and the Andean area from the Neotropical region because of their closest biotic links with Australia, Tasmania, New Guinea, New Zealand and South Africa (Blyth, 1871; Drude, 1884; Allen, 1892; Lydekker, 1896; Diels, 1908; Jeannel, 1938, 1942; Good, 1947; Monrós, 1958; Rapoport, 1968; Cabrera and Willink, 1973; Amorim and Tozoni, 1994; Morrone, 2002, 2006; Moreira-Muñoz, 2007, 2011).

Among the studies recognizing an Austral kingdom (or region) three deserve special comment. Newbigin (1950) recognized five biogeographical regions for the world: Northern Lands (Holarctic), Mediterranean, Northern Paleotropical Desert, Intertropical and Austral. Her Austral region comprises the Andean region, Australia, Tasmania, New Zealand and South Africa. She considered that the biotic relationships among these Austral areas were especially evident for plant taxa. Kuschel (1964) discussed the geographical distribution of plant and animal taxa from the southern continents and, based on the phylogenetic relationships of several Austral taxa and the possibility of Antarctica having behaved in the past as a land-bridge, proposed an Austral biogeographical region to which he assigned the Patagonian (southern South America), South African and Australian subregions. Rapoport (1968) recognized three biogeographical regions for the world: the Holarctic region which comprises the areas of the Northern Hemisphere, including the Nearctic and Palearctic regions of previous studies; the Holotropical region which comprises the

Neotropical region, Africa, southeast Asia and the Pacific islands; and the Holantarctic region (herein treated as the Austral kingdom) which comprises southern South America (Andean region), South Africa, Australia, New Guinea, New Zealand and Antarctica.

Cox (2001) challenged this consensus that recognized an Austral kingdom. Based on formal arguments and without providing an empirical justification, Cox (2001) disintegrated the Austral kingdom, allocating its different parts to the adjacent continents. Morrone (2002) replied to proposal of Cox (2001) and, based on track and cladistic biogeographical analyses, regroups the Austral areas into a single biogeographical kingdom. Studies have noted that the scheme of Morrone (2002) is similar to the proposal of Engler (1879, 1882), with the exception that Morrone (2002) groups all the tropical areas into a single Holotropical kingdom (Moreira-Muñoz, 2007).

Moreira-Muñoz (2007) provided a review of different biogeographical classifications and especially discussed the recognition of an Austral kingdom. Based on a quantitative analysis of plant genera from Chile, New Zealand and South Africa, he concluded that the floristic similarities and the phylogenetic relationships of several plant families clearly supported the recognition of an Austral kingdom including Australasia and southern South America, and possibly including the Cape region (South Africa). Moreira-Muñoz (2007) also traced the relationships among the Austral areas to Gondwanaland, with a history tied to an ancient paleogeography involving geological details not yet fully understood.

I provided a consensus scheme, dividing the world into the Holarctic, Holotropical and Austral kingdoms and nine regions (Morrone, 2015c). In this scheme, the Andean region belongs to the Austral kingdom, which corresponds to the eastern portion of the Gondwanaland paleocontinent (Crisci et al., 1993), known also as temperate Gondwanaland (Amorim et al., 2009). Additionally, the Andean region has a transition zone, known as the South American transition zone, which corresponds to the area of biotic overlap with the Neotropical region (Holotropical kingdom).

Regionalization of the Andean region

Several proposals recognized subregions, dominions, provinces and other units within the Andean region. Initially, Wallace (1876) recognized the Chilean subregion, which he assigned to the Neotropical region, which corresponds to southern or temperate South America. He considered that the Chilean subregion was transitional to the Australian region. Several studies followed Wallace's regions and subregions during the last decades of the nineteenth century and the twentieth century (Heilprin, 1887; Lydakker, 1896; Bartholomew et al., 1911; Mello-Leitão,

1937; Hershkovitz, 1969). Mello-Leitão (1937), based on arachnids, modified slightly the scheme of Wallace (1876) by recognizing an Andean-Patagonian subregion that corresponds to the Andean cordillera from Colombia to Chile.

Cabrera and Yepes (1940) proposed a geographical scheme of South America based on the distribution of mammals. Their Patagonian subregion extended into most of Peru, Bolivia, Argentina and all of Chile. Sclater (1858) and Wallace (1876) previously named this subregion the Chilean subregion, but Cabrera and Yepes (1940) found more appropriate to name it Patagonian. They characterized the Patagonian subregion by the proportion of species belonging to different mammal orders and that its boundary with the Guianan-Brazilian subregion followed an oblique line that goes from northwest to southeast, from northern Peru to central Argentina. They cited characteristic species of Cervidae, Camelidae and Rodentia. Additionally, Cabrera and Yepes (1940) defined 11 smaller divisions, which they named districts (and considered as equivalent to provinces), based on physiographical criteria and the presence of some mammal species. Within the Patagonian subregion they included the Patagonian, Subandean, Chilean, Andean and Incasic districts.

Kuschel (1960) analyzed the zoogeography of southern Chile, specifically the area from the Valdivian forest to Cape Horn. He recognized five zones (Figure 2.2): Valdivian Forest, Magellanic Forest, Magellanic Moorland, Mountain and Patagonian Steppes. Kuschel (1960) considered that the faunas of the two forest zones were rather similar and that the faunas of the three latter zones, characterized by the absence of trees, resembled each other. When addressing the biotic relationships of the area, Kuschel (1960) considered that there were three basic types, namely, Holarctic, Brazilian (=Neotropical) and southern temperate. The two former relationships are less important, whereas the latter is the most important element in the southern Chilean biota. Kuschel (1960) listed several arthropod taxa that show clear relationships between southern Chile and New Zealand, Australia, Tasmania, New Guinea and South Africa, including Pseudoscorpiones, Araneae, Ephemeroptera, Plecoptera, Orthoptera, Heteroptera, Mecoptera, Trchoptera, Diptera, Hymenoptera and Coleoptera.

Ringuelet (1961) published an essay on several issues of the zoogeography of Argentina in which he highlighted the relevance of considering both ecological and historical criteria when regionalizing an area. Ringuelet (1961) considered that the primary divisions within a subregion were the dominions, which he considered equivalent to provinces in Mello-Leitão (1937, 1943) and districts in Cabrera and Yepes (1940). From an historical perspective, he postulated that the fauna of Argentina consisted of 10 cenocrons: primitive Austral-American, Gondwanic, Notogeic, Brazilian,

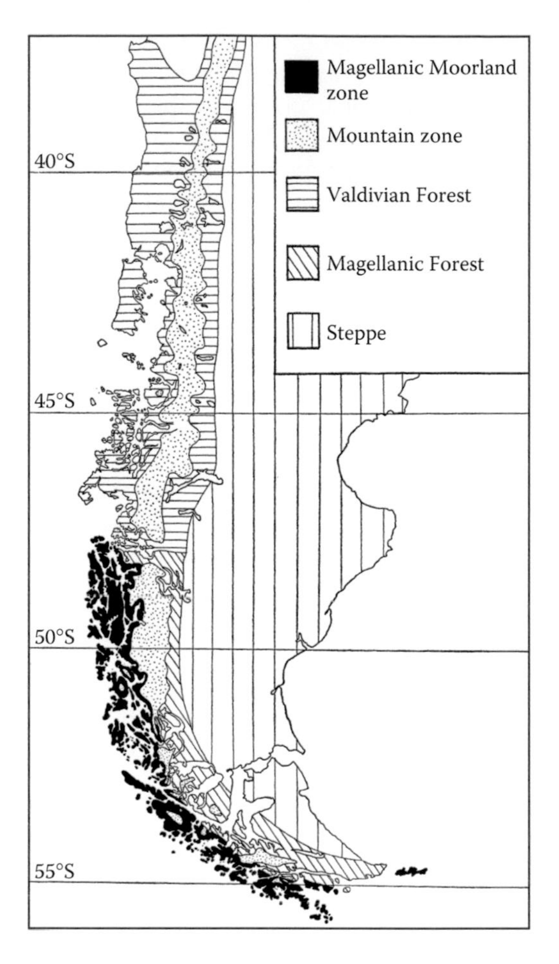

Figure 2.2 Map of the Kuschel (1960) zoogeographical regionalization of Southern Chile. (Modified from Kuschel, G., *Proceedings of the Royal Society of London, series B*, 152, 540–550, 1960. With permission.)

Afro-Brazilian, Palearctic, Nearctic, Pacific, intrusive (in fresh water) and autochthonous. In contrast with previous proposals, Ringuelet (1961) recognized the Andean-Patagonian and Araucanian subregions, the latter corresponding to the *Nothofagus* forests of southern South America. The Andean-Patagonian subregion comprised the Andean, Central and Patagonian dominions (Figure 2.3). Ringuelet (1961) considered that each of these dominions harbored a specific fauna, with taxa belonging to different cenocrons but that were ecologically similar. He also postulated that the Guianan-Brazilian fauna extended in the past southward, reaching northern Patagonia, and that there was a secular

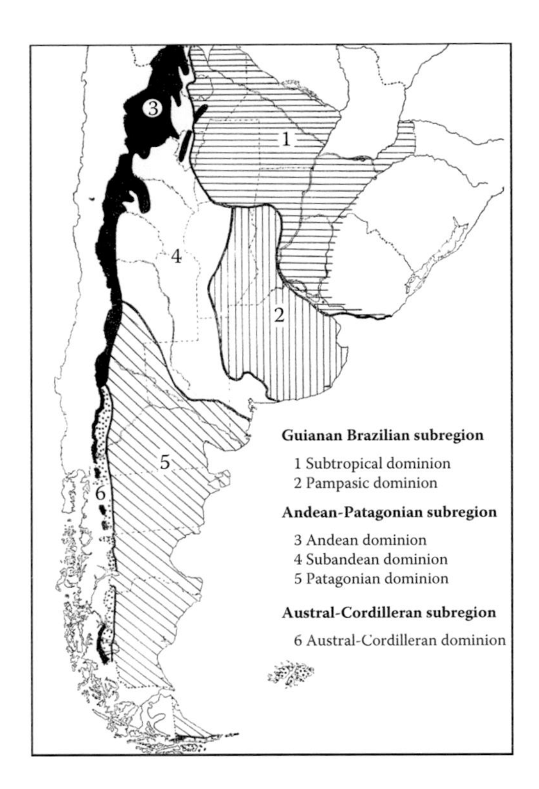

Figure 2.3 Map of the Ringuelet (1961) zoogeographical regionalization of Argentina. (Modified from Ringuelet, R.A., *Physis* [Buenos Aires], 22, 151–170, 1961. With permission.)

dynamism in the boundaries between the Guianan-Brazilian and the Andean-Patagonian subregions.

Peña (1966a,b) provided a zoogeographical regionalization of Chile based mainly on species of the family Tenebrionidae (Coleoptera). His scheme (Figure 2.4) divided the country into 18 regions: High Plateau, Northern Andean Cordillera, Northern Desert, Northern Coast, Intermediate Desert, Coquimban Desert, Central Andean Cordillera, Central Valley, Central Coastal Cordillera, Northern Valdivian Forest, Southern Andean Cordillera, Pehuenar, Valdivian Forest, Valdivian Cordillera, Patagonian Steppe, Aysén Cordillera, Magellan Interoceanic and Southern Pacific. Peña (1966a,b) characterized each of these regions and provided examples of their characteristic plant and insect species.

Rapoport (1968) provided an insightful review of the biogeographical regionalization of the New World, with specific emphasis on the Neotropical region. He divided this region into the Central American (or Mexican), Antillean, Guianan-Brazilian, Andean-Patagonian and

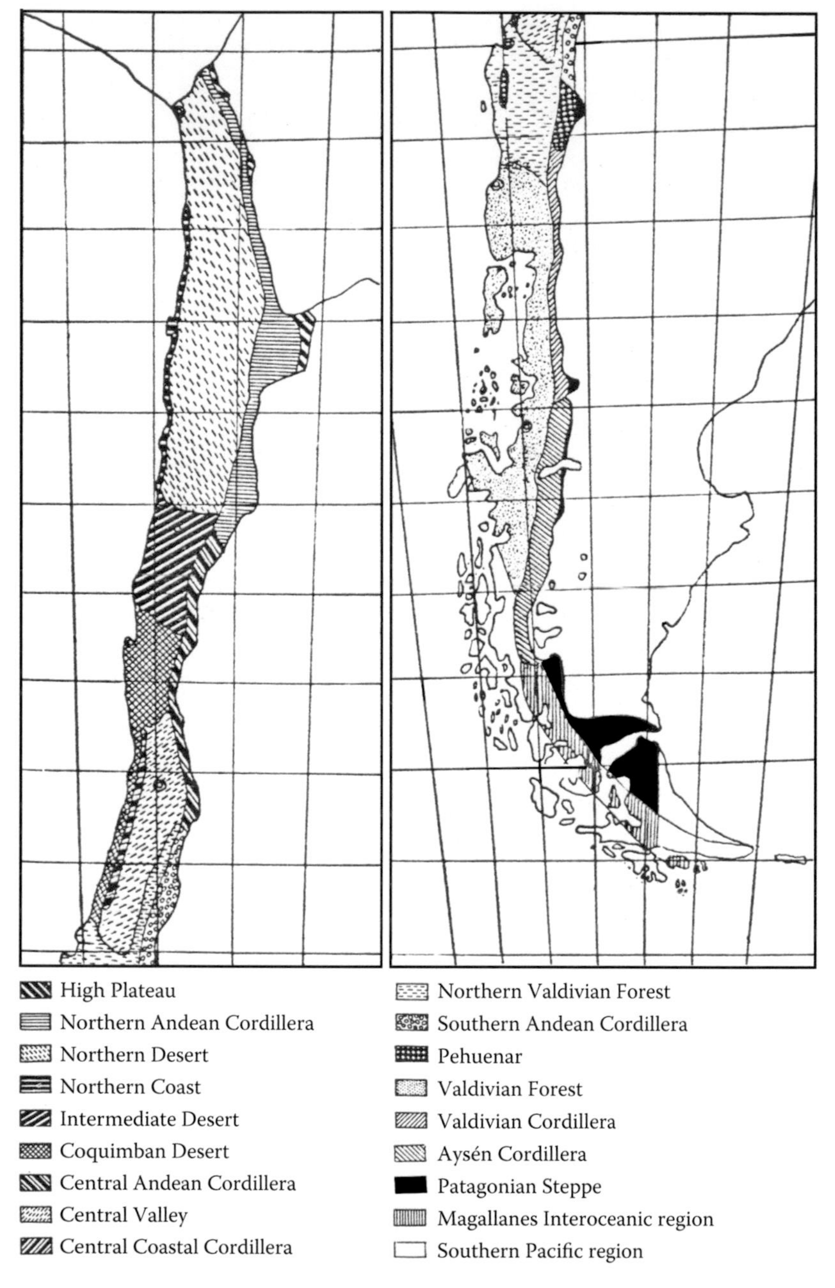

Figure 2.4 Map of the Peña (1966a,b) zoogeographical regionalization of Chile. (Modified from Peña, L.E., *Postilla*, 97, 1–17, 1966. With permission.)

Araucanian subregions, which he characterized in terms of their fauna. Rapoport (1968) considered that the Andean-Patagonian subregion was transitional between the Guianan-Brazilian and the Araucanian subregions. He analyzed specifically the boundary between the Guianan-Brazilian and Andean-Patagonian subregions, which he named the *subtropical line*, discussing the delimitation of previous studies (see also Ruggiero and Ezcurra, 2003).

Fittkau (1969) reviewed the zoogeographical divisions of South America by Mello-Leitão's (1937, 1943) and Cabrera and Yepes' (1940), finding that although they based their divisions on completely different animals (arachnids and mammals, respectively), the areas they delimited were rather similar. Fittkau (1969) accepted the Guianan-Brazilian and Patagonian subregions but considered that their boundaries were difficult to place because of the existence of transition zones between them. He presented a system of 13 zoogeographical provinces, which are similar to the districts of Cabrera and Yepes (1940). His Andean, Subandean, Incasic, Chile and Patagonia provinces correspond to the Andean region.

Kuschel (1969) analyzed the biogeographical regionalization of South America, recognizing the Brazilian and Patagonian subregions (Figure 2.5), based on the distribution of Coleoptera. The Patagonian subregion occupied the Andes above 3000–3500 m and the Patagonian steppe, as well as the Desventuradas, Juan Fernández, Falklands and Tristan da Cunha-Gough islands, corresponding mainly to the Chilean subregion of Wallace (1876). He considered that the area corresponding to the Patagonian subregion has been increasing with the gradual cooling of the Paleogene and Neogene. Additionally, the Pleistocene glaciations forced the plants and animals of the mountain ranges to disperse to the lowlands. This climatic change accompanied a replacement of rainforests to open communities, such as grasslands and steppes. The already adapted southern taxa would have been able to disperse to the newly developed temperate areas in the central and northern Andes. Neotropical taxa also contributed to the northern Andean biota, whereas southern taxa seem to have concentrated towards the southern Andes. Kuschel (1969) also compared the ecology of the beetles in both subregions, referring especially to the ground, forest, litter, soil and aquatic faunas. He highlighted the dominance of ground- and litter-inhabiting beetles in the Patagonian subregion, whereas phloeophagus and xylophagous species are scarce and leaf-mining beetles are completely absent.

Cabrera (1971) provided a detailed phytogeographical regionalization of Argentina, which summarizes his previous studies (Cabrera, 1951, 1953, 1957, 1958) based on extensive floristic data. He recognized 13 provinces (Figure 2.6), which he grouped in two regions and five dominions. The Neotropical region comprised three dominions: Amazonian (Yungas and Parana provinces), Chacoan (Chacoan, Espinal, Prepuna, Monte and

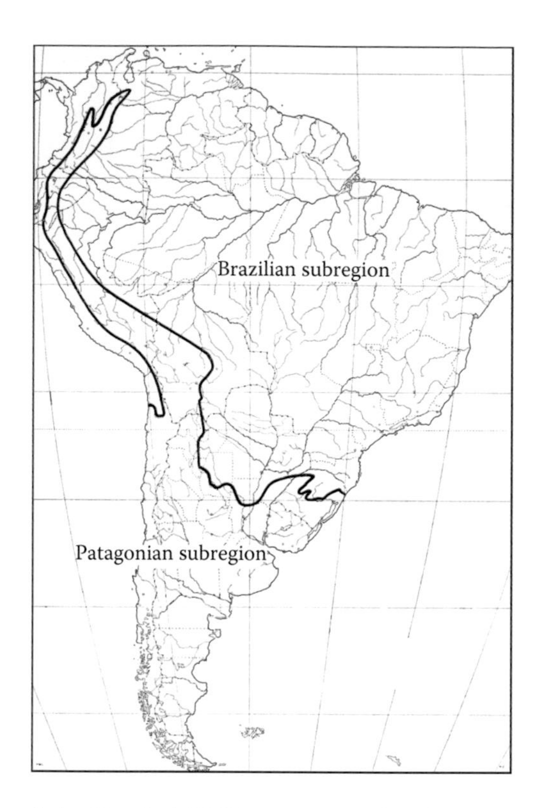

Figure 2.5 Map of the Kuschel (1969) zoogeographical regionalization of South America. (Modified from Kuschel, G., Biogeography and ecology of South American Coleoptera. In: Fittkau, E., J.J. Illies, H. Klinge, G.H. Schwabe and H. Sioli [Eds.], *Biogeography and Ecology in South America, 2.* Junk, The Hague, pp. 709–722, 1969. With permission.)

Pampean provinces) and Andean-Patagonian (High Andean, Puna and Patagonian provinces). The Antarctic region included two dominions: Subantarctic (Subantarctic and Insular provinces) and Antarctic (Antarctic province). The Andean region as recognized in this book corresponds mainly to the Cabrera (1971) Andean-Patagonian and Subantarctic dominions combined. For each phytogeographical province recognized, Cabrera (1971) provided a general description and a characterization of the vegetation, listed the endemic and characteristic species and, in many cases, described districts. Ribichich (2002) analyzed this phytogeographical system, finding some inconsistencies in the hypotheses of Cabrera (1971) concerning the assembly of the flora of some phytogeographical units. Apodaca et al. (2015a) followed closely the scheme of Cabrera (1971).

After decades of separate phytogeographical and zoogeographical regionalizations, Cabrera and Willink (1973) proposed a biogeographical

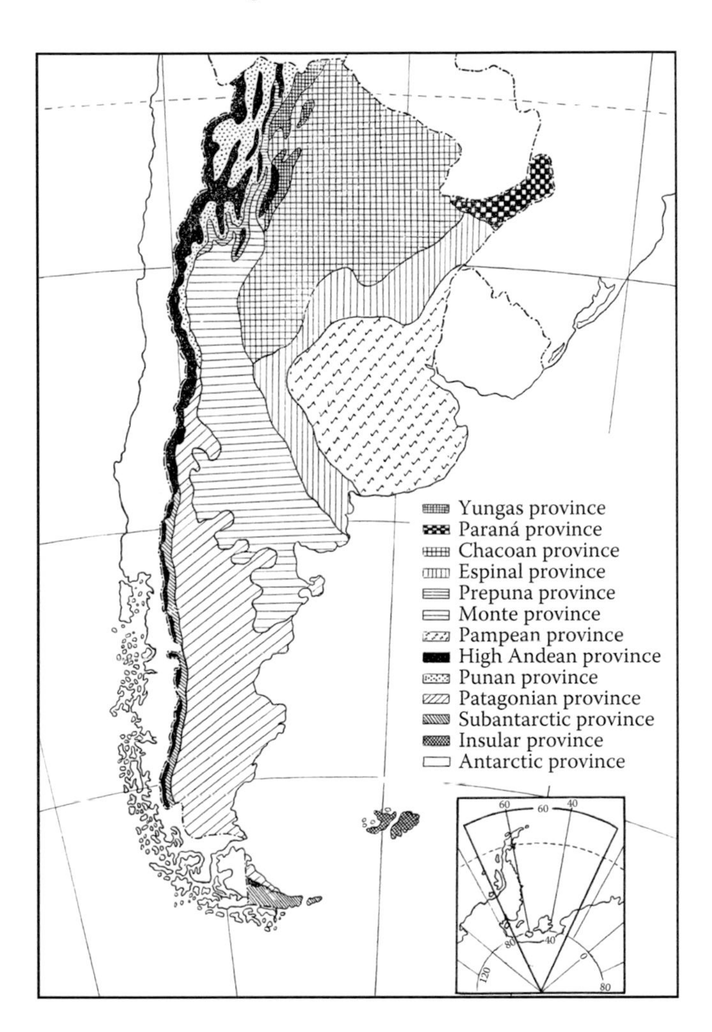

Figure 2.6 Map of the Cabrera (1971) phytogeographical regionalization of Argentina. (Modified from Cabrera, A.L., *Boletín de la Sociedad Argentina de Botánica*, 14, 1–42, 1971. With permission.)

regionalization of Latin America based on plant and animal taxa. Within South America, they recognized 23 provinces for the Neotropical region (Figure 2.7), which they classified into the Caribbean, Amazonian, Guianan, Chacoan and Andean-Patagonian dominions. The Cabrera and Willink (1973) Neotropical region did not include the southernmost area of South America, which they assigned to the Subantarctic dominion of the Antarctic region, but include the Subantarctic and Insular provinces. Many studies adopted this scheme and found it useful for characterizing and naming geographical areas of many plant and animal taxa.

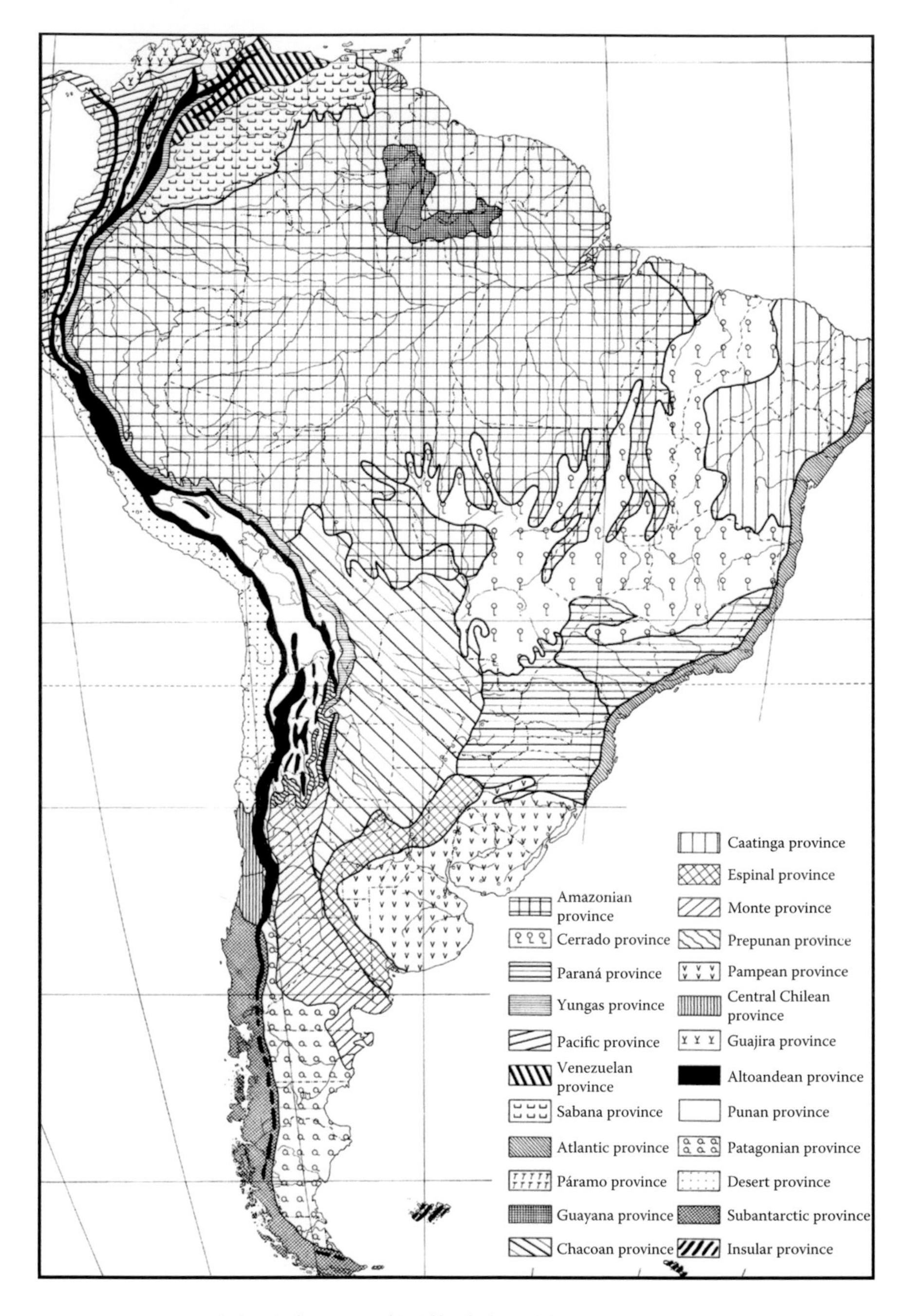

Figure 2.7 Map of the Cabrera and Willink (1973) biogeographical regionalization of South America. (Modified from Cabrera, A.L. and Willink, A., *Biogeografía de América Latina*. Monografía 13, Serie de Biología, OEA, Washington, DC, 1973.)

Müller (1973) analyzed the geographical distribution of Neotropical vertebrate taxa. He identified 40 dispersal centers. The North Andean, Andean Pacific, Puna, Monte, *Nothofagus* and Patagonian centers corresponded to the Andean region. For each center, Müller (1973) mapped the distributional areas of several endemic species. Additionally, for some centers, he recognized nested subcenters. Some of these centers and subcenters are coincident with areas recognized by other studies, whereas other centers represent smaller nested units.

Artigas (1975) conducted a zoogeographical study of Chile based on a quantitative analysis of 903 animal species, including insects, arachnids, crustaceans and vertebrates. He provided an interesting historical review of some previous regionalizations (e.g., Goetsch, 1931; Osgood, 1943; Kuschel, 1960; Mann, 1960; Peña, 1966a; Di Castri, 1968; O'Brien, 1971; Cekalovic, 1974), presenting maps redrawn to the same scale to facilitate comparison. His scheme (Figure 2.8) divided Chile into five areas: Atacaman, Santiagan, Valdivian, Aysenian and Magellanian. Within them, Artigas (1975) further recognized 31 smaller zones, which are treated herein mostly as districts.

Ringuelet (1975) conducted a global analysis of South American freshwater fishes. After discussing several previous schemes, Ringuelet (1975) proposed 20 provinces which he grouped into two subregions and seven dominions. His Austral subregion, equivalent to the Andean region, included the Chilean and Patagonian provinces.

Udvardy (1975) proposed a unified regionalization of the world, intended for biogeographical and conservation purposes in which he recognized the Palearctic, Nearctic, Africotropical, Indomalayan, Oceanian, Australian, Antarctic and Neotropical kingdoms. Within the Neotropical kingdom, Udvardy (1975) recognized 47 provinces. The Valdivian Forest, Chilean *Nothofagus*, Chilean Araucaria Forest, Chilean Sclerophyll, Pacific Desert, Monte, Patagonian, North Andean, Colombian Montane, Puna, Southern Andean and Lake Titicaca provinces corresponded to the Andean region.

Rivas-Martínez and Tovar (1983) proposed that the Andean highlands, the Pacific coast between 5°S and 38°S, and Patagonia constitute a biogeographical unit within the Neotropical kingdom, which they named Andean subkingdom. The other biogeographical units recognized by them for South America were the Caribbean-Amazonian subkingdom, the group of Chacoan regions and the Caatinga and Subantarctic regions (Figure 2.9). Within the Andean subkingdom, they recognized the Páramo, Puna, Pacific Desert, Central Chilean and Patagonian regions, each characterized by different endemic plant genera.

Takhtajan (1986) provided a phytogeographical regionalization of the world. Within the Neotropical kingdom, Takhtajan recognized seven regions: Caribbean, Venezuela-Surinam, Amazon, Central Brazilian,

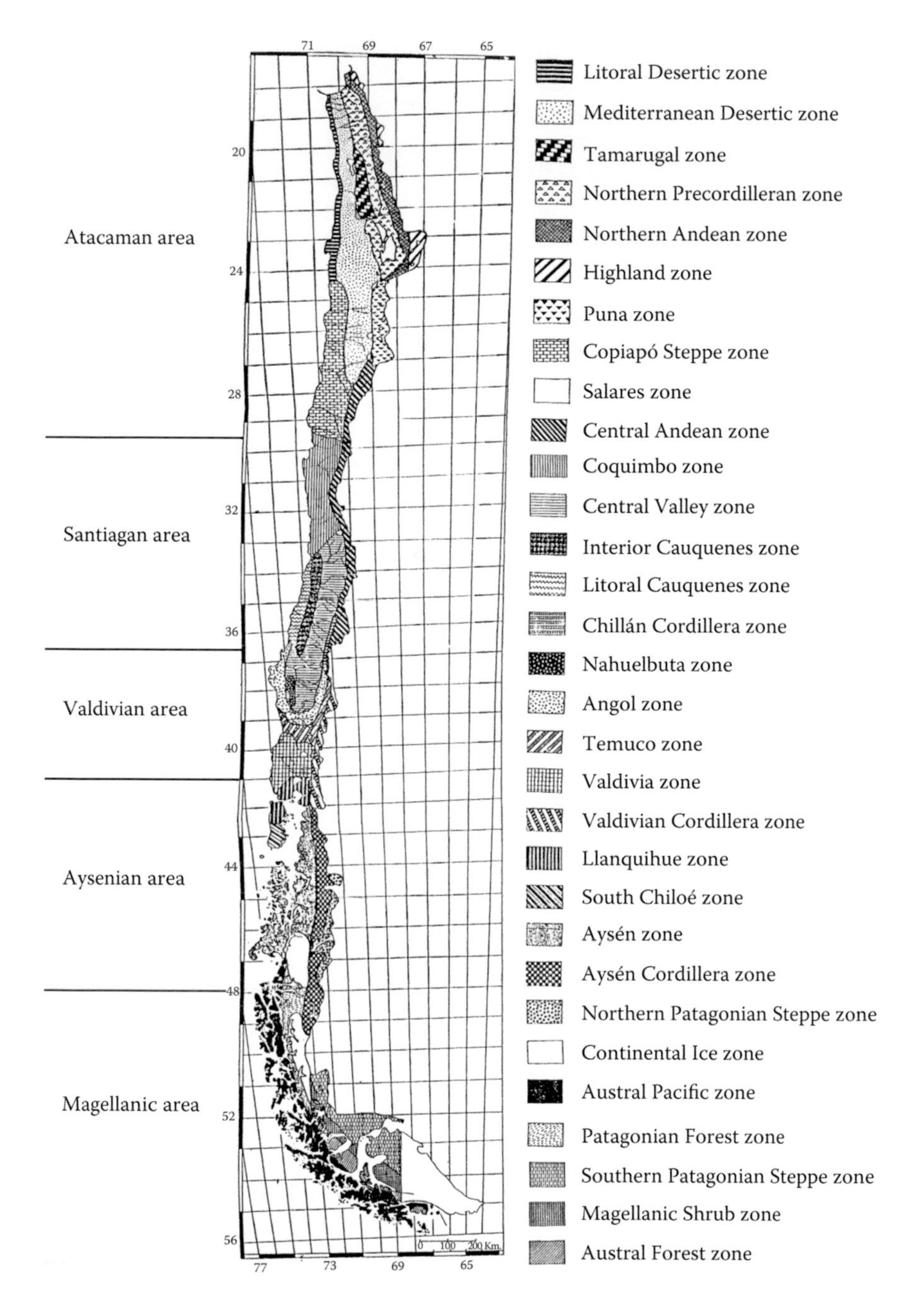

Figure 2.8 Map of the Artigas (1975) zoogeographical regionalization of Chile. (Modified from Artigas, J.N., *Gayana, miscelánea*, 4, 1–25, 1975. With permission.)

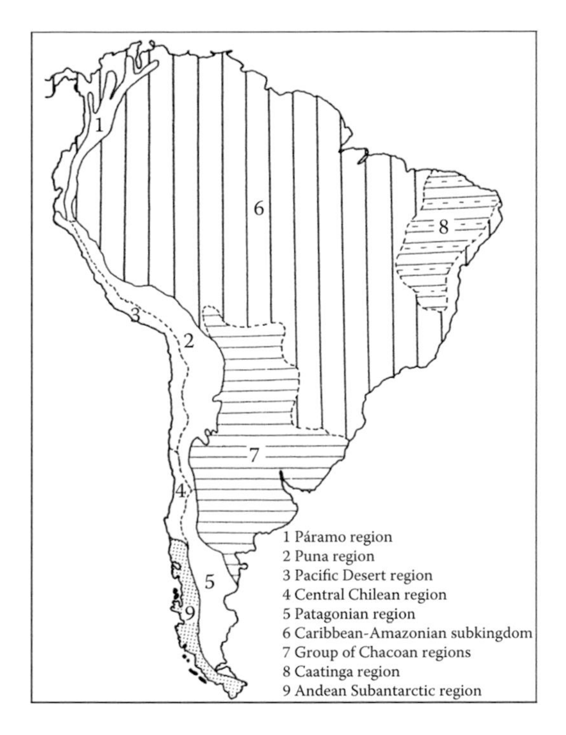

Figure 2.9 Map of the Rivas-Martínez and Tovar (1983) phytogeographical regionalization of South America. (Modified from Rivas-Martínez, S. and Tovar, O., *Collectanea Botanica* [Barcelona], 14, 515–521, 1983. With permission.)

Pampas, Andean and Fernandezian. The Andean region includes the Galapagos and Cocos islands, the flanks of the Andes, the Atacama Desert and the Chilean sclerophyll zone; whereas the Fernandezian region includes the Juan Fernández and Desventuradas islands. The latter is considered herein to belong to the Andean region.

Rivas-Martínez and Navarro (1994) recognized a Neotropical-Austroamerican kingdom. They recognized 49 provinces, which they classified into subkingdoms and regions. The Neotropical subkingdom included the Caribbeo-Mexican region, Colombian-Mesoamerican region, Venezuelan region, Amazonian region, Brazilian-Paraná region, Andean region (Peruvian, Bolivian, Argentinean-Atacaman, Monte, Páramo and Yunga provinces), Chacoan region and Peruvian Pacific Desert region (Peruvian Desert and Atacama Hyperdesert provinces). The Austroamerican subkingdom included the Pampean region, Mesochilean-Patagonian region (Desert Mesochilean and Central Chilean provinces), Andean-Patagonian subregion (Mediterranean Andean, Septentrional Patagonian and Meridional Patagonian provinces) and Valdivian-Magellanic region (Valdivian, Austroandean, Fuegian, Juan Fernández

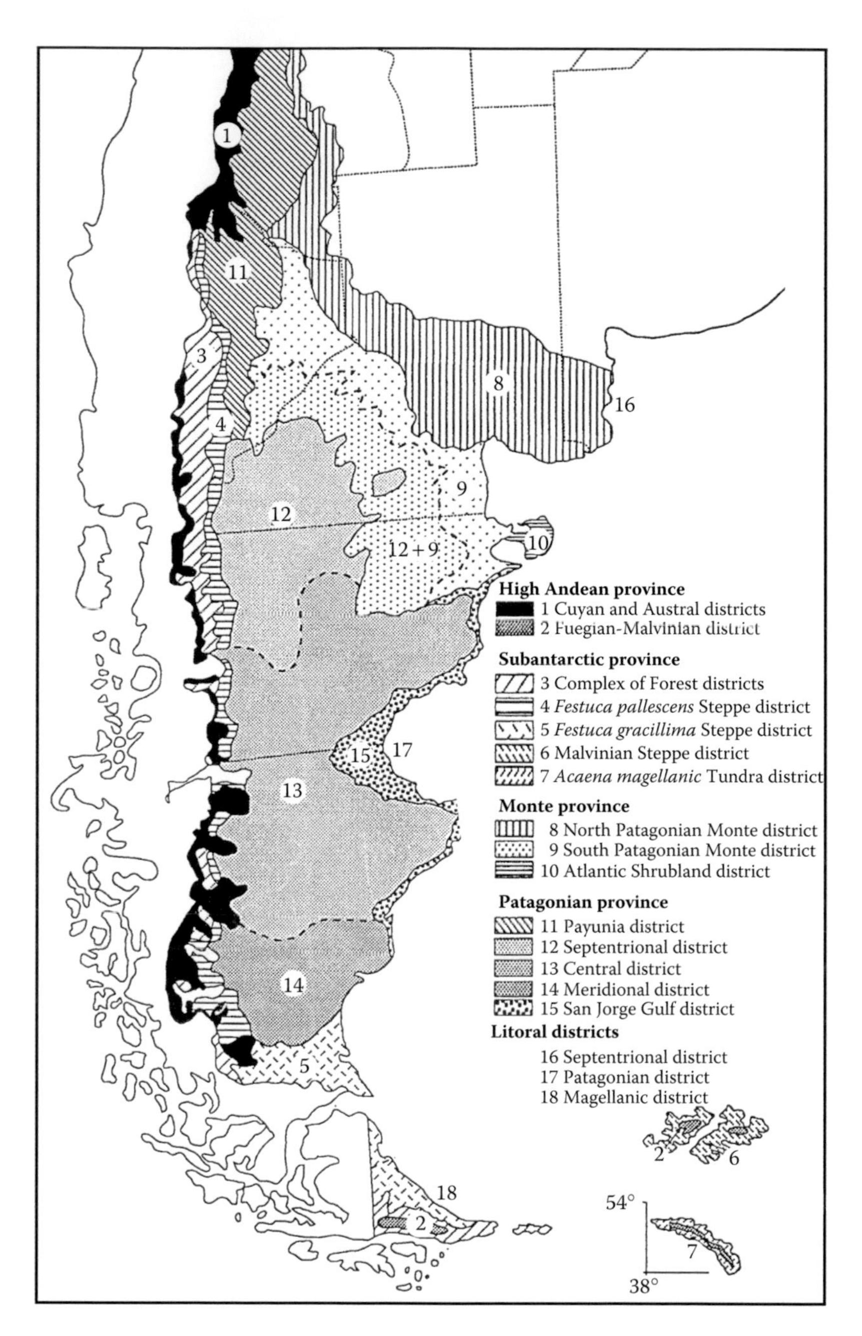

Figure 2.10 Map of the Roig (1998) phytogeographical regionalization of Argentinean Patagonia. (Modified from Roig, F.A., La vegetación de la Patagonia. In: Correa, M.N. [Ed.], *Flora Patagónica, tomo VIII[1]*, INTA, Colección Científica, Buenos Aires, pp. 48–166, 1998. With permission.)

Islands and Antarctic provinces). Rivas-Martínez et al. (2011) presented a revised version of their regionalization with the same provinces but arranged in a slightly different scheme.

Dinerstein et al. (1995) proposed a system of 178 ecoregions for Latin America and the Caribbean, which they classified based on their major ecosystem types into tropical broadleaf forests (tropical moist broadleaf forests and tropical dry broadleaf forests), conifer/temperate broadleaf forests (temperate forests and tropical and subtropical coniferous forests), grasslands/savannas/shrublands (grasslands, savannas and shrublands, flooded grasslands and montane grasslands) and xeric formations (Mediterranean scrub and *restingas*). Dinerstein et al. (1995) grouped the ecoregions into seven major bioregions: Caribbean, Northern Andes, Orinoco, Amazonia, Central Andes, Eastern South America and Southern South America.

Roig (1998) provided a complete phytogeographical regionalization of Argentinean Patagonia (including areas assigned to the Subantarctic and Patagonian subregions as recognized in this book, as well as to the South American transition zone). He characterized several types of temperate forests, grass steppes, arid and semiarid shrub steppes, moorlands, tundra and littoral vegetation. Roig's (1998) phytogeographical scheme (Figure 2.10) divided the area basically into four phytogeographical provinces: Monte, Patagonian, High Andean and Subantarctic.

I synthesized some previous schemes and, mostly based on track analyses of plant and animal taxa, provided a biogeographical regionalization of Latin America and the Caribbean (Morrone, 2001a). My scheme divided the Andean region into four subregions (Paramo-Punan, Central Chilean, Subantarctic and Patagonian) and 14 provinces. Later, I treated the Paramo-Punan subregion as the South American transition zone, adding the Monte province to it (Morrone, 2006, 2015b). More recently, Martínez et al. (2017) recognized the Comechingones province within the South American transition zone. The resulting regionalization of the Andean region, adopted in this book, consisted of three subregions, one transition zone and 16 provinces.

chapter three

The Austral kingdom

The Austral kingdom corresponds to the southern temperate areas of South America, South Africa, Australia, New Zealand, New Guinea, New Caledonia and Antarctica (Newbigin, 1950; Kuschel, 1964; Rapoport, 1968; Udvardy, 1975; Moreira-Muñoz, 2007; Morrone, 2015b,c). It has been treated as a kingdom (Engler, 1882, 1899; Drude, 1890; Diels, 1908; Good, 1947; Udvardy, 1975; Müller, 1986; Takhtajan, 1986; Morrone, 2002, 2015c; Moreira-Muñoz, 2007, 2011) or a region (Newbigin, 1950; Monrós, 1958; Kuschel, 1964; Rapoport, 1968; Cabrera and Willink, 1973). Treviranus (1803) noted the relationships among these widely separated areas and recognized an *Antarctic flora* distributed in Chile, Magallanes, Tierra del Fuego and New Zealand. From a palaeogeographical viewpoint, the Austral kingdom corresponded to eastern Gondwanaland (Crisci et al., 1993) or temperate Gondwanaland (Amorim et al., 2009).

This chapter deals with the Austral kingdom to provide an appropriate context for the classification of the Andean region. I provide examples of taxa endemic and characteristic of the kingdom and discuss the biotic relationships of the regions assigned to it (Andes, Australia, New Zealand, etc.). This chapter provides a geological account of the Austral areas and a geological area cladogram summarizing their relationships. I give specific reference to an event-based biogeographic analysis of the Austral kingdom.

Austral kingdom

Old Oceanic kingdom—Engler, 1879: 346 (regionalization), 1882: 346 (regionalization).

Antarctic kingdom—Drude, 1890: 158 (regionalization); Diels, 1908: 154 (regionalization); Good, 1947: 18 (regionalization); Takhtajan, 1986: 253 (regionalization); Fleming, 1987: 199 (regionalization); Udvardy, 1987: 189 (regionalization); Cox, 2001: 517 (regionalization).

Austral kingdom—Engler, 1899: 149 (regionalization); Morrone, 2002: 150 (regionalization); Glasby, 2005: 241 (regionalization); Moreira-Muñoz, 2007: 1651 (regionalization), 2011: 137 (regionalization); Ebach et al., 2013: 318 (regionalization); Omad, 2014: 573 (systematic revision); Morrone, 2015b: 207 (regionalization), 2015c: 85 (regionalization);

Brion and Ezcurra, 2017: 2 (ecoregionalization); Escalante, 2017: 351 (parsimony analysis of endemicity).
Austral region—Newbigin, 1950: 145 (regionalization); Kuschel, 1964: 443 (regionalization); Blöcher and Frahm, 2002: 81 (floristics).
Neantarctic region—Monrós, 1958: 145 (regionalization).
Holantarctic region—Rapoport, 1968: 88 (regionalization).
Antarctic region—Cabrera and Willink, 1973: 96 (regionalization); Granara de Willink, 2014: 254 (faunistics).
Archinotic kingdom—Müller, 1986: 18 (regionalization).
Holantarctic kingdom—Takhtajan, 1986: 276 (regionalization).

Endemic and characteristic taxa

Moreira-Muñoz (2007, 2011) listed 14 plant families that are endemic to the Austral kingdom: Araucariaceae, Asteliaceae, Atherospermataceae, Berberidopsidaceae, Calceolariaceae, Centrolepidaceae, Corsiaceae, Escalloniaceae, Griseliniaceae, Luzuriagaceae, Nothofagaceae, Proteaceae, Restionaceae and Stylidaceae. Several studies reassessed the biogeography of *Nothofagus*, the single representative of the Nothofagaceae, which studies considered classically a key element of the Austral biota (e.g., van Steenis, 1972). Cook and Crisp (2005) considered that the evolutionary biogeographical history of this genus included several major transoceanic dispersal events. Heads (2006) argued that the distribution of the subgenera of *Nothofagus* was compatible with vicariant differentiation and no dispersal was necessary for explaining their distribution since at least the Upper Cretaceous. It seems that the biogeographical history of *Nothofagus* is much more complex than previously found, and that its biotic history involved both vicariance and dispersal events.

Several animals are endemic to the Austral kingdom (Crisci et al., 1991; Giribet and Edgecombe, 2006; Boyer et al., 2007; Liebherr et al., 2011). Some of them are *Oxelytrum-Ptomaphila* (Coleoptera: Silphidae); Diamesinae and Podonominae (Diptera: Chironomidae); Pseudopsinae (Coleoptera: Staphylinidae); Migadopini, Zolini, Creobiina, Baripina, Nothobroscina, Ceroglossini and Metiini (Coleoptera: Carabidae); *Eriococcus* and *Madarococcus* (Hemiptera: Eriococcidae); *Paralamyctes* (Chilopoda: Henicopidae); and Pettalidae (Opiliones).

Biotic relationships

Some studies provided contrasting proposals to explain the biotic connections between the southern continents assigned to the Austral kingdom, basically favoring vicariance or dispersal explanations. Treviranus (1803) and Hooker (1844–1860) were the first researchers to note that the floras of the southern continents were similar, hypothesizing a former connection between them. Engler (1879, 1882) would later group southern South America, New Zealand's South Island, the Subantarctic islands, most of Australia and South Africa in the Austral kingdom, recognizing an interesting biotic pattern and a vicariant explanation a long time before the acceptance of continental drift (Cox and Moore, 2005).

Amorim and Tozzoni (1994) proposed a general area cladogram of the world based on the cladistic biogeographical analysis of several insect, vertebrate and plant taxa. They replaced the terminal species by pre-defined areas of endemism (Figure 3.1a) and, based on the comparison of the area cladograms, obtained a general area cladogram (Figure 3.1b). This general

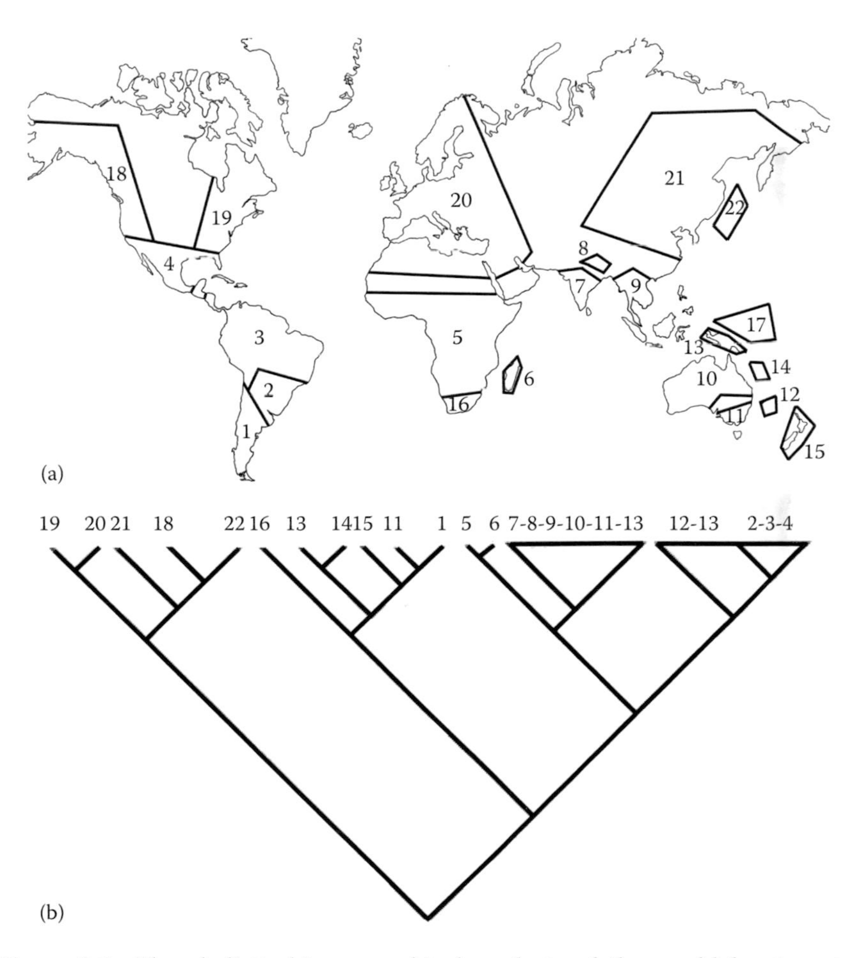

Figure 3.1 The cladistic biogeographical analysis of the world by Amorim and Tozoni (1884). (a) Areas analyzed and (b) general area cladogram obtained. (Modified from Amorim, D.S. and Tozoni, H.S., *Revista Brasileira de Entomologia*, 38, 517–543, 1994. With permission.)

area cladogram differed from previous analyses in the presence of two major Gondwanaland landmasses: one tropical and the other temperate. Tropical Gondwanaland included northern South America, Africa, southeast Asia, New Guinea, northwestern Australia and Oceania. Temperate Gondwanaland included southern South America, southern Africa, southeastern Australia, New Guinea, New Caledonia and New Zealand.

Morrone (2004b) represented the relationships of the biogeographical regions of the world with in a general area cladogram (Figure 3.2). According to it, the Andean region belonged to the Austral kingdom, exhibiting biotic relationships with the Cape, Antarctic and Australian regions. In addition to these relationships with other Austral areas, the Andean region exhibited biotic relationships with the Neotropical region through the South American transition zone.

Moreira-Muñoz (2007) analyzed the floristic similarity of the Austral areas based on genera of vascular plants suggesting the existence of an ancient flora in these areas tied to ancient paleogeographical events. Moreira-Muñoz (2007, 2011) characterized an Australasiatic floristic element comprised of 58 plant genera from Australasia, southern South America and the Pacific islands, which he traced to Gondwanaland. Within this assemblage, Moreira-Muñoz (2011) further characterized three generalized tracks:

1. *Austral Antarctic track*: Genera occurring in southernmost South America, New Zealand, eastern Australia and Tasmania, for example, *Luzuriaga* (Luzuriagaceae), *Eucryphia* (Cunoniacae), *Prumnopitys* (Podocarpaceae), *Jovellana* (Calceolariacae) and *Raukaua* (Araliaceae)
2. *Tropical Pacific track*: Genera occurring in southernmost South America, New Zealand, eastern Australia and/or Tasmania, and

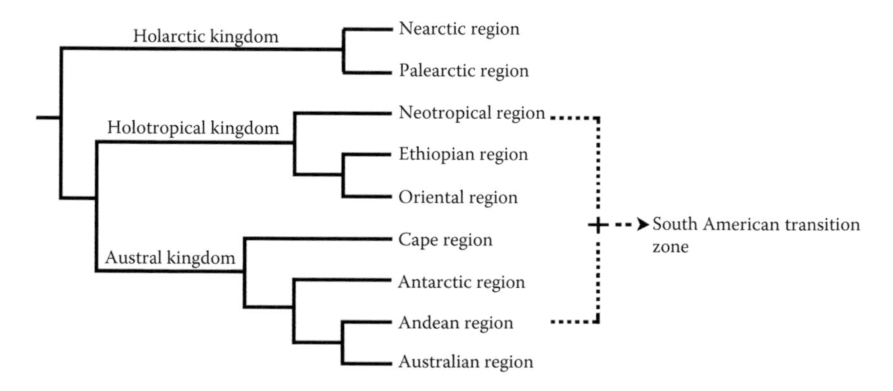

Figure 3.2 General area cladogram depicting the relationships of the biogeographical regions of the world. (Modified from Morrone, J.J., *Neotropica*, 42, 103–114, 1996a. With permission.)

extending to New Guinea and east Asia, for example, *Abrotanella* and *Lagenophora* (Asteraceae), *Araucaria* (Araucariaceae), *Nothofagus* (Nothofagaceae), *Coprosma* (Rubiaceae) and *Hebe* (Plantaginaceae).

3. *Circum-Austral track*: Genera extending their distribution to southern Africa and Madagascar, for example, *Ficinia* (Cyperaceae), *Nertera* (Rubiaceae), *Rumohra* (Dryopteridaceae) and *Wahlenbergia* (Campanulaceae).

Giribet and Edgecombe (2006) analyzed the geographical distribution of the species of the centipede genus *Paralamyctes* (Figure 3.3) distributed in small, relictual areas of southern South America, South Africa, Madagascar, Australia, Tasmania and New Zealand. Based on a phylogenetic hypothesis using both morphological and molecular data, they found relationships between New Zealand, Australia (Queensland) and Tasmania (subgenus *Haasiella*); Australia (New South Wales), New Zealand and Chile (subgenus *Edgecombegdus*); and southern Africa, Madagascar, New Zealand, tropical-warm temperate Australia and southern India

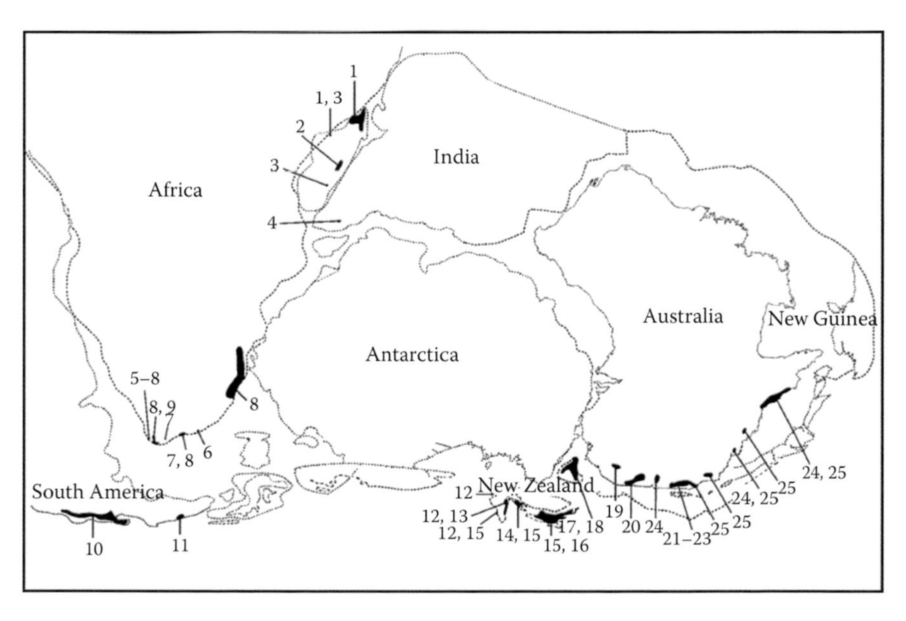

Figure 3.3 Distribution of the species of *Paralamyctes* (Chilopoda: Henicopidae), represented on a Triassic map of Gondwanaland. (Modified from Giribet, G. and Edgecombe, G.D., *Biological Journal of the Linnean Society*, 89, 65–78, 2006. With permission.) 1, *Paralamyctes quadridens*; 2, *P. tridens*; 3, *P. bipartitus*; 4, *P. newtoni*; 5, *P. prendinii*; 6, *P. asperulus*; 7, *P. weberi*; 8, *P. spenceri*; 9, *P. levigatus*; 10, *P. chilensis*; 11, *P. wellingtonensis*; 12, *P. trailli*; 13, *P. halli*; 14, *P. rahuensis*; 15, *P. validus*; 16, *P. harrisi*; 17, *P. subicolus*; 18, *P. mesibovi*; 19, *P. ginini*; 20, *P. grayi*; 21, *P. cassisi*; 22, *P. hornerae*; 23, *P. neverneverensis*; 24, *P. cammooensis*; 25, *P. monteithi*.

(subgenus *Paralamyctes*). In general, Giribet and Edgecombe (2006) noted agreement with the standard hypothesis for Gondwanan relationships in that South Africa resulted basal to other areas, but also found that the relationship between Madagascar and New Zealand was unorthodox. Furthermore, they noted that their analysis questioned the naturalness of Australia because southern New South Wales and Tasmania resolved outside a clade with tropical and warm temperate parts of Australia, Madagascar and New Zealand.

Boyer et al. (2007) analyzed the geographical distribution of a group of Opiliones distributed in all the major continental landmasses based on the results of a molecular phylogenetic analysis. They found that the family Pettalidae is restricted to temperate Gondwanaland, with species occurring in Chile, South Africa, Madagascar, Sri Lanka, western Australia, Queensland and New Zealand. Their estimated age (178–215 m.y.a.) indicates that this family diversified prior to the break-up of Gondwanaland.

Amorim et al. (2009) reviewed the alternative proposals to explain the biogeographical history of the circumantarctic areas, for example, southern South America, southern Africa, southern and southeastern Australia, Tasmania, New Zealand, New Hebrides, New Caledonia and a few other areas of Gondwanan origin. They considered that the vicariance hypothesis of a former Gondwanan biota for explaining southern connections was not completely compatible with different sources of evidence (e.g., fossils and molecular dating). After analyzing the distributional patterns of several taxa, they found that the southern temperate areas host a combination of truly Gondwanan and post-Gondwanan taxa. They classified Austral taxa in four categories, according to the kind of evidence available: category 1, with both Jurassic fossils and South African living taxa; category 2, Jurassic fossils but no South African living taxa; category 3, South African living taxa, but no Jurassic fossils; and category 4, lacking both Jurassic fossils and South African living taxa. Taxa that fit categories 1 and 2 corresponded clearly to Gondwanan groups, taxa in category 3 were candidates for truly Gondwanan taxa but pending confirmation, and taxa in category 4 were less probable to represent Gondwanan groups. Amorim et al. (2009) concluded that sympatrid allochronic taxa characterized the Austral kingdom, meaning that they belonged to different temporal layers (Gondwanan and post-Gondwanan). Further studies will distinguish the old Gondwanan taxa from those that dispersed to Gondwanaland later.

Ribeiro and Eterovic (2011) found several transpacific tracks for species of crane flies (Diptera: Tipulomorpha). They hypothesized that although recent Mesozoic and Cenozoic events of continental breakup drove the biotic differentiation of some Austral taxa at least part of them evolved allopatrically in response to older vicariant events.

Regionalization

The Austral kingdom (Figure 3.4) comprises four regions: Cape (South Africa), Andean, Australian (Australia, Tasmania, New Zealand, New Guinea and New Caledonia) and Antarctic (Morrone, 2015c).

Geological scenario

The following account of the geological evolution of the Austral areas is based on Sanmartín and Ronquist (2004). They summarized the main geological events affecting the Austral areas in terms of a geological area cladogram (Figure 3.5).

During the Triassic, Gondwanaland formed the southern portion of Pangaea. It included two different biotic provinces: Northern Tropical

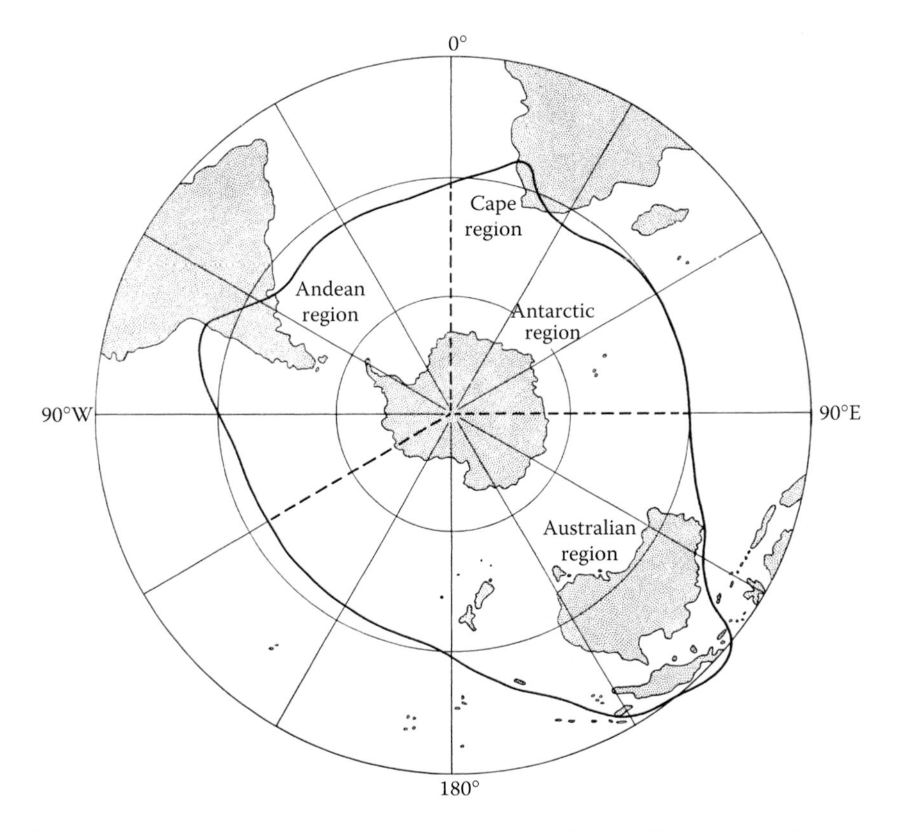

Figure 3.4 Map of the Austral kingdom. (Modified from Kuschel, G., Problems concerning an Austral region. In: Gressitt, J.L., C.H. Lindroth, F.R. Fosberg, C.A. Fleming and E.G. Turbott [Eds.], *Pacific Basin Biogeography: A Symposium, 1963 [1964]*, Bishop Museum Press, Honolulu, pp. 443–449, 1964. With permission.)

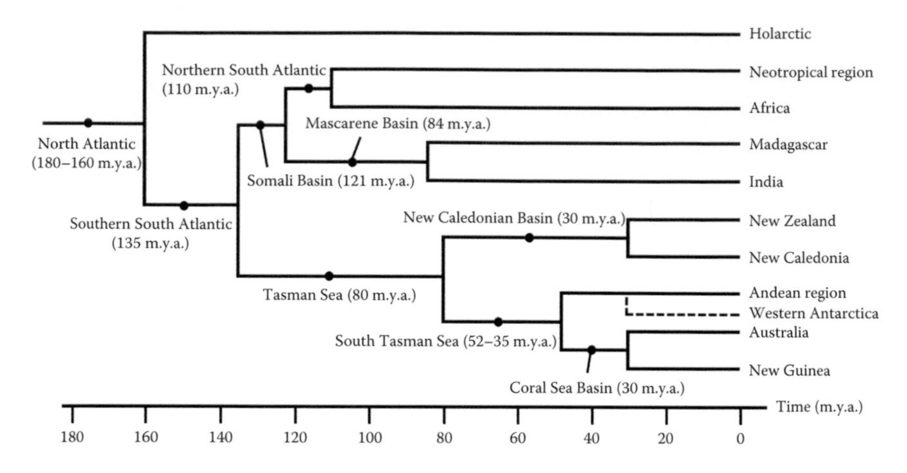

Figure 3.5 Geological area cladogram of the Austral areas. (Modified from Sanmartín, I. and Ronquist, F., *Systematic Biology*, 53, 216–243, 2004. With permission.)

Gondwanaland, including northern South America, Africa, Madagascar, India, northern Australia and New Guinea, with a tropical climate; and Southern Temperate Gondwanaland, including southern South America, Australia, Antarctica, New Zealand and New Caledonia and southern Africa.

Gondwanaland started to breakup in the Jurassic when rifting began between India and Australia-east Antarctica about 165 m.y.a. Then, Madagascar and India broke away from Africa and began moving southeast with Madagascar attaining its present position in the Early Cretaceous (121 m.y.a.). India separated from Madagascar in the Late Cretaceous (88–84 m.y.a.) with the opening of the Mascarene Basin and began drifting northward and finally collided with Asia at about 50 m.y.a. South America began to separate from Africa in the Early Cretaceous (135 m.y.a.) with the opening of the South Atlantic Ocean. Northern South America and Africa remained connected until Mid-Late Cretaceous (110–95 m.y.a.) when a transform fault separated them. As a result, Africa started drifting northeast and collided with Eurasia in the Paleocene (60 m.y.a.), and southern South America drifted southwest into contact with Antarctica.

New Zealand, Australia, South America and Antarctica remained connected until the Late Cretaceous: east Antarctica was adjacent to southern Australia, and New Zealand and southern South America were in contact with west Antarctica. New Zealand was the first to break away from Antarctica in the late Cretaceous, about 80 m.y.a. The continental block of Tasmantis, including New Zealand and New Caledonia, broke away from west Antarctica and moved northwest opening the Tasman Sea. During most of the Paleogene, a marine transgression submerged New Zealand and New Caledonia with probably most of New Zealand beneath water by the Mid-Oligocene.

New Zealand and New Caledonia finally separated in the Paleogene (40–30 m.y.a.), when the Norfolk Ridge foundered opening the New Caledonian Basin. Australia and South America remained in contact across Antarctica until the Eocene. Australia began to separate from Antarctica in the Late Cretaceous (90 m.y.a.), but both continents remained in contact along Tasmania and complete separation did not occur until the late Eocene (35 m.y.a.) with the opening of the South Tasman Sea. Southern South America and Antarctica remained in contact through the Antarctic Peninsula until the Oligocene (30–28 m.y.a.) when the Drake Passage opened allowing the establishment of the Antarctic Circumpolar Current and the onset of the first Antarctic glaciation. Following its separation from Antarctica, Australia began to drift toward Asia. New Guinea then joined to the northern margin of the Australian Plate, although only the southern margin of New Guinea was emergent at that time. The collision of the Australian and Pacific plates in the Oligocene (30 m.y.a.) initiated the uplift of New Guinea, but by the Early Miocene much of southern New Guinea was resubmerged.

Case study: Event-based biogeographical analysis of the Austral areas

Title: "Southern Hemisphere biogeography inferred by event-based models: Plant versus animal patterns" (Sanmartín and Ronquist, 2004).

Goal: To examine the relative roles played by vicariance and dispersal in shaping Southern Hemisphere biotas by analyzing a large data set to test to what extent southern biogeographical patterns fit the break-up sequence of Gondwanaland and to search for the existence of concordant dispersal patterns in the biota of the southern continents.

Location: The world.

Methods: The study analyzed the cladograms of 19 plant and 54 animal taxa using parsimony-based tree fitting in conjunction with permutation tests. To discuss the distributional patterns, the study presented the most commonly discussed patterns postulated for the Southern Hemisphere (Figure 3.6): the southern Gondwanaland pattern (Figure 3.6a) explained by vicariance showed African taxa diverging basally, followed by a New Zealand clade and then a southern South America-Australian clade; the plant southern pattern (Figure 3.6b) and the inverted southern pattern (Figure 3.6c) were incongruent with the geological scenario and explained by dispersal; the northern Gondwanaland pattern (Figure 3.6d) explained by either vicariance or dispersal showed the connection between the areas that were part of tropical Gondwanaland; the tropical Gondwanaland pattern (Figure 3.6e), also explained by vicariance followed the breakup sequence of the western part of northern Gondwanaland; and the

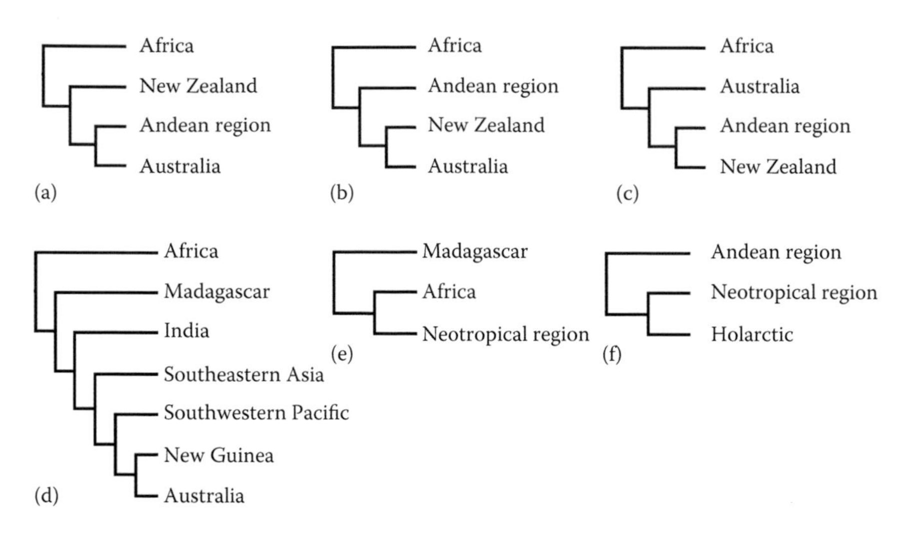

Figure 3.6 Area cladograms representing different patterns of area relationships in the Southern Hemisphere. (Modified from Sanmartín, I. and Ronquist, F., *Systematic Biology*, 53, 216–243, 2004. With permission.) (a) Southern Gondwanaland pattern, (b) plant southern pattern, (c) inverted southern pattern, (d) northern Gondwanaland pattern, (e) tropical Gondwanaland pattern and (f) transamerican pattern.

transamerican pattern (Figure 3.6f) followed earlier studies that suggested dividing South America into two areas with different biogeographical affinities. The areas used in the analysis corresponded to the following historically persistent landmasses: Africa (excluding the region north of the Sahara), Madagascar, India (including Nepal, Tibet and Sri Lanka), Australia (including Tasmania), New Zealand, New Caledonia, New Guinea, southern South America (Andean region), northern South America (Neotropical region), southeast Asia, southwestern Pacific and the Holarctic.

The study used a parsimony-based tree fitting approach under the four-event model (Page, 1995) implemented in TreeFitter 1.2 (Ronquist, 2002). The events considered in the model are vicariance (allopatrid speciation), duplication (speciation within an area), dispersal and extinction. After the study associated each event to a cost, it was possible to fit the cladogram of a taxon and the geographical distributions of its species to any given area cladogram, producing a reconstruction of the biogeographical history of the taxon according to the area cladogram. Each reconstruction specified the ancestral distributions and the biogeographical events that could produce the observed distributional patterns. Then, the study searched the optimal area cladogram. The cost of the most parsimonious cladogram fitting one (or more) taxa to a given area cladogram represented a measure of the fit of that area cladogram to the taxon. The study conducted three different analyses: (1) identify the best area cladogram fitting the animal and plant taxa, (2) identify the optimal cladogram for each taxon in the data matrix and (3) compare the cladogram obtained with each biogeographical pattern and fit the geological cladogram (Figure 3.5) to each taxon analyzed.

Results: Figure 3.7 shows the optimal area cladograms obtained for the animal and plant data sets found. The animal cladogram (Figure 3.7a) has a pattern in agreement with the southern Gondwanaland pattern except for the position of Africa, which appears closely related to other areas. The study obtained the same

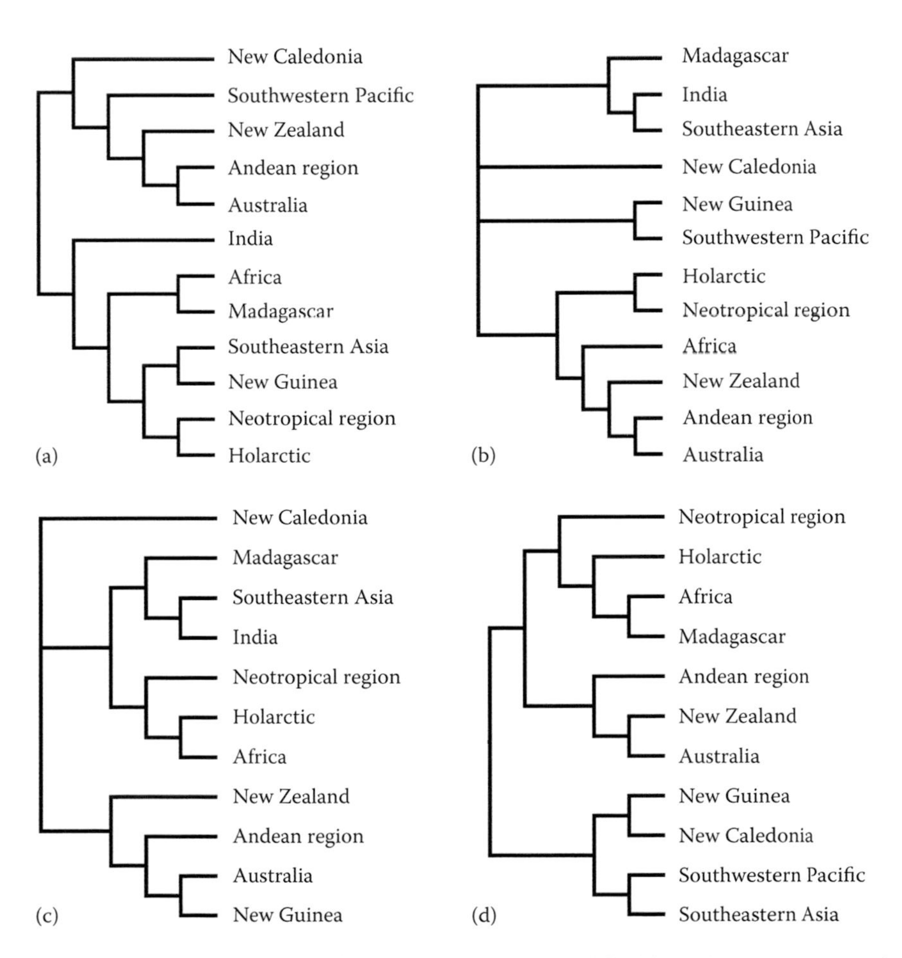

Figure 3.7 Area cladograms of the Austral areas. (Modified from Sanmartín, I. and Ronquist, F., *Systematic Biology*, 53, 216–243, 2004. With permission.) (a) Animals, (b) insects (exc. Eucnemidae), (c) non-insect animals and (d) plants.

area cladogram when analyzing insects separately, but when the study excluded Eucnemidae from the insect data set the southern Gondwanaland pattern became complete, including Africa (Figure 3.7b). Separate analysis of the animal groups excluding insects also reflected the southern Gondwanaland pattern, in this case including New Guinea as sister group to Australia (Figure 3.7c). In contrast to all the animal area cladograms, the plant area cladogram (Figure 3.7d) supported the plant southern pattern. Both the animal and plant area cladograms supported a hybrid origin for South America: northern South America consistently grouped with the Holarctic region, whereas southern South America appeared more closely related to Australia or New Zealand.

When the study compared the plant and animal patterns to the six major patterns (Figure 3.6), the southern Gondwanaland and the northern Gondwanaland patterns dominated the animal data set. Support for the latter is almost exclusively due to the eucnemid beetles. When the study removed Eucnemidae from the analysis, the southern Gondwanaland pattern became the dominant pattern in animals and in insects. Similarly, most of the non-insect animal groups supported the southern Gondwanaland pattern or the tropical Gondwanaland pattern. In contrast, most groups in the plant data set supported either the plant southern pattern or the inverted southern pattern, with almost no support for the southern Gondwanaland pattern. Several groups did not support any clear pattern or supported a pattern different from those represented in Figure 3.6.

The fit to the geological scenario showed that both the animal and plant data sets displayed highly significant, phylogenetically conserved distributional patterns, but being the processes responsible different for plant and animal taxa. In general, animals showed a higher frequency of vicariance, extinction and duplication events than expected by chance, whereas dispersals were significantly rare. Plants also exhibited a higher frequency of duplication and a lower frequency of dispersal than expected, but the frequencies of vicariance and extinction events did not depart significantly from expected values.

The comparison of the frequency of dispersal events between the animal and plant data sets showed that the frequency of terminal dispersals was significantly higher in plants than in animals, but the frequencies of ancestral and total dispersals, although higher in plants, did not differ from expected values. Thus, there was more dispersal in the plant data set, but it appeared to be solely due to dispersal within widespread terminals. The study found several significant concordant dispersal patterns, but these patterns were different in animals and plants. In animals, transantarctic dispersal between Australia and southern South America was significantly more frequent than any of the other dispersal events involving the austral landmasses. In plants, trans-Tasman dispersal (New Zealand-Australia) was significantly more frequent than transantarctic (Australia-southern South America) and transpacific (New Zealand-southern South America) dispersals in plants. New Zealand was more frequently the source area of trans-Tasman dispersal than Australia in animals, whereas the study found the opposite pattern in plants, but none of these differences were significant. There was a highly significant directional asymmetry in transpacific biotic exchange (New Zealand-southern South America), but it followed opposite directions in animals and plants: New Zealand was the most frequent source area of dispersal in animals, whereas southern South America was almost the only source of dispersal in plants. Transamerican exchange (southern South America-northern South America) in animals was significantly lower than the biotic exchange between northern South America and North America (northern South America-Holarctic), and southern South America seemed to be the main source area of these dispersals.

Main conclusions: Significant hierarchical distributional patterns characterized the biota of the Southern Hemisphere, but these patterns differ in animal and plant taxa. Animal distributional patterns conformed to the geological breakup of Gondwanaland, particularly the southern Gondwanaland pattern, and vicariance appeared to have generated it. In plants, the hierarchical patterns (plant southern pattern and inverted southern pattern) were incongruent with the commonly accepted sequence of geological events and concordant dispersal and extinction events may have shaped it. In many of the plant taxa, it appeared that more recent dispersal obscured old, geologically induced vicariance patterns. The resistant dispersal stage of plants (the seed) may have facilitated colonization events, making them less constrained by geological history than animals. When the study fit organism distributions onto a geological cladogram, several types of concordant dispersal patterns emerged. Transantarctic dispersals dominated animals, that is, between South America and Australia, this being apparently due to the long period of geological contact between Australia and South America via Antarctica. In plants there was a lot of exchange between the landmasses not connected at the time, for instance, trans-Tasman dispersal between Australia and New Zealand. Both animal and plant data sets pointed to the hybrid nature of the South American biota, with very few dispersal events between tropical and temperate regions of the continent. New Caledonia and New Zealand appeared to have retained few old Gondwanan lineages, particularly of plants.

chapter four

The Andean region

The Andean region corresponds to southern South America below 26°S, extending through the Andean highlands north of this latitude (Rivas-Martínez and Navarro, 1983; Morrone, 2001b, 2002, 2015b,c). Studies recognized it originally as a subregion of the Neotropical region of the Sclater-Wallace system. Several phytogeographers and zoogeographers working with invertebrates excluded the Andean area from the Neotropical region because of its closest links with Australia, Tasmania, New Guinea, New Caledonia, New Zealand and South Africa (Blyth, 1871; Engler, 1882; Diels, 1908; Good, 1947; Monrós, 1958; Kuschel, 1964; Cabrera and Willink, 1973; Amorim and Tozoni, 1994; Morrone, 2002, 2006, 2014a; Moreira-Muñoz, 2007). The Andean region (also known as Andean-Patagonian, Patagonian, Argentinean, Chilean or Austral) belongs to the Austral kingdom (Morrone, 2002, 2015c). The history of the Andean orogeny has an important role in the biotic evolution of this region (Schaefer, 2011). The uplift of the Andean cordillera during the last 20 m.y. and numerous episodes of tectonism affected different areas at different times, contributing to the dispersal and vicariance of the Andean biota.

In this chapter, I characterize the Andean region, detail some plant and animal taxa (endemic or characteristic) and discuss the biotic relationships with other areas of the world. Within the Andean region, I recognize the Subantarctic, Central Chilean and Patagonian subregions. Additionally, I provide a geological scenario accounting for the evolution of the region and an event-based biogeographical analysis based on weevil taxa.

Andean region

Peruvian subregion—Blyth, 1871: 428 (regionalization).
Chilean subregion—Wallace, 1876: 78 (regionalization).
Andean region—Engler, 1882: 346 (regionalization); Shannon, 1927: 3 (regionalization); Good, 1947: 236 (regionalization); O'Brien, 1971: 198 (regionalization); Morain, 1984: 178 (textbook); Takhtajan, 1986: 251 (regionalization); Morrone, 1999: 11 (regionalization), 2001a: 70 (regionalization), 2001b: 103 (regionalization); Posadas and Morrone, 2001: 267 (cladistic biogeography); Morrone, 2002: 150 (regionalization); Donato et al., 2003: 340 (dispersal-vicariance analysis); Posadas

and Morrone, 2003: 72 (cladistic biogeography); Abrahamovich et al., 2004: 100 (track analysis); Morrone, 2004a: 158 (regionalization), 2004b: 43 (regionalization); Soares and de Carlvalho, 2005: 485 (evolutionary biogeography); Mihoč et al., 2006: 391 (PAE and track analysis); Morrone, 2006: 483 (regionalization); Spinelli et al., 2006: 302 (parsimony analysis of endemicity); Alzate et al., 2008: 1252 (track analysis); Carpintero and Montemayor, 2008: 115 (faunistics); Löwenberg-Neto et al., 2008: 375 (macroecology); Casagranda et al., 2009: 18 (endemicity analysis); Escalante et al., 2009: 379 (endemicity analysis); Löwenberg-Neto and Carvalho, 2009: 1751 (parsimony analysis of endemicity); Chani-Posse, 2010: 5 (faunistics); Lopes-Andrade, 2010: 830 (faunistics); Martin, 2010: 1029 (ecological modelling); Morrone, 2010: 38 (regionalization); Urtubey et al., 2010: 505 (track analysis and cladistic biogeography); Martin, 2011: 122 (ecological modelling); Melo and Faúndez, 2011: 12 (systematic revision); Morrone, 2011: 2085 (evolutionary biogeography); Kutschker and Morrone, 2012: 541 (track analysis); Posadas, 2012: 2 (faunistics); Procheş and Ramdhani, 2012: 263 (regionalization); Alfaro et al., 2013: 243 (faunistics), 2014: 387 (faunistics); Campos-Soldini et al., 2013: 16 (track analysis); Carpintero, 2014: 339 (faunistics); Chani-Posse, 2014: 63 (faunistics); Ferretti et al., 2014b: 1 (parsimony analysis of endemicity); Flores and Cheli, 2014: 285 (faunistics); González, 2014: 515 (faunistics); Morrone, 2014b: 203 (cladistic biogeography); Omad, 2014: 565 (systematic revision); Absolon et al., 2016: 63 (track analysis); Carrara and Flores, 2015: 47 (faunistics); Cione et al., 2015: 48 (biotic evolution); Goin et al., 2015: 132 (biotic evolution); Löwenberg-Neto, 2015: 600 (shapefiles); Morrone, 2015b: 208 (regionalization); Coelho et al., 2016: 26 (endemicity analysis); Ruiz et al., 2016: 385 (track analysis); Agrain et al., 2017: 73 (faunistics); Arana et al., 2017: 421 (shapefiles); Escalante, 2017: 351 (parsimony analysis of endemicity); Martínez et al., 2017: 479 (track analysis and regionalization); Romano, 2017: 230 (shapefiles); Romano et al., 2017: 444 (track analysis); Urra, 2017a: 30 (faunistics), 2017b: 30 (faunistics).

Andean kingdom—Drude, 1890: 158 (regionalization).

Argentinean subarea—Clarke, 1892: 381 (regionalization).

Patagonian subregion—Sclater and Sclater, 1899: 65 (regionalization); Cabrera and Yepes, 1940: 12 (regionalization); Kuschel, 1964: 447 (regionalization); Hershkovitz, 1969: 8 (regionalization); Kuschel, 1969: 712 (regionalization).

Andean dominion—Hauman, 1931: 62 (regionalization); Cabrera, 1951: 48 (regionalization), 1957: 335 (regionalization).

Temperate South America dominion—Hauman, 1931: 62 (regionalization).

Andean-Patagonian subregion—Mello-Leitão, 1937: 232 (regionalization), 1943: 129 (regionalization); Ringuelet, 1961: 156 (biotic evolution);

Rapoport, 1968: 75 (regionalization); Fittkau, 1969: 639 (regionalization); Maury, 1979: 710 (faunistics); José de Paggi, 1990: 330 (regionalization); Sánchez Osés and Pérez-Hernández, 1998: 201 (regionalization), 2005: 168 (regionalization).

Andean-Patagonian dominion—Cabrera, 1971: 29 (regionalization), 1976: 50 (regionalization); Cabrera and Willink, 1973: 83 (regionalization); Willink, 1988: 206 (regionalization), 1991: 138 (regionalization); Zuloaga et al., 1999: 18 (floristics).

Austral subregion—Ringuelet, 1975: 107 (regionalization); Almirón et al., 1997: 23 (regionalization).

Andean subkingdom—Rivas-Martínez and Tovar, 1983: 516 (regionalization).

Argentine subregion—Smith, 1983: 462 (regionalization).

Patagonian region—Takhtajan, 1986: 253 (regionalization).

Austroamerican subkingdom (in part)—Rivas-Martínez and Navarro, 1994: map (regionalization); Rivas-Martínez et al., 2011: 27 (regionalization).

Neotemperate region—Amorim and Pires, 1996: 187 (regionalization).

Andean subregion—Morrone, 1996a: 105 (regionalization); Posadas et al., 1997: 2 (parsimony analysis of endemicity), 2001: 1328 (conservation biogeography); Díaz Gómez, 2009: 2 (dispersal-vicariance analysis); Kotov et al., 2010: 59 (faunistics).

Patagonian-Andean region—Daniels and Veblen, 2000: 225 (climate).

Chile-Patagonian region—Cox, 2001: 519 (regionalization).

Temperate South America region—Kreft and Jetz, 2010: 2044 (regionalization).

Ando-Patagonian region—Stonis et al., 2016: 561 (systematic revision).

Endemic and characteristic taxa

Morrone (2001b, 2006) provided a list of endemic taxa. Some examples of plant taxa include *Oreobolus* (Cyperaceae); *Bowlesia* (Apiaceae); Barnadesioideae, *Leucheria* and *Nassauvia* (Asteraceae); *Gunnera* subgen. *Misandra* (Gunneraceae); *Coriaria* (Coriariaceae); *Epilobium denticulatum* (Figure 4.1a; Onagraceae); *Aretiastrum* (Valerianaceae); and *Calceolaria* (Calceolariaceae). Animal taxa include *Lymnaea cousini* (Gastropoda: Lymnaeidae); *Brachistosternus* (Scorpiones: Bothriuridae); *Parabroteas sarsi* (Copepoda: Centropagidae); Baripina, *Pseudocnides* and *Trechisibus* (Coleoptera: Carabidae); *Cylydrorhinus* and Listroderini (Coleoptera: Curculionidae); Faroninae (Coleoptera: Pselaphidae); *Gigantodax wrighti* (Figure 4.1b; Diptera: Simuliidae); *Dorymyrmex*, *Lasiophanes* and *Nothidris* (Hymenoptera: Formicidae); *Chirodamus* and *Pompilocalus hirticeps* (Hymenoptera: Pompilidae); *Gayella* spp. (Hymenoptera: Vespidae); *Heliothis tergemina* (Lepidoptera: Noctuidae); *Anas georgica*, *A. flavirostris*, *A. specularoides* and *Oxyura ferruginea* (Anseriformes: Anatidae); *Vultur gryphus* (Ciconiiformes: Ciconiidae); *Phalacrocorax gaimardi* (Suliformes: Phalacrocoracidae); *Phoenicopterus chilensis* (Phoenicopteriformes:

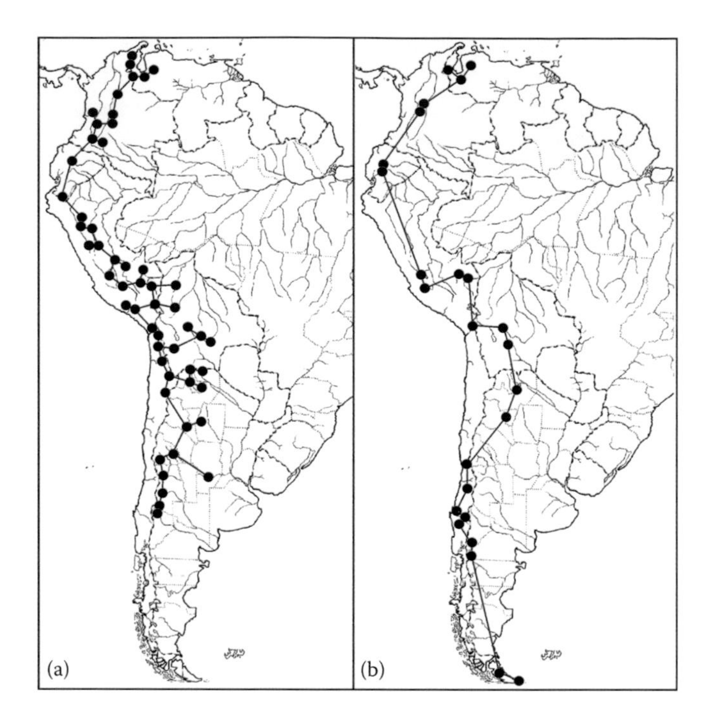

Figure 4.1 Maps with individual tracks in the Andean region. (a) *Epilobium denticulatum* (Onagraceae) and (b) *Gigantodax wrighti* (Diptera: Simuliidae).

Phoenicopteridae); *Podiceps occipitalis* (Podicipediformes: Podicipedidae); *Phrygilus unicolor* (Passeriformes: Fringillidae); *Certhiaxis cinnamonea* and *Cinclodes fuscus* (Passeriformes: Furnariidae); *Anairetes* and *Muscisaxicola maculirostris* (Passeriformes: Tyrannidae); *Rhea pennata* (Rheiformes: Rheidae); *Pseudalopex culpaeus* (Carnivora: Canidae); and *Lontra felina* (Carnivora: Mustelidae). According to Cione et al. (2015), the mammal genera *Vicugna, Hippocamelus, Pudu, Dromiciops, Chinchilla* and *Abrothrix* are endemic to the Andean region.

Andean niche conservatism

Segovia and Armesto (2015) hypothesized the existence of niche conservatism in the warm-temperate biota of the Andean region. They considered that to understand adequately the current distributional patterns in South America it is necessary to consider the influence of an Austral biota that became differentiated during the fragmentation of Gondwanaland. The configuration of this biota would have had a strong impact in the biotic evolution of South America, especially through its interaction with the Neotropical biota in the South American transition zone. The richness of gymnosperms (e.g., Araucariaceae and Podocarpaceae) and of angiosperm taxa that are rare in the Neotropics (e.g., Proteaceae, Myrtaceae, Cunoniaceae and Akamiaceae) is characteristic of the temperate forests of the Andean region. Additionally, the absence of some major Neotropical clades (e.g., Zingiberales, Meliaceae, Sapotacerae, Moraceae and Annonaceae) suggested that the tropical biota could not expand southward during global warming periods (Segovia and Armesto, 2015).

Biotic relationships

Katinas et al. (1999) conducted a track analysis of the Andean region and other areas of the world, based on several species of fungi, gymnosperms, angiosperms, insects and vertebrates. Their analysis identified five basic distributional patterns:

1. *Andean endemic distribution*: Several taxa were endemic to the Andean region, for example, *Chuquiraga* (Asteraceae) and *Hyppocamelus* (Cetartiodactyla: Cervidae). Other Andean taxa were not widespread in the whole region, inhabiting one or more of its subregions and showing basically three generalized tracks: (1) one joining the Central Chilean and Subantarctic subregions with the Puna province of the South American transition zone, (2) another joining the Central Chilean, Subantarctic and Patagonian subregions with the Puna province and (3) a third one joining the Subantarctic and Patagonian subregions.
2. *Austral distribution*: Several tracks related only the Subantarctic subregion with other Austral regions, some of them representing classical examples of intercontinental disjunctions, for example, *Nothofagus* (Nothofagaceae) from southern South America, southeastern Australia, Tasmania, New Zealand, New Guinea and New Caledonia (Figure 4.2a), having a Pacific basin baseline. Other taxa showed similar distributions but excluded New Guinea and/or New Caledonia (Figure 4.2b), for example, *Aristotelia* (Elaeocarpaceae), *Boeckella* (Copepoda: Centropagidae) and *Cyttaria* (Cyttariaceae). A few taxa were absent from New Zealand, for example, Caridae (Coleoptera). Distribution of the plant family Cunoniaceae was in the Subantarctic subregion and Australia (Figure 4.2c). Distribution of some taxa was in the Subantarctic subregion and New Zealand (Figure 4.2d), for example, Donatiaceae, *Kenodactylus* (Coleoptera: Carabidae), *Lepidothamnus* (Podocarpaceae) and *Pseudopanax* (Araliaceae). Distribution of the remaining taxa was also in other subregions of the Andean region and the rest of the Austral kingdom, for example, Aphroteniinae (Diptera: Chironomidae), Araucariini (Coleoptera: Curculionidae), Galaxiidae (Osteichthyes), *Lomatia* (Proteaceae) and Megascolecidae (Oligochaeta). Within the Austral

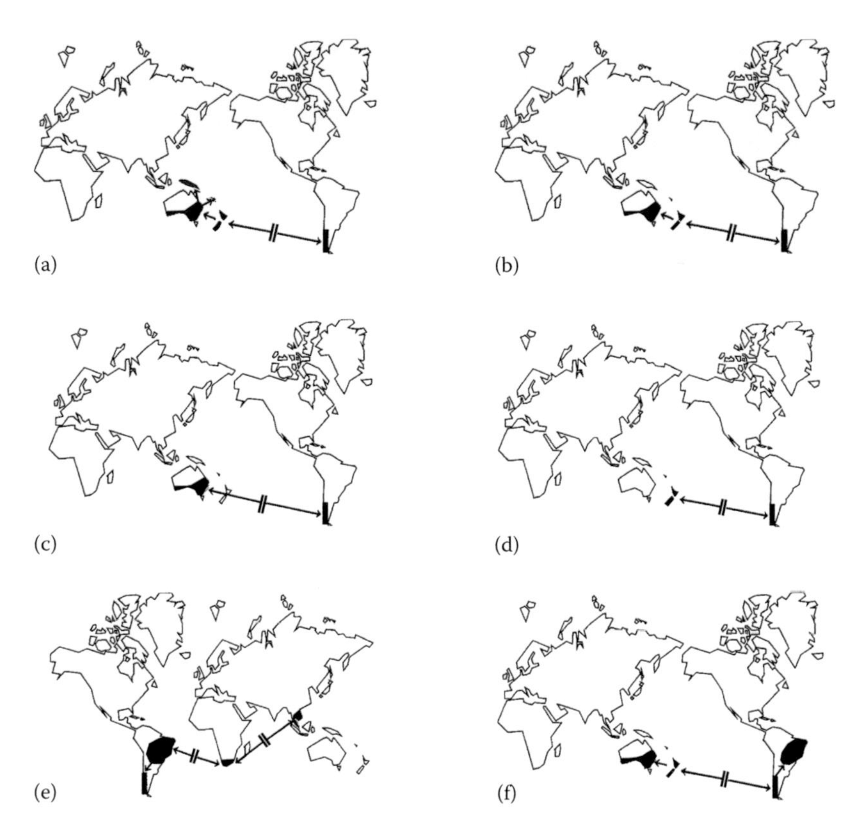

Figure 4.2 Schematic representation of Austral individual tracks. (Modified from Katinas, L. et al., *Aust. Syst. Bot.*, 47, 111–130, 1999. With permission.) (a) *Nothofagus* (Nothofagaceae), (b) *Aristotelia* (Elaeocarpaceae), (c) Cunoniaceae, (d) Donatiaceae, (e) Oxycorinini (Belidae) and (f) *Araucaria* (Araucariaceae).

pattern, two generalized tracks were evident: (1) one joining the Subantarctic subregion, Australia and New Zealand and (2) another joining the Subantarctic subregion and New Zealand.

3. *Tropical distribution*: Taxa distributed in both the Andean region and the Holotropical kingdom fell into two groups: those whose distribution was in the whole Holotropical kingdom, or those restricted to the Neotropical region. One example of Holotropical distribution was the tribe Oxycorinini (Coleoptera: Belidae), which comprised the genera *Oxycraspedus* from the Subantarctic subregion; the Neotropical *Hydnorobius, Alloxycorynus, Oxycorynus* and *Parallocorynus; Hispodes* and *Afrocorynus* from South Africa; and *Metrioxena* from southeastern Asia, having its individual track along both Atlantic and Indian Ocean baselines (Figure 4.2e). In

comparison with the Austral pattern, there was a low level of relationship of the Subantarctic subregion with the Neotropical region, for example, *Diposis* (Apiaceae; Subantarctic, Central Chile and Neotropical region), *Listroderes* (Coleoptera: Curculionidae; Subantarctic, Central Chile, South American transition zone and Neotropical), *Lucilia* (Asteraceae; Subantarctic, Central Chile, South American transition zone and Neotropical) and *Myrceugenia* (Myrtaceae; Subantarctic and Neotropical region). Some taxa, for example, *Araucaria* (Araucariaceae), Belini (Coleoptera: Belidae), Centrolepidaceae, *Colobanthus* (Caryophyllaceae), *Discaria* (Rhamnaceae), Epacridaceae, *Gunnera* (Gunneraceae) and Winteraceae, inhabited the Subantarctic subregion, the Neotropical region and one or more of the remaining Austral areas (Australasia, South Africa and Antarctica) showing a Pacific basin baseline (Figure 4.2f). Most of these taxa inhabited different subregions of the Andean region and other Austral regions, and so no generalized tracks. The distribution of the remaining taxa analyzed constitute three generalized tracks: (1) one joining the Andean and Neotropical regions (Figure 4.3a), for example, *Azorella* (Apiaceae), Barnadesioideae (Asteraceae) and *Tropaeolum* (Tropaeolaceae); (2) another joining the South American transition zone and the Neotropical region (Figure 4.3b), for example, *Dunalia* (Solanaceae), *Malvaviscus* (Malvaceae) and *Proctoporus* (Squamata: Gymnophthalmidae) and (3) a third one joining the Patagonian subregion with the Neotropical region (Figure 4.3c), for example, *Acicarpha* (Calyceraceae), *Diplolaemus* and *Leiosaurus* (Squamata: Leiosauridae), *Ophiodes* (Squamata: Anguidae) and *Teius* (Squamata: Teiidae).

4. *Amphitropical distribution*: Taxa distributed in both the Andean region and the Holarctic kingdom were rather uncommon and found mainly in plant taxa. For example, members of the plant family Empetraceae grew in the Subantarctic subregion and in the north temperate areas of Eurasia and America (Figure 4.3d). There were plant taxa growing in the South American transition zone, the Central Chilean, Subantarctic and Patagonian subregions and the Holarctic kingdom (*Gutierrezia*, Asteraceae; Figure 4.3e). A few of the amphitropical taxa of the Andean region were also present in other areas of the Austral kingdom, for example, *Epilobium* sect. *Boisduvalia* (Onagraceae), Keroplatidae (Diptera) and Rhinorhynchini (Coleoptera: Nemonychidae).
5. *Cosmopolitan distribution*: Taxa shared by the Andean region and two or three kingdoms (cosmopolitan taxa) were uncommon, for example, *Adenocaulon* (Asteraceae) and *Agalinis* (Orobanchaceae). They did not show any generalized track.

The intersection of the Andean endemic, Austral and Tropical generalized tracks led Katinas et al. (1999) to identify three nodes (Figure 4.3f): one was

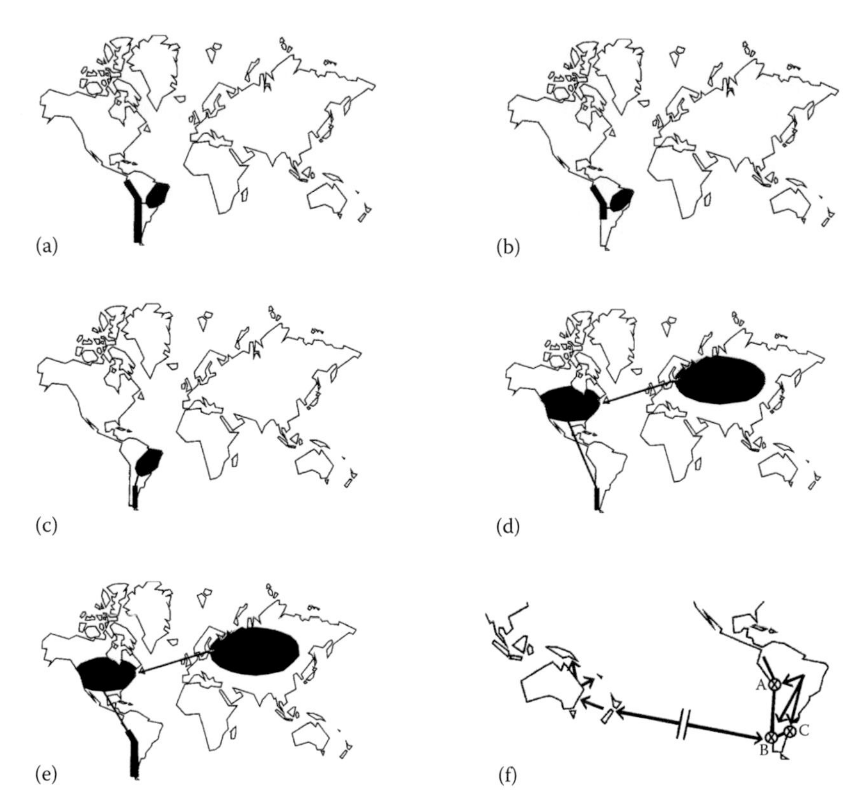

Figure 4.3 Schematic representation of Austral individual tracks, generalized tracks and nodes. (Modified from Katinas, L. et al., *Aust. Syst. Bot.*, 47, 111–130, 1999. With permission.) (a) *Azorella* (Apiaceae), (b) *Dunalia* (Solanaceae), (c) *Acicarpha* (Calyceraceae), (d) Empetraceae, (e) *Gutierrezia* (Asteraceae) and (f) representation of the most consistently found generalized tracks and the nodes delimited by them. A: Puna node, B: Subantarctic node and C: Patagonian node.

situated in the Subantarctic subregion, in the intersection of the Andean endemic, Austral and Tropical generalized tracks; and two others were situated in the Puna and Patagonian provinces, in the intersection of the Andean endemic and Tropical generalized tracks. The Subantarctic was the only subregion of the Andean region where the three patterns appeared. The Andean endemic and Tropical patterns predominated in the Puna province of the South American transition zone and the Patagonian province. Katinas et al. (1999) concluded that the biota of the Andean region has a composite origin. Cretaceous events in Gondwanaland relating it with other austral areas, mainly Australia and New Zealand principally shaped the evolution of a part of this biota, represented by the Subantarctic subregion. The biota of the remaining subregions and the South American

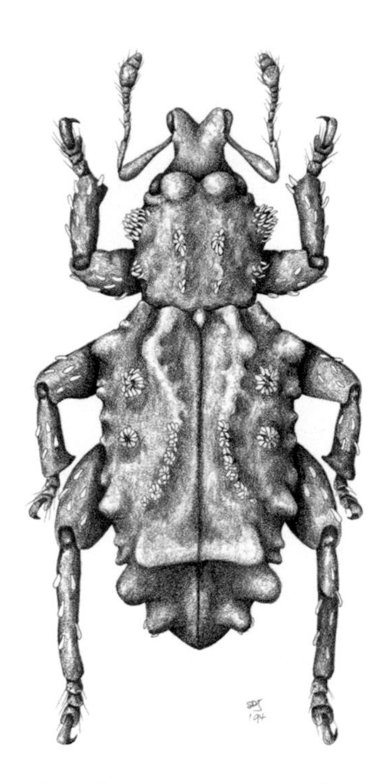

Figure 4.4 *Trachodema tuberculosa*, a representative species of the Andean Listroderini (Coleoptera: Curculionidae). (Courtesy of Sergio Roig-Juñent.)

transition zone appeared to link to that of the Neotropical region, probably due to more recent events like the uplift of the Andes in the Neogene and Pleistocene glaciations.

Morrone (1994a) analyzed the biotic relationships within the Andean region based on a track analysis and a parsimony analysis of endemicity of species belonging to 21 genera of South American Listroderini (Coleoptera: Curculionidae) that include *Antarctobius, Germainilellus, Trachodema* (Figure 4.4), *Falklandius* and *Telurus*, among others. He concluded from his analysis the restriction of the ancient Austral biota originally to the Subantarctic and Central Chilean subregions. Then, this biota dispersed north and east to occupy all the Andean region. Finally, the uplift of the Andes and the development of open-country communities (Figure 4.5) affected this biota. Donato et al. (2003) reanalyzed the Listroderini using a dispersal-vicariance approach and found similar dispersal events, although they postulated a much more widespread ancestral biota that occupied the whole Andean region.

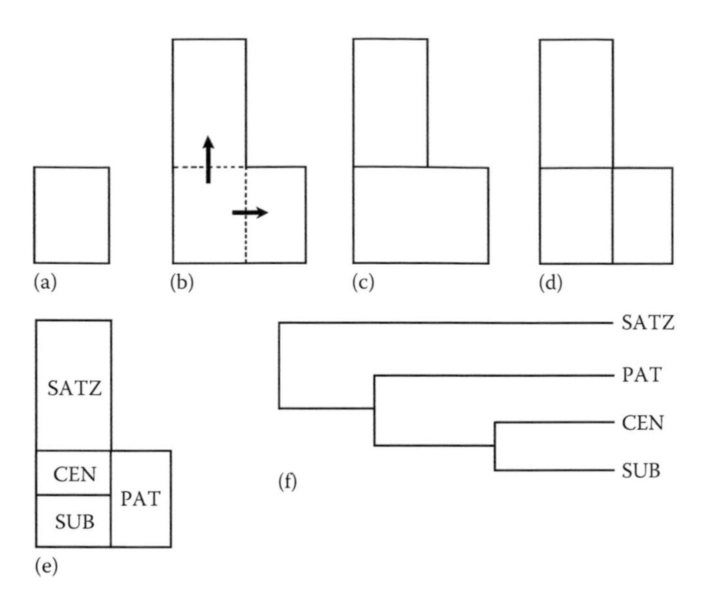

Figure 4.5 Morrone's (1994) biogeographical analysis of the Andean region based on a track analysis and a parsimony analysis of endemicity of species of Curculionidae. (Modified from Morrone, J.J., *Global Ecol. Biogeogr. Lett.*, 4, 188–194, 1994a. With permission.) (a) Original biota in the southern part of the region, (b) dispersal north and east, (c)–(e) successive isolation of the subregions and transition zone and (f) area cladogram showing the vicariant events and the progressive depauperation of the Andean biota. (CEN) Central Chilean subregion, (PAT) Patagonian subregion, (SATZ) South American transition zone and (SUB) Subantarctic subregion.

Regionalization

Several studies recognized dominions, provinces and districts within the Andean region (Cabrera and Willink, 1973; Müller, 1973; Udvardy, 1975; Rivas-Martínez and Navarro, 1994; Dinerstein et al., 1995; Rivas-Martínez et al., 2011). In addition to these general schemes there were regionalizations referred in particular to Argentina (Cabrera, 1951, 1953, 1957, 1971; Ringuelet, 1961; Roig, 1998; Burkart et al., 1999; Apodaca et al., 2015a), Chile (Reiche, 1905; Goetsch, 1931; Mann, 1960; Kuschel, 1960; Peña, 1966a,b; O'Brien, 1971; Cekalovic, 1974; Artigas, 1975), Colombia (Hernández Camacho et al., 1992a,b) and Peru (Lamas, 1982). I synthesized these previous schemes and, mostly based on track analyses of plant and animal taxa, regionalized this region (Morrone, 2006, 2015b) considering the Central Chilean, Subantarctic and Patagonian subregions, as well as the South American transition zone. The assignment of the Subantarctic subregion, from an ecoregional perspective, was to the Humid Temperate domain, whereas the Central Chilean and Patagonian subregions, as well as the South American transition zone, belonged to the Dry domain (Bailey, 1998).

The Andean region (Figure 4.6) comprises the Subantarctic, Central Chilean and Patagonian subregions, and it overlaps with the Neotropical region in the South American transition zone (Morrone, 2004a, 2006, 2010, 2015b), which belongs simultaneously to both regions. I consider that the Subantarctic subregion represent the core of the Andean region (Figure 4.7a), harboring its most characteristic biota. The combination of the Subantarctic with the Central Chilean and Patagonian subregions constitutes the Andean region in the strict sense (Figure 4.7b). Finally, the addition of the South American transition zone to them becomes the Andean region *sensu lato* (Figure 4.7c).

Geological scenario

The following account of the geological evolution of the Andean area during the last 83 m.y.a. (Late Cretaceous–Cenozoic) is based on Donato et al. (2003). During the Mesozoic and Cenozoic, control of the tectonic development of South America was through a complex subduction regime along its western margin and the evolving spreading center of the Mid-Atlantic Ridge along its eastern margin, combined with sea-level variations. The main geological events that have shaped the Andean evolution are as follows:

Late Cretaceous–Early Paleocene: During the Late Cretaceous–Early Paleocene (*ca.* 70–64 m.y.a.), the absence of large continental barriers allowed an Atlantic transgression that covered southern South America, from Patagonia to Bolivia and Peru (Figure 4.8). This shallow Salamancan Sea divided South America into the northeastern and southwestern regions, which corresponded basically to the Neotropical and Andean regions, respectively.

Late Paleocene–Oligocene: During the Palaeogene (Paleocene–Oligocene; *ca.* 64–24 m.y.a.), there was a pause in the Andean magmatism and a marine retreat. During the Late Paleocene, transformation of the Salamancan Sea into a series of broad alluvial plains and large lake basins occurred. During the Late Paleocene–Eocene (*ca.* 64–36 m.y.a.), with southern Patagonia still covered by the sea, central and northern Patagonia turned into vast loess plains, whose sediments were pyroclastic in origin. Most of the Oligocene (36.6–23.7 m.y.a.) was a time of relative tectonic quiescence, but during the Late Oligocene (*ca.* 25 m.y.a.) the modification of the convergence direction between the Nazca and South American plates induced important modifications in the Andean arc. Igneous activity expanded over large areas and invaded west-central Argentina,

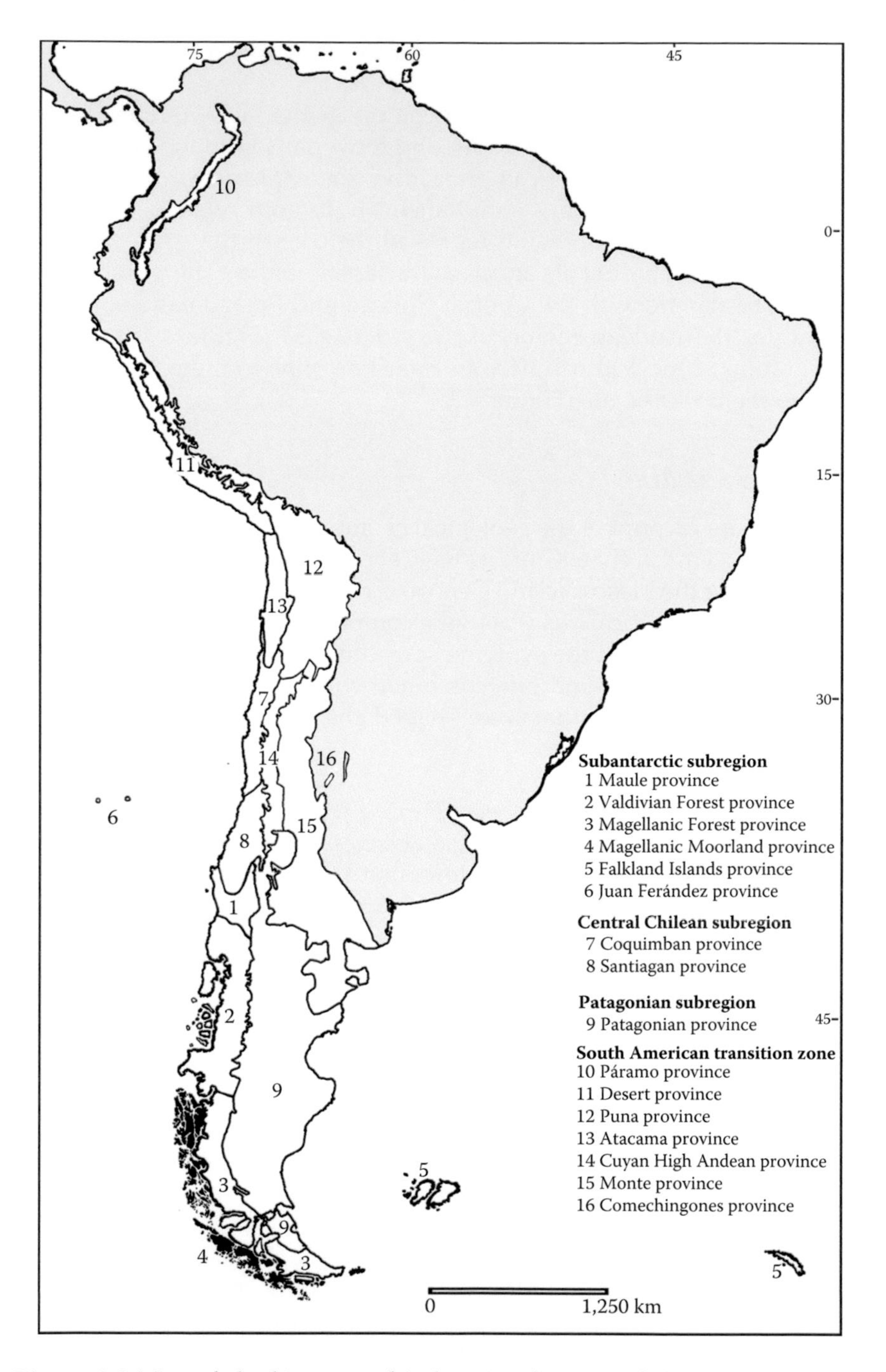

Figure 4.6 Map of the biogeographical regionalization of the Andean region. (Modified from Morrone, J.J., *Zootaxa*, 3936, 207–236, 2015b. With permission.)

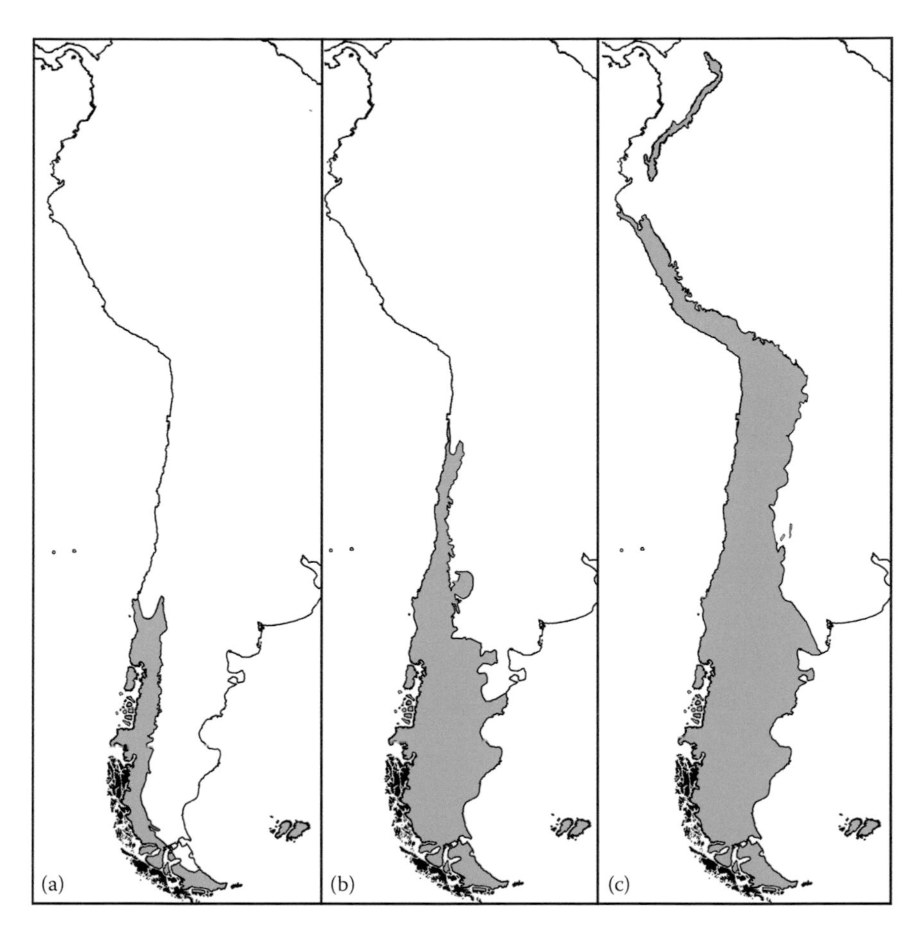

Figure 4.7 Three perspectives of the Andean region. (a) Subantarctic subregion, (b) Andean region *sensu stricto* and (c) Andean region *sensu lato* (also including the South American transition zone).

Bolivia and Peru, while in southern Argentina and Chile the magmatic activity was modest.

Late Oligocene–Early Miocene: During the Late Oligocene–Early Miocene (*ca.* 26–20 m.y.a.), the present configuration of the Southern Andes began to develop. Also, a new Atlantic transgression (Patagonian Sea) occurred in the same general areas as had occurred in the Late Cretaceous–Early Paleocene, whereas a generalized Pacific transgression occurred on the west margin of South America.

Middle Miocene–Late Miocene: In the Middle Miocene (*ca.* 15 m.y.a.), domination of southern South American landscapes by the processes leading to the present configuration of the Andean tectonic-magmatic

Figure 4.8 Main South American areas affected by the Late Cretaceous–Early Paleocene marine transgressions. (Modified from Donato, M. et al., *Biol. J. Linn. Soc.*, 80, 339–352, 2003. With permission.)

belt began. During the Middle Miocene, a small Pacific transgression covered Central Chile. Also, during the Middle and Late Miocene, three successive Atlantic marine transgressions occurred in southern South America (Paranean Sea) and an open seaway again separated the terrestrial environments of southern South America from those farther north. The northwestern part of this Paranean Sea connected with the Tethys Waterspout (Figure 4.9) that covered widespread areas of the Andean Cordillera and the Guayanian and Brazilian bedrocks.

Late Miocene–Late Pliocene: Widespread plains succeeded the widespread surface flooded by the Paranean Sea and extended from northern

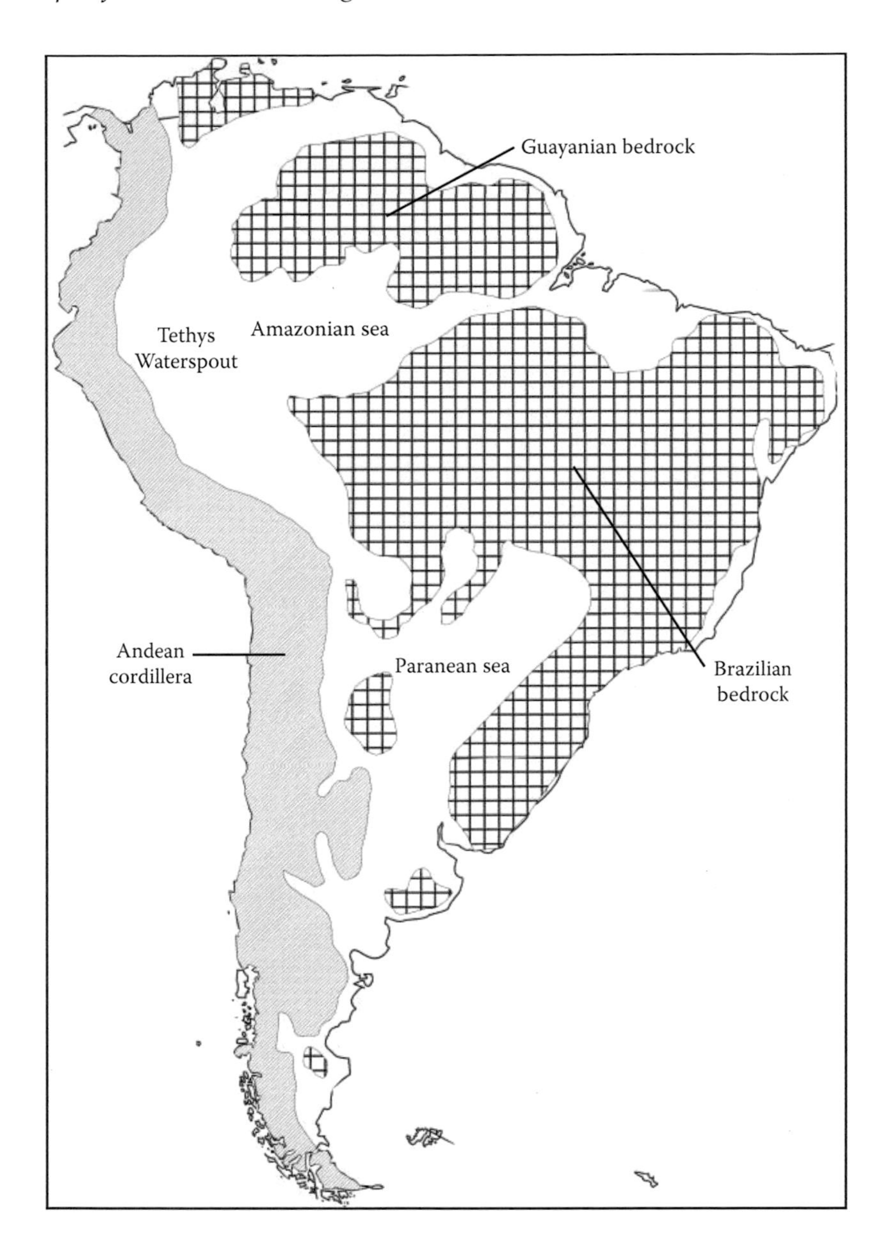

Figure 4.9 Main South American areas affected by the Middle Miocene–Late Miocene marine transgressions. (Modified from Donato, M. et al., *Biol. J. Linn. Soc.*, 80, 339–352, 2003. With permission.)

Patagonia northward. The development of these new habitats correlated to the Quechua Phase of the Andean diastrophism, where the Andean Cordillera was successively uplifted, progressively forming a major barrier to moisture-laden South Pacific winds. The resulting rain-shadow effect on the eastern Patagonian landscapes led to the first stages on the differentiation of the Subantarctic and Patagonian subregions. Additionally, at the beginning of the Age of the Southern Plains (*ca.* 13–11 m.y.a.), the final opening of the Drake Passage occurred, separating South America from Antarctica, finally establishing the cool Circum-Antarctic Current and initiating the formation of an ice sheet in West Antarctica. Thus, during the Age of the Southern Plains the climate was cooler and seasonally more marked than it was in the Middle Miocene. The end of the Age of the Southern Plains correlated to the elevation of the Central Cordillera of Argentina and Chile, the Puna and other orographic systems. The final uplift of the Central Andes and the Puna had marked ecological consequences, producing a rain-shadow effect that resulted in extremely xeric conditions east of the Andes.

Case study: Event-based biogeographical analysis of the Andean region

Title: "Historical biogeography of the Andean region: Evidence from Listroderina (Coleoptera: Curculionidae: Rhytirrhinini) in the context of the South American geobiotic scenario" (Donato, M., P. Posadas, D. R. Miranda-Esquivel, E. Ortiz Jaureguizar and G. Cladera, *Biological Journal of the Linnean Society*, 80: 339–352, 2003).

Goal: To analyze the distributional patterns of the weevils of the subtribe Listroderina applying a dispersal-vicariance analysis and to interpret their biogeographical evolution in the context of the geobiotic evolution of South America.

Location: Andean region.

Methods: To reconstruct the ancestral distributions of the genera of Listroderina the study used software DIVA 1.1 (Ronquist, 1996), applying an exact search according to the dispersal-vicariance optimization proposed by Ronquist (1997). This software allowed inference of the ancestral distribution of a taxon and calculated the frequencies of vicariance and dispersal events among different areas under consideration. The input information was the phylogenetic and distributional information encoded on the taxon-area cladogram. The study analyzed the following areas (Figure 4.10): Central Chile, Neotropics, Nearctic, Páramo, Patagonia, Puna and Subantarctic. The cladogram proposed by Morrone (1997) included a total of 25 genera (Figure 4.11).

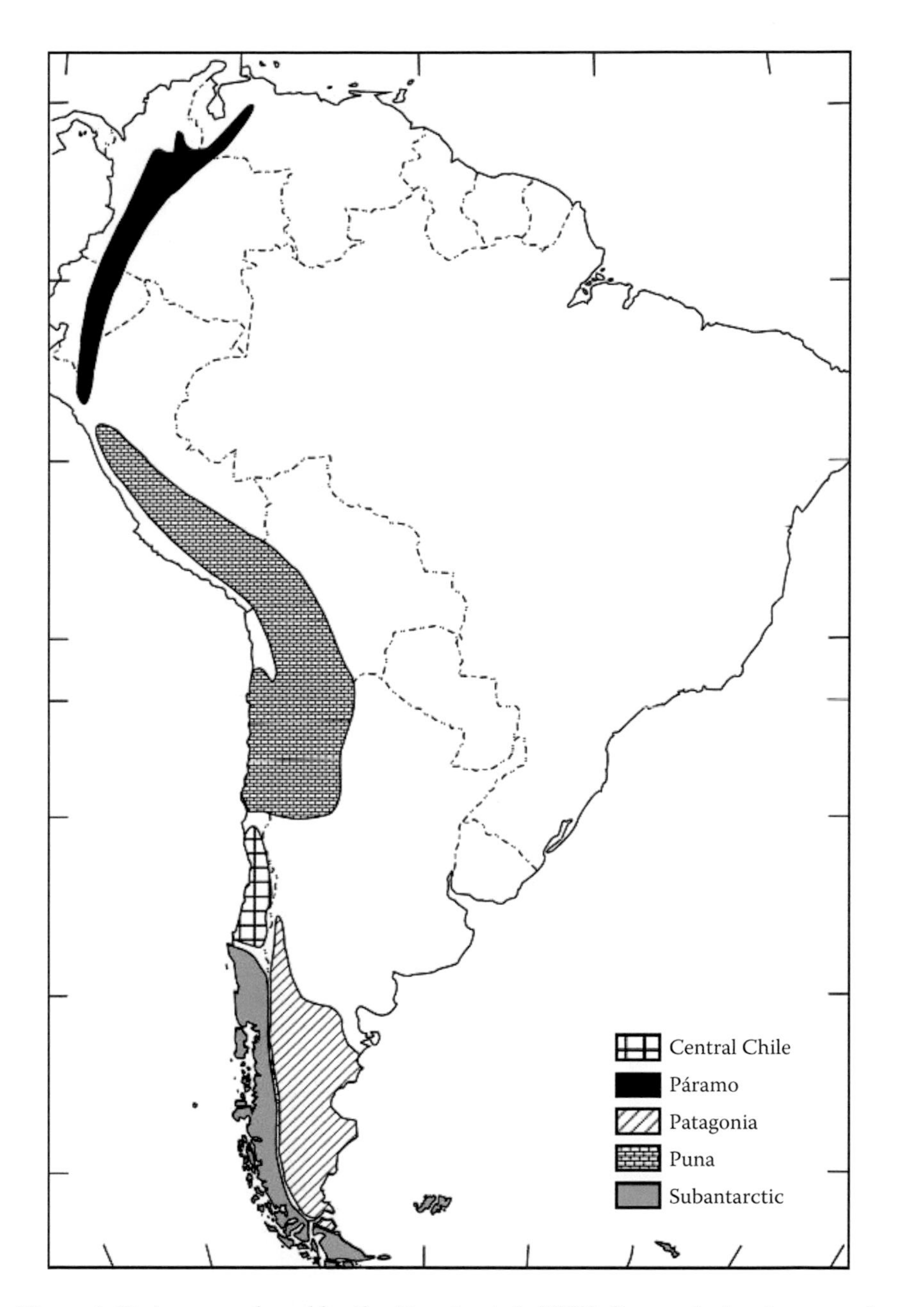

Figure 4.10 Areas analyzed by the Donato et al. (2003) dispersal-vicariance analysis of the Andean region. (Modified from Donato, M. et al., *Biol. J. Linn. Soc.*, 80, 339–352, 2003. With permission.)

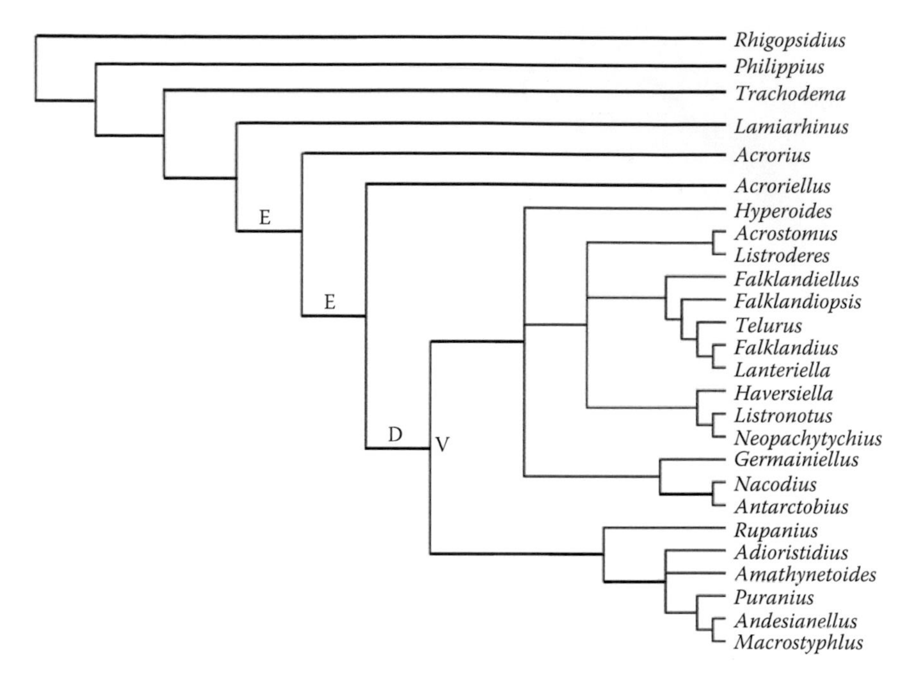

Figure 4.11 Cladogram of the subtribe Listroderina used in the Donato et al. (2003) dispersal-vicariance analysis of the Andean region. (Modified from Donato, M. et al., *Biol. J. Linn. Soc.*, 80, 339–352, 2003. With permission.) D = Dispersal event, E = extinction events and V = vicariant event.

Results: DIVA produced 84 alternative, equally optimal reconstructions requiring 30 dispersal events. The study assigned the highest frequencies for vicariant events to the Páramo + Puna related to the Subantarctic (28.69%), followed by Central Chile + Páramo related to the Subantarctic (16.80%), Neotropics + Puna related to Central Chile + Páramo + Subantarctic (14.75%), and the Páramo related to Central Chile + Subantarctic (13.11%). Dispersal events observed usually occurred between contiguous areas. Two dispersal events had the highest scores: the first occurred from Puna to Páramo (45.69%) and the second was from Subantarctic to Central Chile (21.89%). The study assigned the third and fourth scores of dispersals from Subantarctic to Patagonia and from Puna to the Neotropics (both with 8.84%).

Main conclusions: The first ancestral distribution showed that Listroderina inhabited the areas today occupied by Central Chile, Páramo, Puna and Subantarctic, which would reflect the Andean origin of the group during the Late Cretaceous–Early Palaeocene. During the Late Cretaceous–Early Palaeocene biotic interchange,

Listroderina immigrated from South America to North America. Evidence for this event was a Listroderina fossil record from the Oligocene of North America, which the study interpreted as a relict of an older lineage. Validation for this interpretation is the lower diversity reached by this taxon in North America, and the isolation of South America after the South and North American connection, which ended in the Pliocene when the Panamanian bridge connected both continents. Later, restriction of the subtribe to the areas presently occupied by the Páramo and Puna occurred, probably due to local extinction events. This restricted distribution could link to the climatic and geological events that occurred during the Eocene (58–37 m.y.a.). The explanation for the return of the subtribe to the area presently occupied by the Subantarctic subregion could be the relative tectonic quiescence that characterized most of the Oligocene and the low magmatic activity in southern Argentina and Chile at this time. After that, a vicariant event separated the Subantarctic subregion from the Puna and Páramo. The development of a sea barrier represented by the Paranean Sea and a new phase of the Andean uplift, beginning in the Middle Miocene, could be the cause of this vicariant event. The vicariant event, which split Subantarctic from Páramo + Puna, confined the *Macrostyphlus* generic group to the Páramo and Puna subregions and from there dispersed it to other areas. The restriction of the *Antarctobius, Falklandius, Listronotus* and *Listroderes* generic groups to the Subantarctic occurred. The interpretation of the widespread distributions of some terminal taxa pointed to the result of recent dispersal events. Finally, in the Diaguita Phase, the uplift of the Puna created an isolated and climatic extreme habitat and resulted in the local diversification and high endemism exhibited by Listroderina.

chapter five

The Subantarctic subregion

The Subantarctic subregion represents the core of the Andean region and corresponds to Austral Chile from 37°S to Cape Horn, the archipelago of southern Chile and Argentina, and the Falkland Islands (Malvinas), South Georgia Island and Juan Fernández Islands (Ringuelet, 1955a, b; Cabrera and Willink, 1973; Morrone, 2000b, 2001b, 2006, 2015b). Forests with Austral genera (e.g., *Nothofagus, Dacrydium, Saxegothaea, Austrocedrus, Pilgerodendron* and *Fitzroya*) are predominant in the northern part of the subregion; however, there are also moorlands and areas lacking *Nothofagus* completely, like the Falkland and Juan Fernández islands (Morrone, 2001b). Areas with xeric vegetation surround the Subantarctic subregion, thus behaving as an island-like biota (Vuilleumier, 1985).

The Subantarctic subregion is well characterized in terms of numerous plant and animal taxa. This chapter briefly discusses the subregion's biotic relationships with the Australian region and the Central Chilean subregion. Within the Subantarctic subregion, this chapter recognizes six provinces: Maule, Valdivian Forest, Magellanic Forest, Magellanic Moorland, Falkland Islands and Juan Fernández Islands. The chapter characterizes each of these provinces and details their endemic and characteristic taxa, describes their vegetation types and discusses their relationships. In addition, this chapter recognizes and characterizes 11 districts and provides five case studies.

Subantarctic subregion

Subantarctic dominion—Skottsberg, 1905: 415 (regionalization); Cabrera, 1951: 57 (regionalization), 1971: 36 (regionalization); Cabrera and Willink, 1973: 96 (regionalization); Cabrera, 1976: 71 (regionalization); Willink, 1988: 206 (regionalization), 1991: 138 (regionalization); Morrone, 1993b: 122 (track analysis and cladistic biogeography); Morrone et al., 1994: 105 (cladistic biogeography).

Subantarctic Forests area—Hauman, 1920: 52 (regionalization); Parodi, 1934: 171 (regionalization).

Moist Andean zone—Shannon, 1927: 3 (regionalization).

Forests region—Goetsch, 1931: 2 (regionalization).

Subantarctic Andes province—Hauman, 1931: 62 (regionalization).

Subantarctic Forests province—Hauman, 1931: 62 (regionalization).

Subantarctic province—Cabrera, 1951: 58 (regionalization), 1953: 107 (regionalization), 1971: 37 (regionalization), 1976: 72 (regionalization); Cabrera and Willink, 1973: 97 (regionalization); Willink, 1988: 206 (regionalization), 1991: 138 (regionalization); Morrone, 1994a: 191 (parsimony analysis of endemicity), 1996a: 106 (regionalization); Posadas et al., 1997: 2 (parsimony analysis of endemicity); Carpintero, 1998: 148 (faunistics); Roig, 1998: 140 (regionalization); Posadas et al., 2001: 1328 (conservation biogeography); Sérsic et al., 2011: 477 (phylogeographical analysis); Apodaca et al., 2015a: 97 (regionalization).

Austral-cordilleran dominion—Ringuelet, 1955a: 84 (biotic evolution), 1961: 160 (biotic evolution).

Subantarctic forests region—Hueck, 1957: 40 (regionalization).

Araucanan subregion—Monrós, 1958: 145 (regionalization); Ringuelet, 1961: 156 (biotic evolution); Rapoport, 1968: 75 (regionalization).

Chile province (in part)—Fittkau, 1969: 642 (regionalization).

Southern Andes area (in part)—Sick, 1969: 465 (regionalization).

Nothofagus center—Müller, 1973: 155 (regionalization); Cracraft, 1985: 36 (regionalization).

Andean Subantarctic region—Rivas-Martínez and Tovar, 1983: 521 (regionalization).

Neotropic-Archinotic transition zone—Müller, 1986: 24 (regionalization).

Chilean subregion (in part)—Flint, 1989: 1 (regionalization).

Valdivian-Magellanic region—Rivas-Martínez and Navarro, 1994: map (regionalization).

Subantarctic area—Coscarón and Coscarón-Arias, 1995: 726 (areas of endemism); Apodaca et al., 2015b: 5 (dispersal-vicariance analysis).

Patagonian Forests ecoregion—Burkart et al., 1999: 35 (ecoregionalization).

Subantarctic subregion—Morrone, 1999: 13 (regionalization); Ocampo and Morrone, 1999: 21 (faunistics); Morrone, 2000b: 1 (regionalization), 2001b: 118 (regionalization); Posadas and Morrone, 2001: 267 (cladistic biogeography); Donato et al., 2003: 340 (dispersal-vicariance analysis); Posadas and Morrone, 2003: 72 (cladistic biogeography); Morrone, 2004a: 158 (regionalization); Posadas and Morrone, 2004: 353 (faunistics); Soares and de Carvalho, 2005: 485 (evolutionary biogeography); Morrone, 2006: 484 (regionalization); Spinelli et al., 2006: 302 (parsimony analysis of endemicity); Alzate et al., 2008: 1252 (track analysis); Casagranda et al., 2009: 19 (endemicity analysis); Escalante et al., 2009: 379 (endemicity analysis); Chani-Posse, 2010: 37 (faunistics); Martin, 2010: 1029 (ecological modelling); Morrone, 2010: 38 (regionalization); Urtubey et al., 2010: 506 (track analysis and cladistic biogeography); Martin, 2011: 122 (ecological modelling); Morrone, 2011: 2085 (evolutionary biogeography); Kutschker and Morrone, 2012: 541 (track analysis); Posadas, 2012: 2 (faunistics); Alfaro et al., 2013: 244 (faunistics); Fergnani et al., 2013: 296 (cluster analysis); Vivallo, 2013: 529

(systematic revision); Alfaro et al., 2014: 387 (faunistics); Marvaldi and Ferrer, 2014: 535 (faunistics); Carrara and Flores, 2015: 47 (faunistics); Curler et al., 2015: 116 (faunistics); Goin et al., 2015: 133 (biotic evolution); Morrone, 2015b: 213 (regionalization); Monckton, 2016: 125 (systematic revision and map); Ruiz et al., 2016: 383 (track analysis); Arana et al., 2017: 421 (shapefiles); Barriga and Cepeda, 2017: 19 (faunistics); Romano et al., 2017: 444 (track analysis).

Valdivean [sic]-Magellanian province—Rivas-Martínez et al., 2011: 27 (regionalization).

Endemic and characteristic taxa

Morrone (2000b, 2015b) provided a list of endemic and characteristic taxa. Some examples included *Cyttaria* (Cyttariaceae); *Pilgerodendron* (Cupressaceae); *Prumnopitys andina* (Podocarpaceae); *Oreobolus* spp. and *Schoenus* spp. (Cyperaceae); *Marsippospermum* spp. (Juncaceae); *Azorella filamentosa* and *A. selago* (Apiaceae); *Abrotanella* and *Perezia nutans* (Asteraceae); *Nothofagus* (Nothofagaceae); *Gomorteca* (Gomortecaceae); *Aristotelia chilensis* (Elaeocarpaceae); *Embothrium* (Proteaceae); *Misodendrum* (Misodendraceae); *Glossiphonia mesembrina* (Hirudinea: Glossiphoniidae); *Americobdella* (Hirudinea: Americobdellidae); *Gnolus* (Araneae: Mimetidae); *Austropsopilio* (Opiliones: Caddidae); *Austrochthonius chilensis* (Pseudoscorpiones: Chthoniidae); *Ilyocryptus brevidentatus* (Branchiopoda: Ilyocryptidae); *Atractuchus* and *Trichophtalmus* (Coleoptera: Belidae); *Antarctiola, Antarctonomus, Bembidarenas, Creobius, Cascellius* (Figure 5.1a; Roig-Juñent, 1995), *Merizodus* and *Pseudomigadops*

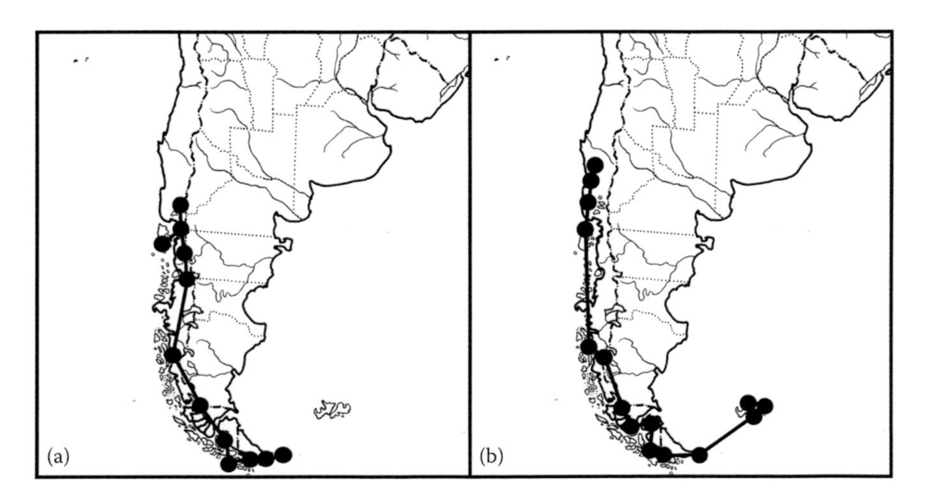

Figure 5.1 Maps with individual tracks in the Subantarctic subregion. (a) *Cascellius* (Coleoptera: Carabidae) and (b) *Germainiellus* (Coleoptera: Curculionidae).

(Coleoptera: Carabidae); *Stenomela* (Coleoptera: Chrysomelidae); *Antarctobius, Falklandius, Germainiellus, Germainius* (Figure 5.1b) and *Sinophloeus* (Coleoptera: Curculionidae); *Nannomacer* (Coleoptera: Nemonychidae); *Frickius* (Coleoptera: Scarabaeidae); *Notiocoenia pollinosa, Scatella neglectus* and *S. sturdeeanus* (Diptera: Ephydridae); *Mitrodetus* (Diptera: Mydidae); *Gigantodax antarcticus* and *Parastrosimulium* (Diptera: Simuliidae); *Aposycorax chilensis* (Diptera: Psychodidae; Curler et al., 2015); *Microtomus gayi* (Heteroptera: Reduviidae); *Gladicauda* (Hymenoptera: Diapriidae); *Cosila chilensis* (Hymenoptera: Tiphiidae); *Hypsochila argyrodile* (Lepidoptera: Pieridae); *Conchopterella, Hemerobius chilensis* and *Megalomus nigratus* (Neuroptera: Hemerobiidae); *Rialla villosa* (Odonata: Cordulidae); *Neopetalia punctata* (Odonata: Neopetalidae); Tropidostethini (Ortoptera: Tristiridae); *Neofulla* and *Udamocercia* (Plecoptera: Notonemouridae); *Batrachyla leptopus* (Leptodactylidae); *Rhinoderma* (Rhinodermatidae); *Chloephaga hybrida, C. poliocephala* and *Tachyeres patachonicus* (Anseriformes: Anatidae); *Buteo ventralis* (Accipitriformes: Accipitridae); *Carduelis barbata* (Passeriformes: Fringillidae); *Aphrastura spinicauda, Cinclodes antarcticus, Pygarrhichus albogularis* and *Sylviorhynchus desmursii* (Passeriformes: Furnariidae); *Turdus falcklandii* (Passeriformes: Muscicapidae); *Enicognathus ferrugineus* and *E. leptorhynchus* (Psittaciformes: Psittacidae); *Dromiciops australis* (Microbiotheria: Microbiotheriidae); *Hippocamelus bisulcus* and *Pudu puda* (Cetartiodactyla: Cervidae); *Felis guigna* (Carnivora: Felidae); and *Abrothrix longipilis, Auliscomys micropus* and *Irenomys tarsalis* (Rodentia: Cricetidae).

Biotic relationships

Several studies (Monrós, 1958; Kuschel, 1960; Cabrera and Willink, 1973; Cabrera, 1976) emphasized the distinctive character of the Subantarctic biota and its relationships with the biota of the Australian region. There were several plant genera shared by the Subantarctic and Australian regions, for example, *Abrotanella, Aristotelia, Astelia, Austrocedrus, Boquila, Carpha, Donatia, Drapetes, Eucryphia, Fitzroya, Gaimardia, Lardizabala, Laurelia, Laureliopsis, Lebetanthus, Lomatia, Luzuriaga, Marsippospermum, Nothofagus, Oreobolus, Orites, Phyllachne, Pilgerodendron, Pseudopanax, Rostkowia, Schoenus, Selliera* and *Tetrachondra* (Cabrera, 1976; Zuloaga et al., 1999). Villagrán and Hinojosa (1997) analyzed the plant taxa of this subregion and concluded that the low number of species per genus and the high number of monotypic genera suggested a great age for the biota.

Within the Andean region, current distributional patterns of plant and animal taxa suggested a close relationship between the Subantarctic and Central Chilean subregions (Morrone, 1994a). The weevils (Coleoptera: Curculionidae) of the tribe Aterpini (Morrone, 1996b) and the genus *Rhyephenes* (Figure 5.2; Morrone, 1996c) clearly showed this connection.

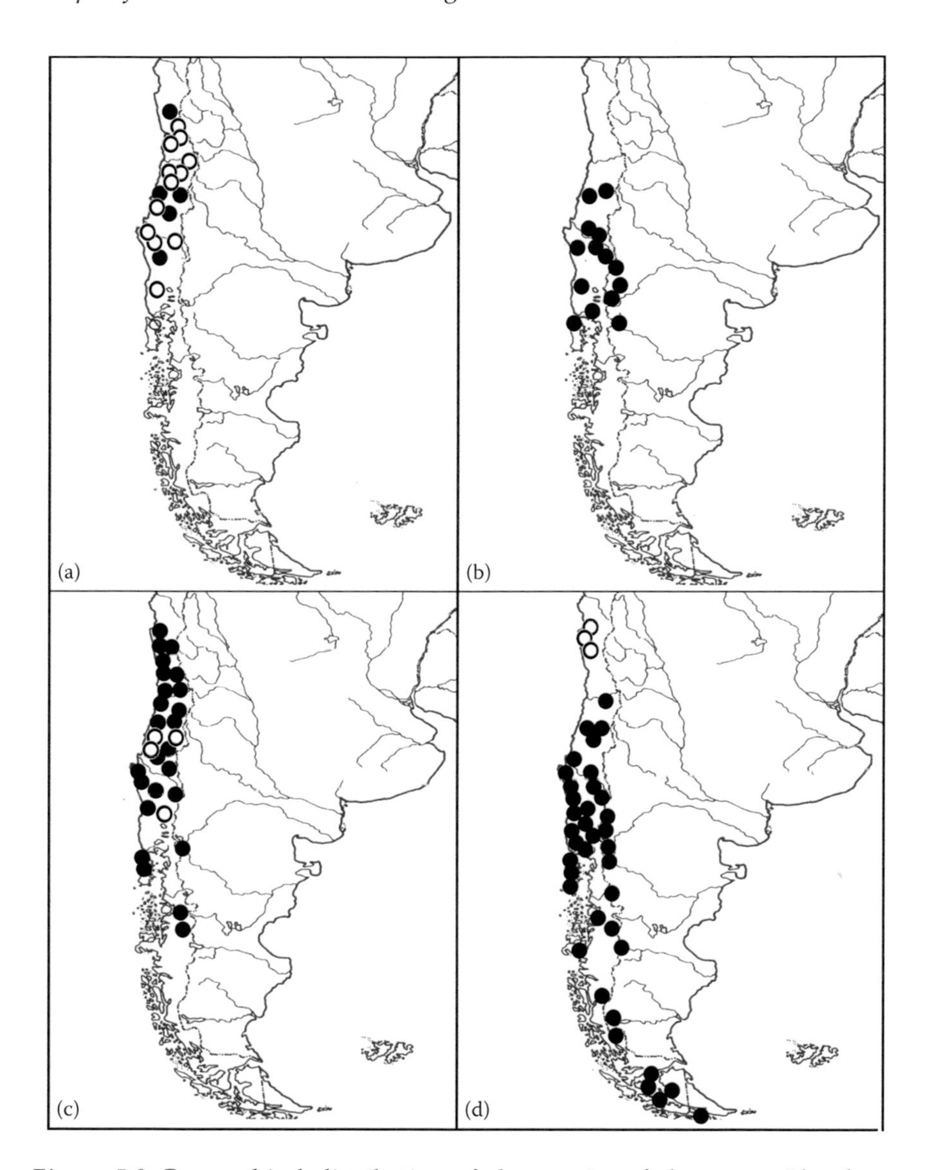

Figure 5.2 Geographical distribution of the species of the genus *Rhyephenes* (Coleoptera: Curculionidae) in the Subantarctic and Central Chilean subregions. (Modified from Morrone, J.J., *J. N. Y. Entomol. Soc.*, 104, 1–20, 1996c. With permission.) (a) *Rhyephenes clathratus* (black circles) and *R. gayi* (open circles), (b) *R. goureaui,* (c) *R. humeralis* (black circles) and *R. lateralis* (open circles) and (d) *R. maillei* (black circles) and *R. squamiger* (open circles).

Morrone (1993b) analyzed the distributional patterns of the genera *Germainiellus, Puranius, Antarctobius, Falklandius, Lanteriella, Telurus, Falklandiellus, Trachodema, Philippius* and *Haversiella* (Coleoptera: Curculionidae), combining the results of a track analysis and a cladistic biogeographical analysis. The track analysis identified a generalized track beginning in the Valdivian Forest, then joining the Magellanic Forest, then the Falkland Islands and finally the Magellanic Moorland. The general area cladogram showed the following sequence of vicariant events: (Magellanic Moorland [Valdivian Forest [Magellanic Forest, Falkland Islands]]). Based on these results, Morrone (1993b) inferred the biotic history of the areas, which he represented as a generalized track joining the four areas, and their splitting according to the general area cladogram (Figure 5.3).

Regionalization

The Subantarctic subregion comprises six provinces (Figure 5.4): Maule, Valdivian Forest, Magellanic Forest, Magellanic Moorland, Falkland Islands and Juan Fernández Islands (Morrone, 2000b, 2006, 2015b).

Case study: Cladistic biogeographical analysis of the Subantarctic subregion

Title: "Biogeografía histórica de la familia Curculionidae (Insecta: Coleoptera) en las subregiones Subantártica y Chilena Central" (Posadas, P. and J. J. Morrone, *Revista de la Sociedad Entomológica Argentina*, 62, 71–80, 2003.)

Goal: To analyze the historical relationships of the provinces of the Subantarctic subregion and the Central Chilean subregion.

Location: Andean region.

Methods: The units of the analysis were five provinces of the Subantarctic subregion (excluding the Juan Fernández province) and the Central Chilean subregion (Figure 5.5). The taxa analyzed comprised six genera and one generic group of weevils (Coleoptera: Curculionidae) for which phylogenetic analyses were available, for example, *Aegorhinus, Alastoropolus, Puranius, Germainiellus, Antarctobius, Rhyephenes* and *Falklandius* generic group (Figure 5.6). The study applied three methods: reconciled tree analysis, Brooks parsimony analysis and dispersal-vicariance analysis. The study applied reconciled tree analysis (Page, 1994) with Component 2.0 (Page, 1993), performing three tree searches minimizing duplications, extinctions and losses + extinctions, respectively. The study provided analysis of the data matrix for Brooks parsimony analysis (Wiley, 1988) with Hennig86 (Farris, 1988). The study applied dispersal-vicariance analysis (Ronquist, 1997) with DIVA 1.1 (Ronquist, 1996). The first two methods led to general

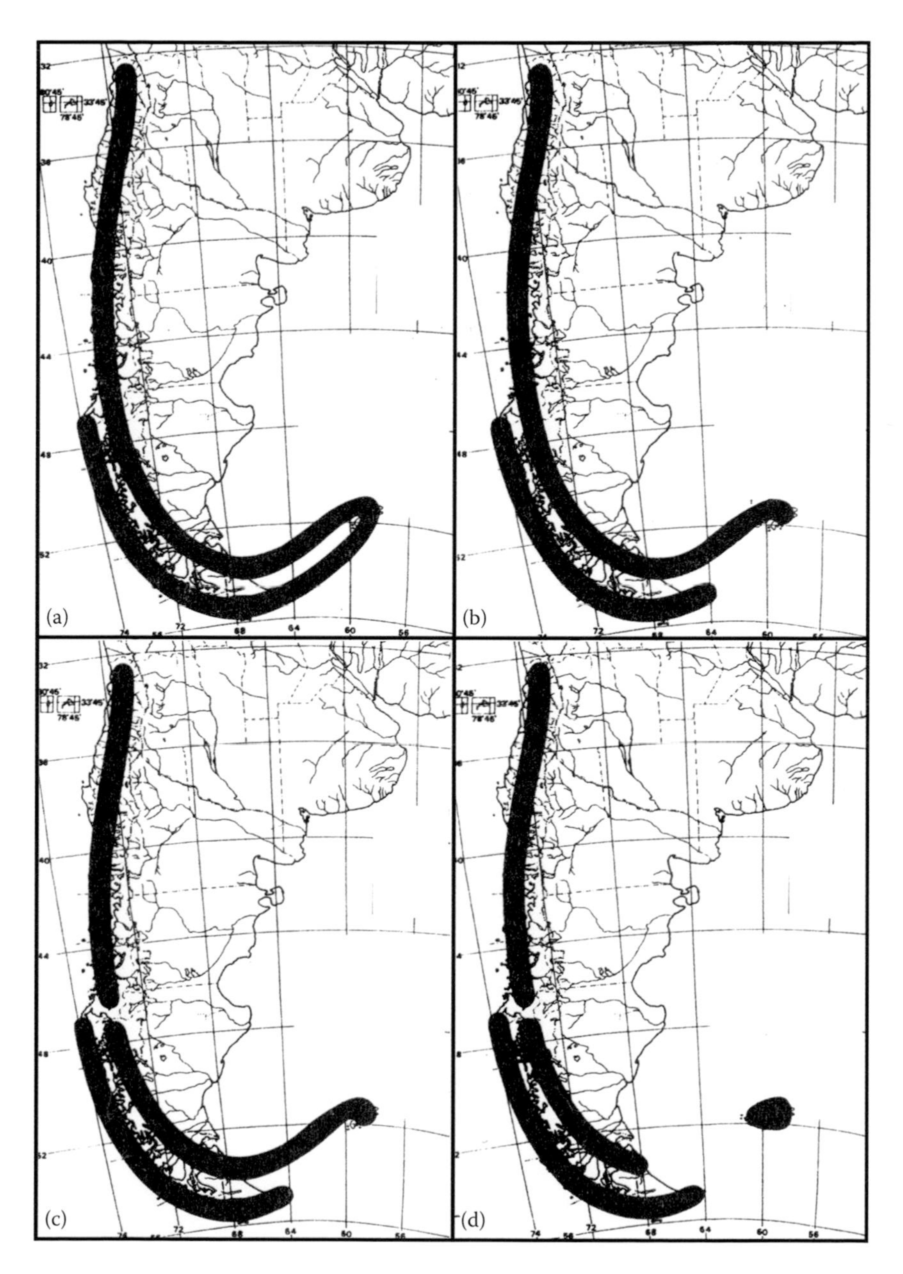

Figure 5.3 Maps showing the splitting of the Subantarctic generalized track based on species of Curculionidae. (Modified from Morrone, J.J., *Boletín de la Sociedad de Biología de Concepción*, 64, 121–145, 1993b. With permission.) (a) Subantarctic generalized track, (b) initial vicariance of the Magellanic Moorland, (c) vicariance of the Valdivian Forest and (d) final vicariance of the Magellanic forest–Falkland Islands.

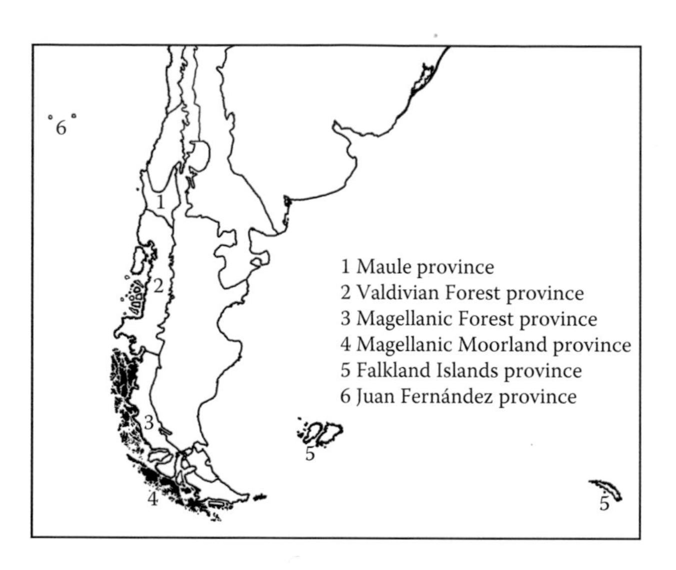

Figure 5.4 Map of the provinces of the Subantarctic subregion.

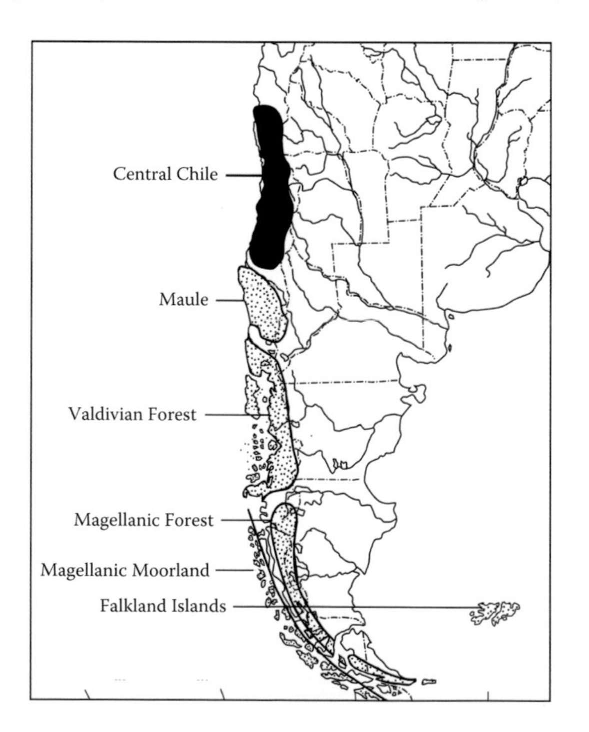

Figure 5.5 Areas analyzed in the Posadas and Morrone (2003) cladistic biogeographical analysis of the Subantarctic and Central Chilean subregions. (Modified from Posadas, P. and Morrone, J.J., *Revista de la Sociedad Entomológica Argentina*, 62, 71–80, 2003. With permission.)

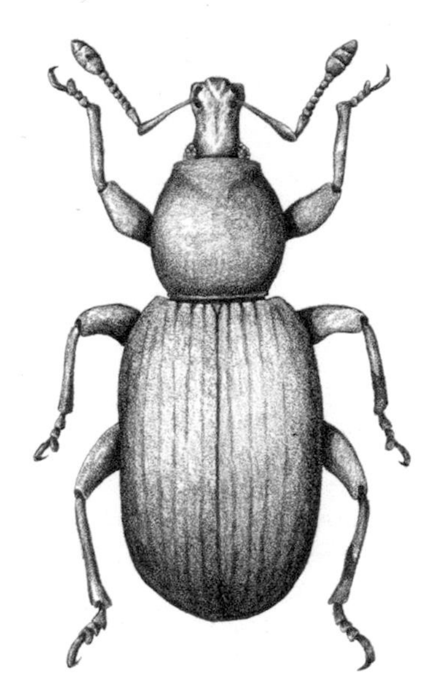

Figure 5.6 Falklandius antarcticus, a representative species of the *Falklandius* generic group (Coleoptera: Curculionidae) distributed in the Subantarctic subregion. (Courtesy of Sergio Roig-Juñent.)

area cladograms showing hierarchic relationships among the areas, whereas DIVA reconstructed ancestral areas, indicating the frequencies of postulated dispersal and vicariant events.

Results: The reconciled tree analysis and Brooks parsimony analysis identified the same general area cladogram (Figure 5.7a), which showed the Central Chilean subregion as the sister area to the two northern Subantarctic provinces (Maule and Valdivian Forest), and these areas sister to the clade including the Falkland Islands, Magellanic Forest and Magellanic Moorland. For the DIVA analysis (Figure 5.7b), the most frequent dispersal events involved: Maule–Valdivian Forest (19.65%), Valdivian Forest-Maule (12.80%), Maule-Central Chilean (16.73%) and Central Chilean-Maule (9.55%). Whereas the most frequent vicariant event implied the separation of the Falkland Islands and Magellanic Forest–Magellanic Moorland.

Main conclusions: The complex relationships between the northern part of the Subantarctic subregion and the Central Chilean subregion might be due to dispersal, especially from the Maule biogeographical province to both the Central Chilean subregion and the Valdivian Forest. This could be related to a stronger capacity of dispersal of the Maule biota. The relationships of the southern Subantarctic provinces, in turn, might be clearly due to vicariance events.

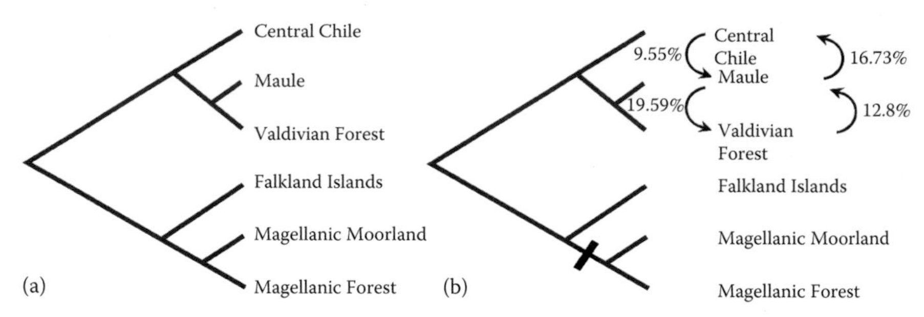

Figure 5.7 General area cladograms obtained in the Posadas and Morrone (2003) cladistic biogeographical analysis of the Subantarctic and Central Chilean subregions. (a) BPA and reconciled tree analysis and (b) DIVA analysis, showing the four most frequent dispersal events and on vicariant event. (Modified from Posadas, P. and Morrone, J.J., *Rev. Soc. Entomol. Argent.*, 62, 71–80, 2003. With permission.)

Maule province

Maule to Southern Concepción Littoral region—Reiche, 1905: map (regionalization).

South Chilean-Fuegian province (in part)—Skottsberg, 1905: 416 (regionalization).

Valdivian district (in part)—Osgood, 1943: 27 (regionalization).

Forest zone (in part)—Mann, 1960: 33 (regionalization).

Northern Valdivian Forest region (in part)—Peña, 1966a: 11 (regionalization), 1966b: 217 (regionalization).

Valdivian region (in part)—O'Brien, 1971: 203 (regionalization).

Maule district—Cabrera and Willink, 1973: 98 (regionalization); Posadas et al., 2001: 1328 (conservation biogeography).

Valdivian area (in part)—Artigas, 1975: map (regionalization).

Maule area—Morrone et al., 1994: 107 (cladistic biogeography); Roig-Juñent, 1994b: 181 (cladistic biogeography); Marvaldi and Ferrer, 2014: 535 (faunistics).

Valdivian province (in part)—Rivas-Martínez and Navarro, 1994: map (regionalization).

Maule subregion—Solervicens and Elgueta, 1994: 140 (faunistics).

Preandean Xeric Forest district—Roig, 1998: 141 (regionalization).

Maule province—Morrone, 1999: 14 (regionalization); Ocampo and Morrone, 1999: 24 (faunistics); Morrone, 2000b: 3 (regionalization), 2001b: 122 (regionalization); Posadas and Morrone, 2001: 267 (cladistic biogeography), 2003: 72 (cladistic biogeography); Morrone, 2004a: 158 (regionalization); Soares and de Carvalho, 2005: 485 (biotic evolution); Morrone, 2006: 485 (regionalization); Spinelli et al., 2006: 302 (parsimony analysis of endemicity); Alzate et al., 2008: 1252 (track analysis); Casagranda et al., 2009: 22 (endemicity analysis); Escalante et al., 2009: 379 (endemicity analysis); Chani-Posse, 2010: 45 (faunistics); Morrone, 2010: 38 (regionalization); Kutschker and Morrone, 2012: 543 (track analysis); Alfaro et al., 2014: 387 (faunistics); Morrone, 2015b: 214 (regionalization); Monckton, 2016: 125 (systematic revision and map); Ruiz et al., 2016: 389 (track analysis); Arana et al., 2017: 421 (shapefiles); Barriga and Cepeda, 2017: 19 (faunistics); Martínez et al., 2017: 479 (track analysis and regionalization).

Araucanian region—Roig-Juñent and Domínguez, 2001: 557 (faunistics).

Valdivean [sic] province (in part)—Rivas-Martínez et al., 2011: 27 (regionalization).

Definition

The Maule province covers southern Chile and Argentina, between 37°S and 39°S (Cabrera and Willink, 1973; Morrone, 2000b, 2001b, 2006, 2015b).

Endemic and characteristic taxa

Morrone (2000b, 2015b) provided a list of endemic and characteristic taxa. Some examples included *Araucaria araucana* (Araucariaceae); *Chaetanthera serrata* (Figure 5.8a) and *Triptilion achilleae* (Figure 5.8b; Asteraceae); *Acanthogonatus brunneus*, *A. hualpen*, *A. mulchen*, *A. nahuelbuta*, *A. recinto* and *A. tolhuaca* (Araneae: Nemesiidae); *Hylotribus signatipes* (Coleoptera: Anthribidae); *Atractuchus argus*, *Callirhynchinus exquisitus*, *Dicordylus*

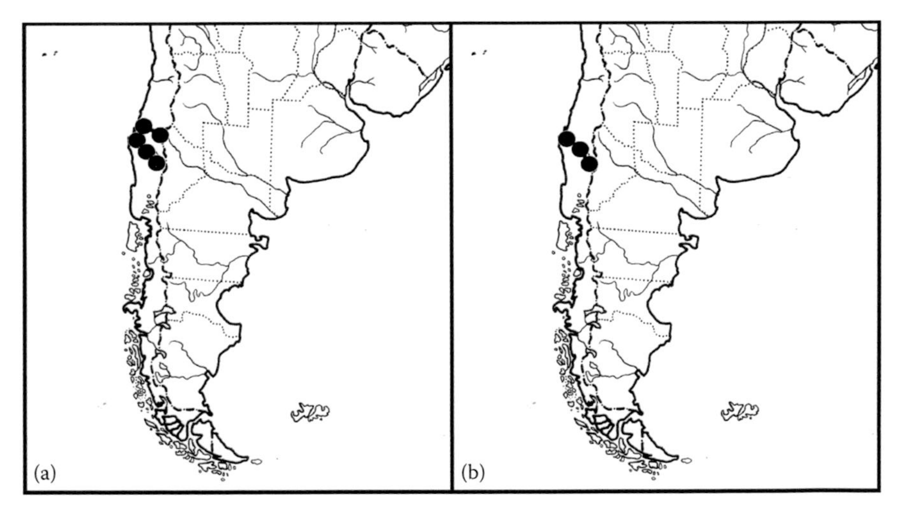

Figure 5.8 Maps with individual tracks in the Maule province. (a) *Chaetanthera serrata* (Asteraceae) and (b) *Triptilion achilleae* (Asteraceae).

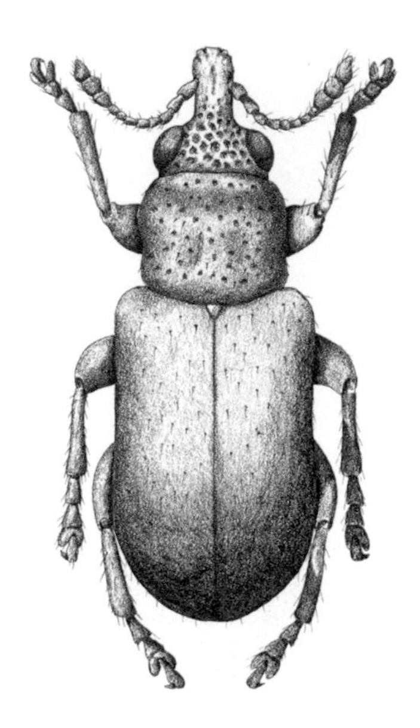

Figure 5.9 *Nannomacer germaini,* a Nemonychidae (Coleoptera) endemic to the Maule province. (Courtesy of Sergio Roig-Juñent.)

balteatus and *Oxycraspedus* (Coleoptera: Belidae); *Ceroglossus chilensis, C. chilensis temucensis, Cnemalobus germaini* and *Cylindera chilensis* (Coleoptera: Carabidae); *Caenominurus* (Coleoptera: Caridae); *Aegorhinus albolineatus, A. nitens, A. silvicola, A. suturalis, Anthonomus araucanus, Araucarietus, Araucarius, Berberidicola carinatus, Calvertius tuberosus, Dasydema annucella, Eisingius, Lamiarhinus horridus, Megalometides, Megalometis andigena, Nothofaginoides, Nothofagius fimbriatus, Omoides validus, Planus* and *Tartarisus perforatipennis* (Coleoptera: Curculionidae); *Palophagoides* (Coleoptera: Megalopodidae); *Mecomacer, Nannomacer germaini* (Figure 5.9) and *Rhynchitomacerinus* (Coleoptera: Nemonychidae); *Ctenomys maulinus* (Rodentia: Ctenomyidae); and *Aconaemys sagei* and *Octodon bridgesi* (Rodentia: Octodontidae).

Vegetation

The Maule province has transitional temperate forests (Figure 5.10; Luebert and Pliscoff, 2006) with some elements of the Central Chilean subregion. According to Smith (2017), the forests of the Maule province are transitional between the Mediterranean sclerophyllous forests of Central Chile and the wet temperate Valdivian forest (Figure 5.11). Dominant plant species include *Acaena pinnatifida, Alstroemeria aurantiaca, Aristotelia chilensis, Baccharis concava, Berberis buxifolia, Boquila trifoliata, Chusquea*

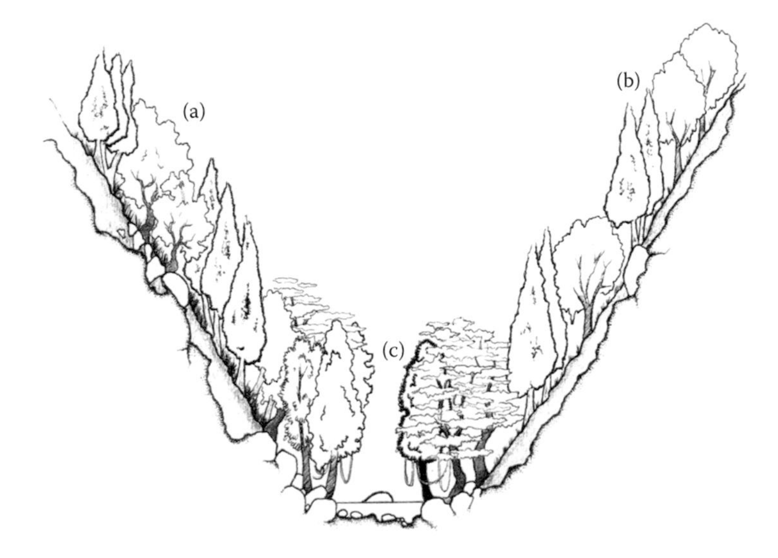

Figure 5.10 Vegetation in the Maule province. (Modified from Luebert, F. and Pliscoff, P., *Sinopsis bioclimática y vegetacional de Chile*. Editorial Universitaria, Santiago de Chile, 2006.) (a) Forest with *Austrocedrus chilensis*, (b) forest with *Quillaja saponaria* and *Austrocedrus chilensis* and (c) forest with *Nothofagus dombeyi*.

Figure 5.11 Forest of the Maule province, Reserva Nacional Los Ruiles, Chile. (Courtesy of Mario Elgueta.)

couleu, Cortaderia pilosa, Cryptocaria mammosa, Escallonia virgata, Lapageria rosea, Laurelia sempervicens, Myrtus luma, Nothofagus alexandri, N. dombeyi, N. leonii, N. obliqua, N. pumilio, Pernetya mucronata, Persea lingue, Podocarpus nubigena, Rhaphithammus spinosum and *Ribes magellanicum* (Cabrera, 1971, 1976; Cabrera and Willink, 1973). There are small *Araucaria araucana* forests in the Pehuén district where this species coexists with *Nothofagus dombeyi, N. pumilio, Austrocedrus chilensis* and *Chusquea coleou* (Peña, 1966a; Cabrera and Willink, 1973; Roig, 1998).

Biotic relationships

A track analysis based on species of weevils (Coleoptera: Curculionidae; Morrone, 1996b) and another based on species of Valerianaceae (Kutschker and Morrone, 2012) postulated that the Maule province was a node between the Subantarctic and Central Chilean subregions. Smith (2017) combined the Maule and Valdivian Forest provinces in a single ecoregion. A cladistic biogeographical analysis based on species of Coleoptera (Morrone et al., 1994) showed that the Maule province is the sister area to the remaining Subantarctic provinces. Another cladistic biogeographical analysis, based on species of Diptera (Soares and de Carvalho, 2015), showed a close relationship of this province with the Valdivian Forest and Magellanic Forest provinces.

Regionalization

Morrone (2015b) treated four areas previously recognized by Cabrera (1951, 1971, 1976), Peña (1966a, b), Artigas (1975) and Roig (1998) as the Angol, Chillán Cordillera, Pehuén and Temuco districts. Their circumscription is not clear (Mario Elgueta, pers. comm.).

Angol district

Angol zone—Artigas, 1975: map (regionalization).

Angol district—Morrone, 2015b: 215 (regionalization); Monckton, 2016: 125 (systematic revision and map).

Definition: The Angol district extends between 37° 30′ S and 39° 15′ S (Artigas, 1975).

Vegetation: Within this district, Luebert and Pliscoff (2006) identified different vegetation types. In the western slopes of the Cordillera de la Costa, between sea level and 800 m there is a sclerophyllous forest dominated by *Lithraea caustica* (Anacardiaceae), *Cryptocarya alba* (Lauraceae) and *Azara integrifolia* (Salicaceae); other species are *Adiantum chilense, Alstroemeria revoluta, Azara integrifolia, Bomarea salsilla, Chusquea cummingii, Colletia hystrix, Escalonia revoluta, Lomatia hirsuta, Myrceugenia obtusa, Peumus boldus, Proustia pyrifolia* and *Triptilion spinosum*. Dominant in the coastal deciduous forest are *Nothofagus obliqua* (Nothofagaceae), *Gomortega keule* (Gomortegaceae), *Podocarpus saligna* (Podocarpaceae), *Lomatia dentata* (Proteaceae) and *Aextoxicon punctatum* (Aextoxicaceae); other species are *Aristotelia chilensis, Berberidopsis coralina, Caldluvia paniculata, Gevuina avellana, Lapageria rosea, Pitavia punctata* and *Teline monspessulana*. Dominant in the mixed coastal temperate forest are *Nothofagus dombeyi, N. obliqua* (Nothofagaceae) and *Eucryphia cordifolia* (Cunoniaceae); other species are *Dasyphyllum diacanthoides, Drimys winteri, Gevuina avellana, Laureliopsis philippiana, Luma apiculata, Persea lingue, Pseudopanax laetevirens, Sophora microphylla* and *Weinmannia trichosperma*. Dominant on Mocha island and the western slopes of the Nahuelbuta cordillera in the temperate forest are *Aextoxicon punctatum* (Aextoxicaceae) and *Laurelia sempervivens* (Atherospermataceae); other species are *Amomyrtus luma, Azara lanceolata, Boquila trifoliolata, Chusquea quila, Cissus striata, Eucryphia cordifolia, Lapageria rosea, Luma apiculata, Persea lingue, Peumus boldus* and *Rhaphithamnus spinosus*.

Chillán Cordillera district

Cordilleras from Chillán to Tierra del Fuego region (in part)—Reiche, 1905: map (regionalization).

Chillán Cordillera zone—Artigas, 1975: map (regionalization).

Chillán Cordillera district—Morrone, 2015b: 215 (regionalization); Monckton, 2016: 125 (systematic revision and map).

Definition: The Chillán Cordillera district extends in the eastern part of the province, between 35° 30′ S and 37° 30′ S (Artigas, 1975).

Vegetation: Luebert and Pliscoff (2006) characterized the interior Mediterranean deciduous forest dominated by *Nothofagus obliqua* (Nothofagaceae), *Cryptocarpa alba* (Lauraceae) and *Peumus boldus* (Monimiaceae); other species are *Aextoxicon punctatum, Azara dentata, Blechnum hastatum, Escallonia pulverulenta, Gevuina avellana, Lapageria rosea, Persea lingue, Podocarpus saligna, Quillaja saponaria* and *Uncinia phleoides*.

Pehuén district

Pehuén district—Cabrera, 1951: 61 (regionalization), 1971: 37 (regionalization); Cabrera and Willink, 1973: 99 (regionalization); Cabrera 1976: 73 (regionalization); Morrone, 2015b: 215 (regionalization); Monckton, 2016: 125 (map); Barriga and Cepeda, 2017: 19 (faunistics).

Araucaria forests subregion—Hueck, 1957: 40 (regionalization).

Pehuenar region—Peña, 1966a: 12 (regionalization), 1966b: 217 (regionalization).
Nahuelbuta zone—Artigas, 1975: map (regionalization).
Chilean *Araucaria* Forest province—Udvardy, 1975: 41 (regionalization).
Araucaria araucana Forest subdistrict—Roig, 1998: 141 (regionalization).
West Pehuenar range—Flores and Vidal, 2000: 193 (systematic revision).
Araucanía region—Roig-Juñent and Domínguez, 2001: 557 (faunistics).
Definition: The Pehuén district corresponds to the *Araucaria araucana* forests, which extend in two disjunct areas: coastal from 37° 30′ to 38°S, and Andean from 37° 30′ to 39°S (Peña, 1966a).
Endemic taxa: *Araucaria araucana* (Araucariaceae); *Oxycraspedus* (Coleoptera: Belidae); *Paraholopterus nahuelbutensis* (Coleoptera: Cerambycidae); *Araucarietus* (Coleoptera: Curculionidae); *Stenomela viviane* (Coleoptera: Chrysomelidae; Barriga and Cepeda, 2017); *Palophagoides vargasorum* (Coleoptera: Megalopodidae; Kuschel and May, 1996); *Telmatobufo bullocki* (Anura: Calyptocephalellidae); and *Alsodes barrioi* and *A. vanzolinii* (Anura: Alsodidae; Charrier, 2012).
Vegetation: Luebert and Pliscoff (2006) characterized the temperate forest dominated by *Araucaria araucana* (Figure 5.12; Araucariaceae) and *Nothofagus pumilio* (Nothofagaceae); other species are *Adenocaulon chilense, Alstroemeria aurea, Azara alpina, Berberis serrato-dentata, Chusquea culeou, Festuca scabriuscula, Maytenus disticha, Pernettya myrtilloides, Senecio angustissimus, S. chilense* and *Viola magellanica*.

Figure 5.12 *Araucaria araucana*, endemic to the Pehuén district, Parque Nacional Los Paraguas, Chile. (Courtesy of Sergio Roig-Juñent.)

Temuco district

Temuco zone—Artigas, 1975: map (regionalization).

Temuco district—Morrone, 2015b: 215 (regionalization); Monckton, 2016: 125 (systematic revision and map).

Definition: The Temuco district extends in the southernmost part of the province between 38° 30′ and 40°S (Artigas, 1975).

Vegetation: Luebert and Pliscoff (2006) characterized the temperate deciduous forest as dominated by *Nothofagus obliqua* (Nothofagaceae), *Laurelia sempervivens* (Atherospermataceae), *Podocarpus saligna* (Podocarpaceae) and *Eucryphia cordifolia* (Cunoniaceae); other species are *Aextoxicon punctatum, Azara dentata, Blechnum hastatum, Escallonia pulverulenta, Gevuina avellana, Lapageria rosea, Persea lingue, Podocarpus saligna, Quillaja saponaria* and *Uncinia phleoides*. Dominant in the interior temperate forest are *Nothofagus dombeyi* (Nothofagaceae) and *Eucryphia cordifolia* (Cunoniaceae); other species are *Amomyrtus luma, A. meli, Azara lanceolata, Blechnum blechnoides, Caldluvia paniculata, Chusquea quila, Fuchsia magellanica, Hydrangea serratifolia, Hymenophyllum caudiculatum, Lomatia hirsuta, Luzuriaga radicans, Pseudopanax laetevirens, Rhaphithamnus spinosus, Sarmienta repens* and *Saxegothaea conspicua*.

Valdivian Forest province

Austral Littoral region (in part)—Reiche, 1905: map (regionalization).

South Chilean-Fuegian province (in part)—Skottsberg, 1905: 416 (regionalization).

Valdivian Forest area—Hauman, 1920: 53 (regionalization); Soriano, 1950: 33 (regionalization); Morrone et al., 1994: 108 (cladistic biogeography); Roig-Juñent, 1994b: 181 (cladistic biogeography); Marvaldi and Ferrer, 2014: 535 (faunistics).

Valdivian district—Osgood, 1943: 27 (regionalization); Cabrera, 1951: 60 (regionalization), 1971: 38 (regionalization); Cabrera and Willink, 1973: 98 (regionalization); Cabrera, 1976: 75 (regionalization).

Subantarctic Forests subregion—Hueck, 1957: 40 (regionalization).

Mountain zone (in part)—Kuschel, 1960: 545 (regionalization).

Valdivian Forest zone—Kuschel, 1960: 541 (regionalization).

Forest zone (in part)—Mann, 1960: 33 (regionalization).

Valdivian Cordillera region (in part)—Peña, 1966a: 15 (regionalization), 1996b: 218 (regionalization).

Valdivian Forest region (in part)—Peña, 1966a: 14 (regionalization), 1996b: 218 (regionalization).; Roig-Juñent and Domínguez, 2001: 557 (faunistics).

Deciduous Forest district—Cabrera, 1971: 37 (regionalization), 1976: 73 (regionalization); Cabrera and Willink, 1973: 100 (regionalization).
Valdivian region (in part)—O'Brien, 1971: 203 (regionalization).
Aysenian area—Artigas, 1975: map (regionalization).
Valdivian area (in part)—Artigas, 1975: map (regionalization).
Valdivian Forest province—Udvardy, 1975: 41 (regionalization); Morrone, 2001b: 123 (regionalization); Posadas and Morrone, 2001: 267 (cladistic biogeography), 2003: 72 (cladistic biogeography); Morrone, 2004a: 158 (regionalization); Soares and de Carvalho, 2005: 485 (evolutionary biogeography); Morrone, 2006: 485 (regionalization); Spinelli et al., 2006: 302 (parsimony analysis of endemicity); Casagranda et al., 2009: 22 (endemicity analysis); Escalante et al., 2009: 379 (endemicity analysis); Morrone, 2010: 38 (regionalization); Posadas, 2012: 5 (faunistics); Alfaro et al., 2013: 243 (faunistics); Curler et al., 2015: 116 (faunistics); Morrone, 2015b: 215 (regionalization); Monckton, 2016: 125 (systematic revision and map); Ruiz et al., 2016: 385 (track analysis); Arana et al., 2017: 421 (shapefiles); Martínez et al., 2017: 479 (track analysis and regionalization); Romano et al., 2017: 445 (track analysis).
Valdivian province—Rivas-Martínez and Navarro, 1994: map (regionalization); Morrone, 1999: 14 (regionalization); Ocampo and Morrone, 1999: 24 (faunistics); Morrone, 2000b: 4 (regionalization); Chani-Posse, 2010: 45 (faunistics); Escalante, 2017: 351 (parsimony analysis of endemicity).
Valdivian Forest subregion—Solervicens and Elgueta, 1994: 141 (faunistics).
Valdivian Temperate Forests ecoregion—Dinerstein et al., 1995: 101 (ecoregionalization); Martin, 2010: 1025 (ecological modelling), 2011: 122 (ecological modelling); Smith, 2017: 1 (ecoregionalization).
Austrocedrus chilensis Forest subdistrict—Roig, 1998: 141 (regionalization).
Valdivian Forest district—Roig, 1998: 140 (regionalization); Posadas et al., 2001: 1328 (conservation biogeography).
Valdivean [sic] province (in part)—Rivas-Martínez et al., 2011: 27 (regionalization).
Valdivian Rainforest province—Alfaro et al., 2014: 387 (faunistics).

Definition

The Valdivian Forest province covers southern Chile south of the Maule province between 39°S and 47°S, and Argentina in the western margins of the Río Negro, Neuquén and Chubut provinces (Cabrera and Willink, 1973; Morrone, 2000b, 2001b, 2006, 2015b). Altitude ranges from sea level to 1500 m (Lücking et al., 2003). Annual precipitation can reach more than 6000 mm in the southern part of the province in Chile; precipitation decreases significantly on the eastern slope of the Andes in Argentina (Smith, 2017).

Endemic and characteristic taxa

Morrone (2000b, 2015b) provided a list of endemic and characteristic taxa. Some examples included *Austrocedrus* and *Fitzroya* (Cupressaceae); *Podocarpus nubigena* and *Saxegothaea* (Podocarpaceae); *Crinodendron hookerianum* (Figure 5.13a; Elaeocarpaceae); *Misodendrum angulatum* (Figure 5.13b), *M. brachystachyum* and *M. gayanum* (Misodendraceae); *Aemalodera limbata, Allendia chilensis, Bembidiomorphum, Ceroglossus valdiviae chiloensis, Chaetauchenium loki, Creobius eudouxy, Crossonychus viridis, Monolobus ovalipennis, Nothanillus luisae, Nothobroscus chilensis, Thalassobius testaceus, Trachysarus antarcticus, Trechisibus obtusiusculus, Tropopsis biguttata* and *T. marginicollis* (Coleoptera: Carabidae); *Chilecar pilgerodendri* (Coleoptera: Caridae); *Dorymolpus elizabethae* (Coleoptera: Chrysomelidae); *Antichthonidris bidentatus* (Hymenoptera: Formicidae); *Osornogyndes tumifrons* (Opiliones: Gonyleptidae); *Chilenodes* (Araneae: Malkaridae); *Apophatus parvus, Palaephatus albiterminus, P. amplisaccus, P. fusciterminus, Sesommata albimaculata, S. leuroptera* and *S. paraplatysaris* (Diptera: Palaephatidae); *Pseudopsis adustipennis* (Coleoptera: Staphylinidae); *Diasia* (Opiliones: Triaenonychidae); *Rhyncholestes raphanurus* (Paucituberculata: Caenolestidae); *Dromiciops gliroides* (Microbiotheria: Microbiotheriidae); *Pseudalopex fulvipes* (Carnivora: Canidae); and *Abrothrix sanborni* and *Pearsonomys annectens* (Rodentia: Cricetidae). Smith (2017) characterized this province as extraordinary endemism with a high number of phylogenetically isolated genera belonging to monogeneric families (e.g., Aextoxicaceae, Gomortecaceae, Desfontainaceae, Eucryphiaceae and Misodendraceae).

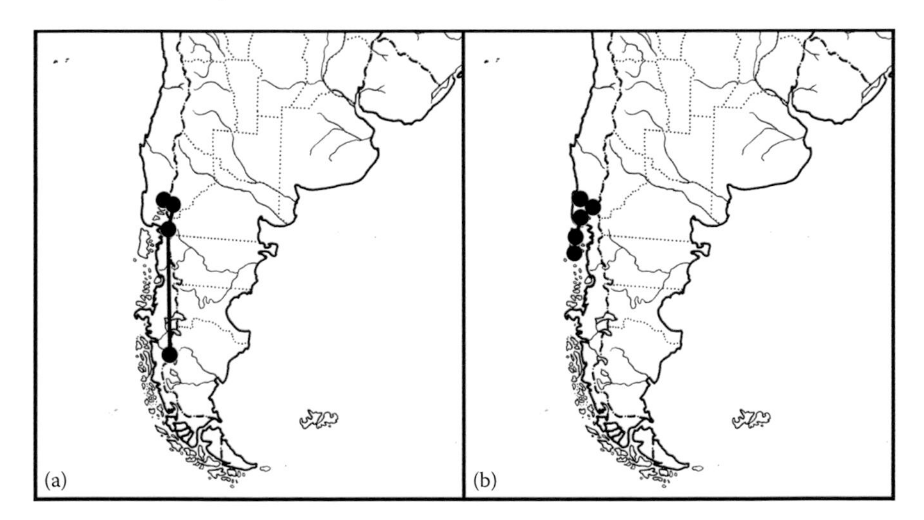

Figure 5.13 Maps with individual tracks in the Valdivian Forest province. (a) *Crinodendron hookerianum* (Elaeocarpaceae) and (b) *Misodendrum angulatum* (Misodendraceae).

Vegetation

The Valdivian Forest province has temperate forests (Figures 5.14 and 5.15; Kuschel, 1960; Luebert and Pliscoff, 2006; Smith, 2017). Dominant plant species include *Aextoxicon punctatum, Anemone multifida, Austrocedrus chilensis, Berberis* spp., *Blechnum chilense, Chusquea couleu, Coriaria ruscifolia, Dasyphyllum diacanthoides, Diostea juncea, Drimys andina, D. winteri, Eucryphia cordifolia, Fabiana imbricata, Festuca scabriuscula, Fragaria chiloensis, Fitzroya cupressoides, Gevuina avellana, Laureliopsis philippiana, L. sempervivens, Lomatia hirsuta, Maytenus boaria, Mutisia decurrens, M. spinosa, Myrceugenella apiculata, Nothofagus alpina, N. antarctica, N. betuloides, N. dombeyi, N. obliqua, N. pumilio, Podocarpus nubigena, Senecio microcephalus* and *Weinmannia trichosperma* (Soriano, 1950; Cabrera and Willink, 1973; Roig, 1998; Lücking et al., 2003; Luebert and Pliscoff, 2006;

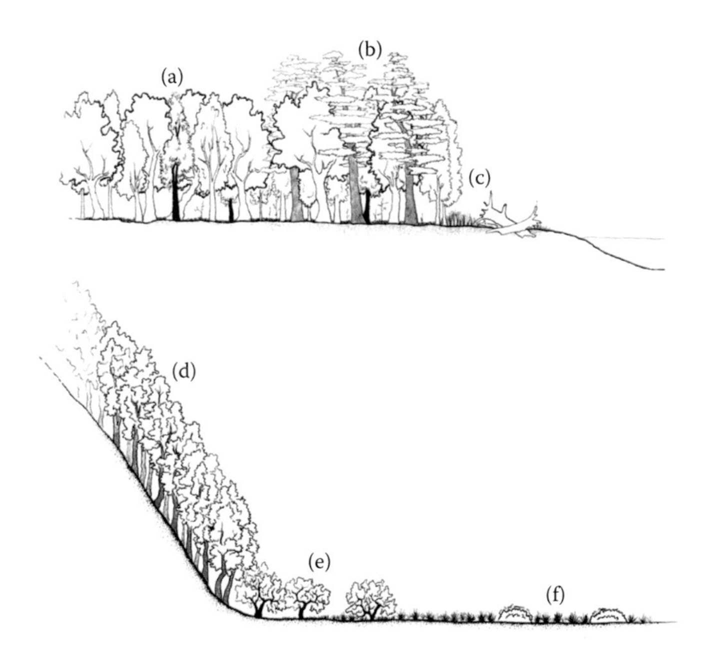

Figure 5.14 Vegetation in the Valdivian Forest province. (Modified from Luebert, F. and P. Pliscoff, *Sinopsis bioclimática y vegetacional de Chile*. Editorial Universitaria, Santiago de Chile, 2006.) (a)–(c) Temperate forest with *Nothofagus obliqua* and *Laurelia sempervivens* and (d)–(f) forest-scrubland-steppe transition. (a) Sclerophyllous forest with *Nothofagus obliqua*, (b) more dense forest, (c) herbs, (d) forest with *Nothofagus pumilio* and *Berberis ilicifolia*, (e) shrubland with *Nothofagus antarctica* and *Berberis microphylla* and (f) steppe with *Festuca pallescens* and *Mulinum spinosum*.

Figure 5.15 *Nothofagus* forest of the Valdivian Forest province, Lake Fagnano, Argentina. (Courtesy of Sergio Roig-Juñent.)

Moreira-Muñoz, 2011). Forests with *Fitzroya cupressoides* (Cupressaceae) are in areas with higher humidity (Roig, 1998).

Biotic relationships

Kuschel (1960) considered that the Valdivian Forest province showed faunistic relationships with the Magellanic Forest province. Lücking et al. (2003) analyzed the lichen taxa of the Valdivian Forest province, detecting clear biotic affinities with taxa from similar forests in New Zealand and Tasmania (Australian region), and characterized these lichens as belonging clearly to an Austral element. Smith (2017) combines the Maule and Valdivian Forest provinces in a single ecoregion.

A cladistic biogeographical analysis based on species of Coleoptera (Morrone et al., 1994) showed that the Valdivian Forest province is intermediate between the Maule province and the remaining Subantarctic provinces. Another cladistic biogeographical analysis, based on species of Diptera (Soares and de Carvalho, 2005), found a closer relationship of this province with the Maule and Magellanic Forest provinces.

Regionalization

Morrone (2015b) treated five areas recognized by Peña (1966a, b) and Artigas (1975) as the Aysén Cordillera, Llanquihue, South Chiloé, Valdivian and Valdivian Cordillera districts.

Aysén Cordillera district

Cordilleras from Chillán to Tierra del Fuego region (in part)—Reiche, 1905: map (regionalization).

Aysén Cordillera region—Peña, 1966a: 15 (regionalization), 1966b: 219 (regionalization).

Aysén Cordillera zone—Artigas, 1975: map (regionalization).

Aysén Cordillera district—Morrone, 2015b: 216 (regionalization).

Definition: The Aysén Cordillera district extends from 45°S southward to Cerro Payne in Magallanes, Chile (Peña, 1966a).

Vegetation: Luebert and Pliscoff (2006) characterized different vegetation types. Dominant in the temperate deciduous forest are *Nothofagus pumilio* (Nothofagaceae) and *Berberis ilicifolia* (Berberidaceae); other species are *Acaena ovalifolia, Berberis serrato-dentata, Chiliotrichum diffusum, Codonorchis lessonii, Embothrium coccineum, Leucheria thermarum, Nothofagus betuloides, Osmorhiza chilensis, Poa alopecurus* and *Rubus geoides.* Dominant in the arborescent scrubland are *Nothofagus antarctica* (Nothofagaceae) and *Berberis microphylla* (Berberidaceae); other species are *Acaena splendens. Baccharis patagonica, Berberis darwini, Festuca pallescens, Geranium berterianum, Mulinum spinosum, Ovidia andina, Ranunculus peduncularis* and *Ribes magellanicum.* Also Dominant in the temperate Mediterranean steppes are *Festuca pallescens* (Poaceae) and *Mulinum spinosum* (Apiaceae); other species are *Adesmia boroniodes, Azorella caespitosa, Cerastium arvense, Colliguaya integerrima, Elymus patagonicus, Galium antarcticum, Geranium patagonicum, Poa ligularis, Nassauvia abbreviata, Quinchamalium chilense* and *Vicia bijuga.*

Llanquihue district

Llanquihue zone—Artigas, 1975: map (regionalization).

Llanquihue district—Morrone, 2015b: 216 (regionalization).

Definition: The Llanquihue district extends from 41°S to 43°S, extending over continental Chile and the northern portion of the Chiloé island (Artigas, 1975).

Vegetation: Luebert and Pliscoff (2006) characterized the interior temperate forest as dominated by *Nothofagus dombeyi* (Nothofagaceae) and *Eucryphia cordifolia* (Cunoniaceae); other species were *Amomyrtus luma, A. meli, Azara lanceolata, Blechnum blechnoides, Caldluvia paniculata, Chusquea quila, Fuchsia magellanica, Hydrangea serratifolia, Hymenophyllum caudiculatum, Lomatia hirsuta, Luzuriaga radicans, Pseudopanax laetevirens, Rhaphithamnus spinosus, Sarmienta repens* and *Saxegothaea conspicua.*

South Chiloé district

South Chiloé zone—Artigas, 1975: map (regionalization).
South Chiloé district—Morrone, 2015b: 216 (regionalization).
Definition: The South Chiloé district corresponds to the southern portion of the Chiloé island (Artigas, 1975).
Vegetation: Luebert and Pliscoff (2006) characterized the coastal temperate forest as dominated by *Pilgerodendron uviferum* (Cupressaceae) and *Tepualia stipularis* (Myrtaceae); other species are *Berberis ilicifolia, Blechnum chilense, Campsidium valdividanum, Drimys winteri, Embothrium coccineum, Gleichenia quadripartita, Lebetanthus myrsinites, Lomatia ferruginea, Myrceugenia parvifolia, Pernettya mucronata, Podocarpus nubigena, Pseudopanax laetevirens, Saxegothaea conspicua* and *Tepualia stipularis*.

Valdivian district

Valdivian zone—Artigas, 1975: map (regionalization).
Valdivian district—Morrone, 2015b: 216 (regionalization); Monckton, 2016: 125 (systematic revision and map).
Definition: The Valdivian district extends from 39°S to 48°S in Chile and reaching small parts of Argentina (Peña, 1966a).
Vegetation: Luebert and Pliscoff (2006) characterized different vegetation types. Dominant in the temperate deciduous forest were *Nothofagus obliqua* (Nothofagaceae), *Laurelia sempervivens* (Atherospermataceae), *Podocarpus saligna* (Podocarpaceae) and *Eucryphia cordifolia* (Cunoniaceae); other species are *Aextoxicon punctatum, Azara dentata, Blechnum hastatum, Escallonia pulverulenta, Gevuina avellana, Lapageria rosea, Persea lingue, Podocarpus saligna, Quillaja saponaria* and *Uncinia phleoides*. Dominant in the coastal temperate forest are *Weinmannia trichosperma* (Cunoniaceae), *Laureliopsis phillippiana* (Monimiaceae) and *Eucryphia cordifolia* (Cunoniaceae); other species are *Anomomyrtus luma, Azara lanceolata, Chusquea quila, Crinodendrom hookerianum, Fascicularia bicolor, Gevuina avellana, Griselinia ruscifolia, Mitraria coccinea, Myrceugenia ovata, Saxegothaea conspicua, Vestia foetida* and *Weinmmania trichosperma*. Dominant in the interior evergreen temperate forest are *Nothofagus nitida* (Nothofagaceae) and *Podocarpus nubigena* (Podocarpaceae); other species are *Asplenium dareoides, Asteranthera ovata, Crinodendron hookerianum, Desfontainia spinosa, Eucryphia cordifolia, Luma apiculata, Philesia magellanica, Pseudopanax laetevirens, Sarmienta repens, Tepualia stipularis* and *Weinmmania trichosperma*.

Valdivian Cordillera district

Cordilleras from Chillán to Tierra del Fuego region (in part)—Reiche, 1905: map (regionalization).

Valdivian Cordillera region—Peña, 1966a: 15 (regionalization), 1966b: 218 (regionalization).

Valdivian Cordillera zone—Artigas, 1975: map (regionalization).

Valdivian Cordillera district—Morrone, 2015b: 216 (regionalization); Monckton, 2016: 125 (systematic revision and map).

Definition: The Valdivian Cordillera district extends from near 39° to 45° at altitudes above 700 m with several breaks where the Valdivian district crosses it (Peña, 1966a).

Vegetation: Luebert and Pliscoff (2006) characterized the Andean temperate deciduous forest as dominated by *Nothofagus pumilio* (Nothofagaceae) and *Berberis ilicifolia* (Berberidaceae); other species are *Acaena ovalifolia, Adenocaulon chilense, Chiliotrichum diffusum, Escallonia alpina, Gunnera magellanica, Maytenus disticha, Myoschilos oblonga, Osmorhiza chilensis, Pernettya myrtilloides, Ribes magellanicum, Rubus geoides* and *Valeriana lapathifolia.*

Cenocrons

The Valdivian Forest together with the Maule and Magellanic Forests have been isolated from other South American forests since the Neogene (Axelrod et al., 1991). During the Quaternary, cooling cycles induced contractions of temperate forests followed by expansions caused by warmer periods (Villagrán and Hinojosa, 1997). Some areas in the coastal range may have remained free of ices and represented the sources for the recovery of the forest biota (Smith, 2017).

Case study: Post-glacial recolonization of marsupial Dromiciops gliroides *in the Valdivian Forest*

Title: "Historical biogeography and post-glacial recolonization of South American temperate rain forest by the relictual marsupial Dromiciops gliroides." (Himes, C. M. T., M. H. Gallardo and G. J. Kenagy, *Journal of Biogeography*, 35: 1415–1424, 2008).

Goal: To examine hypotheses concerning dispersal and vicariance of the biogeographically relictual marsupial *Dromiciops gliroides* (Figure 5.16) in response to long-term climatic variation and potential isolation produced by mountains and water barriers.

Location: Valdivian forest, Chile and Argentina.

Methods: The study conducted an intraspecific phylogeographical analysis of *D. gliroides* based on field samples and museum skins from 21 localities. Bayesian and maximum-likelihood phylogenetic

Figure 5.16 *Dromiciops gliroides*, the single living species of Microbiotheriidae (Microbiotheria), endemic to the Valdivian Forest province. (Courtesy of Mariano A. Rodríguez-Cabal.)

analyses on sequences of mitochondrial DNA allowed evaluation of the influence of two major barriers, the Andean cordillera and the waterway between the mainland and the large island of Chiloé, and the responses of populations to historical north–south shifts of habitat associated with glacial history and sea-level change. The study determined the model of DNA evolution using the hierarchical likelihood ratio test implemented through Modeltest ver. 3.7 (Posada and Crandall, 1998). The study reconstructed the phylogenetic relationships among haplotypes using maximum likelihood (Felsenstein, 1981) and Bayesian methods in PAUP (Swofford, 2003) and MrBayes ver. 3.1.2 (Ronquist and Huelsenbeck, 2003).

Results: The study obtained a phylogenetic hypothesis with three main clades (Figure 5.17). Two clades containing forms from both sides of the Andes are widespread, containing nearly all the haplotypes: a northern clade (36°S–39°S) and a southern clade (40°S–43°S). Differentiation within the southern clade, island and mainland forms was not significant. Tests of recent demographic change reveal that southern populations experienced recent expansion, whereas northern populations exhibited long-term stability. The direction of recent gene flow and range expansion seems to be predominantly from Chile to Argentina with a modest reciprocal exchange across the Andes. The study also supported recent gene flow from the island of Chiloé to the mainland.

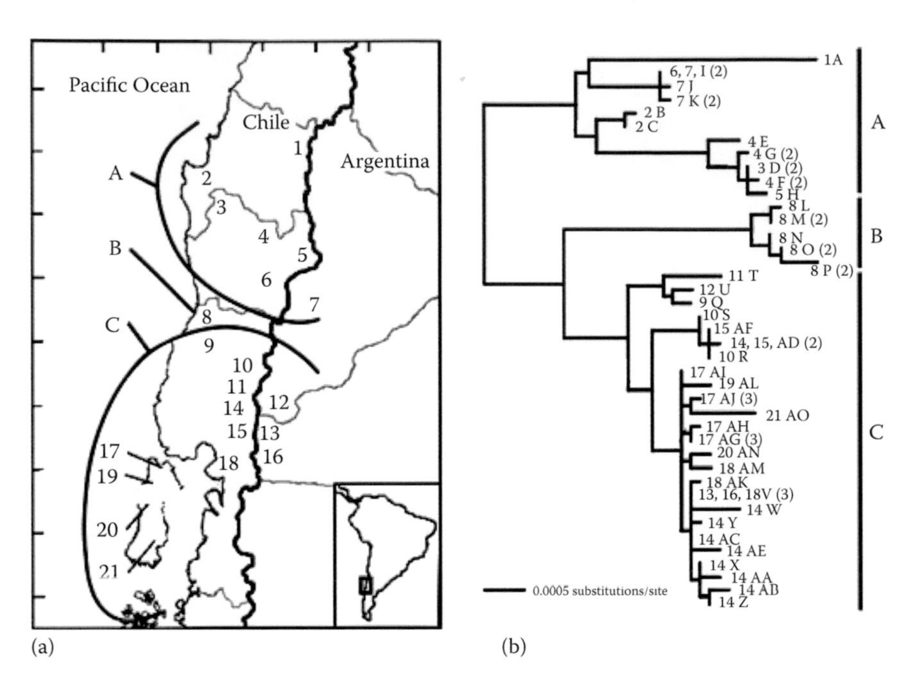

Figure 5.17 Phylogenetic relationships of *Dromiciops gliroides* haplotypes over the geographical range of the species. (Modified from Himes, C.M.T. et al., *J. Biogeogr.*, 35, 1415–1424, 2008. With permission.) (a) Map with 21 localities in Southern Chile and Argentina and (b) unrooted maximum-likelihood tree for the combined mtDNA data set.

Main conclusions: The genetic structure of contemporary *D. gliroides* populations suggested recent gene flow across the Andes and between the mainland and the island of Chiloé. Populations across the entire species range in Chile and Argentina divided at about 40°S into a large northern and a large southern clade. Results suggested that northern populations persisted in ice-free areas during the most recent glacial maximum, whereas southern populations expanded upon glacial retreat. The differences in demographic history detected between the northern and southern populations resulted from historical southward shifts of habitat associated with glacial recession in South America. The study considered that these results demonstrated the value of genetic data to illuminate how environmental history shaped species range and population structure.

Magellanic Forest province

Austral Littoral region (in part)—Reiche, 1905: map (regionalization).
Magellanic region—Reiche, 1905: map (regionalization).

South Chilean-Fuegian province (in part)—Skottsberg, 1905: 416 (regionalization).
Magellanic region (in part)—Goetsch, 1931: 2 (regionalization); O'Brien, 1971: 204 (regionalization).
Fuegian district (in part)—Osgood, 1943: 27 (regionalization).
Magellanic district—Cabrera, 1951: 61 (regionalization), 1971: 39 (regionalization); Cabrera and Willink, 1973: 98 (regionalization); Cabrera, 1976: 77 (regionalization); León et al., 1998: 133 (regionalization).
Magellanic Forest subregion—Hueck, 1957: 40 (regionalization).
Magellanic Forest area—Holdgate, 1960: 560 (regionalization); Morrone et al., 1994: 108 (cladistic biogeography).
Magellanic Forest zone—Kuschel, 1960: 543 (regionalization).
Forest zone (in part)—Mann, 1960: 33 (regionalization).
Magallanes Interoceanic region—Peña, 1966a: 15 (regionalization), 1966b: 219 (regionalization).
Austral Forest zone—Cekalovic, 1974: 305 (regionalization); Artigas, 1975: map (regionalization).
Shrub zone—Cekalovic, 1974: 306 (regionalization).
Magellanic area (in part)—Artigas, 1975: map (regionalization).
Magellanic Shrub zone—Artigas, 1975: map (regionalization).
Chilean *Nothofagus* province—Udvardy, 1975: 41 (regionalization).
Austroandean province (in part)—Rivas-Martínez and Navarro, 1994: map (regionalization).
Eastern forest area—Roig-Juñent, 1994b: 182 (cladistic biogeography).
Subpolar *Nothofagus* Forests ecoregion (in part)—Dinerstein et al., 1995: 96 (ecoregionalization); Brion and Ezcurra, 2017: 1 (ecoregionalization).
Atlantic Forest district—Roig, 1998: 140 (regionalization).
Evergreen Magellanic Forest and Bogs district—Roig, 1998: 140 (regionalization).
Magellanic Forest province—Morrone, 1999: 14 (regionalization), 2000b: 6 (regionalization), 2001b: 124 (regionalization); Posadas and Morrone, 2001: 267 (cladistic biogeography), 2003: 72 (cladistic biogeography); Morrone, 2004a: 158 (regionalization); Soares and de Carvalho, 2005: 485 (evolutionary biogeography); Morrone, 2006: 485 (regionalization); Spinelli et al., 2006: 302 (parsimony analysis of endemicity); Escalante et al., 2009: 379 (endemicity analysis); Morrone, 2010: 38 (regionalization); Posadas, 2012: 2 (faunistics); Alfaro et al., 2013: 244 (faunistics); Ferretti et al., 2014b: 2 (parsimony analysis of endemicity); Jerez and Muñoz-Escobar, 2015: 207 (faunistics); Morrone, 2015b: 220 (regionalization); Ruiz et al., 2016: 385 (track analysis); Arana et al., 2017: 421 (shapefiles); Martínez et al., 2017: 479 (track analysis and regionalization).

Magellanic Forest district—Posadas et al., 2001: 1328 (conservation biogeography).
Magallanic Forest region—Roig-Juñent and Domínguez, 2001: 557 (faunistics).
Magallanic Paramo province—Chani-Posse, 2010: 50 (faunistics).
Temperate Magellanian province—Rivas-Martínez et al., 2011: 27 (regionalization).

Definition

The Magellanic Forest region covers southern Chile from 47°S to Cape Horn, and southern Argentina in a small longitudinal area west of the Patagonian steppe in Santa Cruz and southern Tierra del Fuego (Cabrera and Willink, 1973; Morrone, 2000b, 2001b, 2006, 2015b; Brion and Ezcurra, 2017). It is colder and drier than the Valdivian Forest, with mean annual temperature between 6°C in the north and 3°C in the south, and mean annual precipitation decreasing from west to east from 4000 to 700 mm, respectively (Brion and Ezcurra, 2017).

Endemic and characteristic taxa

Morrone (2000b, 2015b) provided a list of endemic and characteristic taxa. Some examples included *Lepidothamnus fonkii* (Podocarpaceae); *Epilobium conjugens* (Figure 5.18a; Onagraceae); *Scotinoecus fasciatus* (Araneae: Dipluridae); *Peryphus rufoplagiatus, Pseudomigadops ovalis* and *Trechisibus antarcticus* (Coleoptera: Carabidae); *Aegorhinus delfini, Alastaropolus strumosus, Antarctobius germaini, A. hyadesii, A. lacunosus, Cylydrorhinus ursinus, C. vittatus, Falklandiopsis magellanica, Germainiellus laevirostris* and *G. lugens* (Figure 5.18b; Coleoptera; Curculionidae); *Parahelops quadricollis* (Coleoptera: Perimylopidae); *Megachile australis* (Hymenoptera: Megachilidae; Durante and Abrahamovich, 2002); and *Abrothrix lanosus* (Rodentia: Cricetidae).

Vegetation

The Magellanic Forest region consists of humid evergreen forests with abundant *Nothofagus betuloides*, deciduous *Nothofagus pumilio* and *N. antarctica* forests (Figure 5.19) and swamp forests (Kuschel, 1960; Cabrera and Willink, 1973; Cekalovic, 1974; Dinerstein et al., 1995; Brion and Ezcurra, 2017). Luebert and Pliscoff (2006) characterized the Andean antiboreal deciduous forest as dominated by *Nothofagus pumilio* (Nothofagaceae) and *Maytenus disticha* (Celastraceae); other species are *Acaena magellanica, Berberis ilicifolia, Draba magellanica, Empetrum rubrum, Escallonia alpina, Gunnera magellanica, Osmorhiza chilensis, Pernettya mucronata, Ribes*

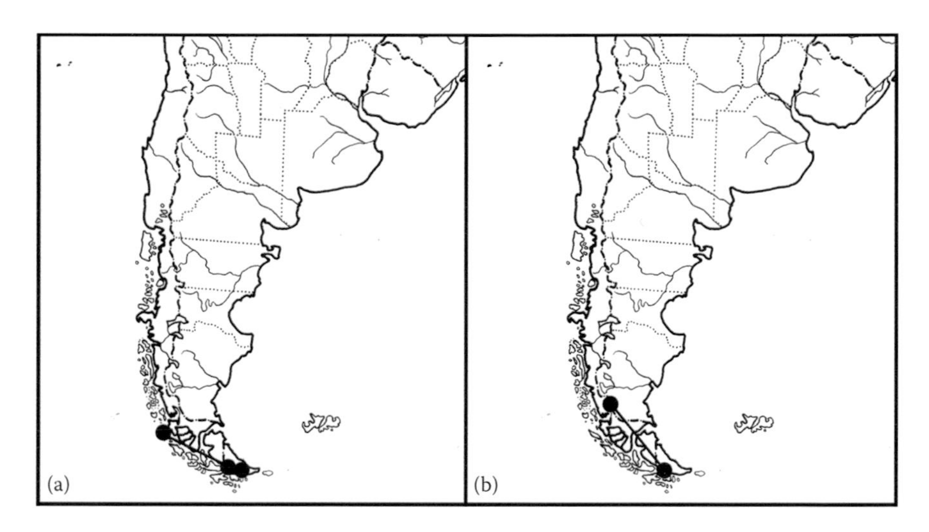

Figure 5.18 Maps with individual tracks in the Magellanic Forest province. (a) *Epilobium conjugens* (Onagraceae) and (b) *Germainiellus lugens* (Coleoptera: Curculionidae).

Figure 5.19 *Nothofagus* forest in the Magellanic Forest province, Southern Chile. (Courtesy of Marcelo Arana.)

magellanicus, Senecio acanthifolius, Thlaspi magellanicum and *Viola magellanica*. Luebert and Pliscoff (2006) also recognized the Andean temperate antiboreal deciduous arborescent scrubland as dominated by *Nothofagus antarctica* (Nothofagaceae) and *Chiliotrichum diffusum* (Asteraceae); other species are *Acaena pinnatifida, Baccharis magellanica, Cotula scariosa, Elymus agropyroides, Galium aparine, Gamochaeta spiciformis, Hieracium antarcticum, Osmorhiza chilensis, Poa patagonica, Ribes magellanicum* and *Vicia magellanica*.

Biotic relationships

Kuschel (1960) considered the fauna of the Magellanic Forest province to represent a subset of the fauna of the Valdivian Forest province. Peña (1966a) considered that the Magellanic Forest was transitional between the Magellanic Moorland and Patagonian provinces.

A track analysis based on plants, insects, crustaceans and mollusks (Morrone, 1992) showed that the southern part of Tierra del Fuego (Magellanic Forest and Magellanic Moorland provinces) was a node with generalized tracks connecting it with the Falklands and Campbell (New Zealand) islands. Two cladistic biogeographical analyses based on species of Coleoptera (Morrone, 1993b; Morrone et al., 1994) showed that the Magellanic Forest province was related to the Magellanic Moorland, Valdivian Forest and Falkland Islands provinces. Another cladistic biogeographical analysis based on species of Diptera (Soares and de Carvalho, 2005) showed a closer relationship of this province with the Valdivian Forest and Maule provinces.

Case study: Integrative biogeographic analysis of the fly genus Palpibracus

Title: "Biogeography of *Palpibracus* (Diptera: Muscidae): An integrative study using panbiogeography, parsimony analysis of endemicity, and component analysis." (Soares, E. D. G. and C. J. B. de Carvalho, In: Llorente, J. and J. J. Morrone (Eds.), *Regionalización biogeográfica en Iberoamérica y tópicos afines,* Las Prensas de Ciencias, UNAM, Mexico City, pp. 485–494, 2005.)

Goal: To identify areas of endemism of the fly genus *Palpibracus* and analyze their relationships appling three different biogeographic methods.

Location: Andean region.

Methods: This study used three methods: track analysis (Craw et al., 1999), parsimony analysis of endemicity (Rosen, 1988) and component analysis (Page, 1993). Biogeographical units used were grid-cells of five degrees of latitude by five degrees of longitude for the parsimony analysis of endemicity and the study used NONA 2.0 (Goloboff, 1998) and WinClada version 0.9.9 beta (Nixon, 1999) to carry out the analysis. The study used the provinces of the Andean region (Figure 5.20) as units for the component analysis. This analysis used phylogenetic and distributional data of *Palpibracus, Germainiellus, Apsil* and *Reynoldsia* (Figure 5.21). The study built the general area cladogram with Component 2.0 (Page, 1993).

Results: The study found eleven generalized tracks and five nodes in the track analysis (Figure 5.22a). The parsimony analysis of endemicity resulted in 10 most parsimonious cladograms (Figure 5.22b),

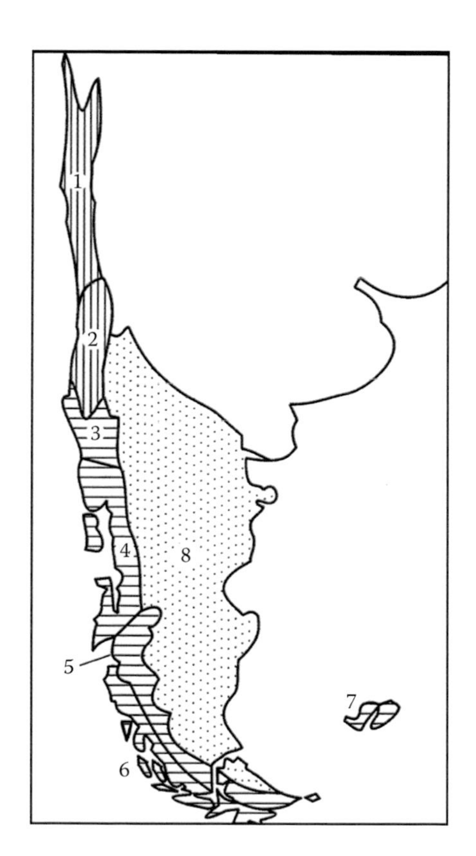

Figure 5.20 Map with the provinces of the Subantarctic and Central Chilean subregions analyzed by Soares and de Carvalho (2005). (Modified from Soares, E.D.G. and de Carvalho, C.J.B., Biogeography of *Palpibracus* (Diptera: Muscidae): An integrative study using panbiogeography, parsimony analysis of endemicity and component analysis, In: Llorente, J. and Morrone, J.J. [Eds.], *Regionalización biogeográfica en Iberoamérica y tópicos afines,* Las Prensas de Ciencias, UNAM, Mexico City, pp. 485–494, 2005. With permission.) (1) Coquimbo, (2) Santiago, (3) Maule, (4) Valdivian Forest, (5) Magellanic Forest, (6) Magellanic Moorland, (7) Falkland Islands and (8) Patagonia.

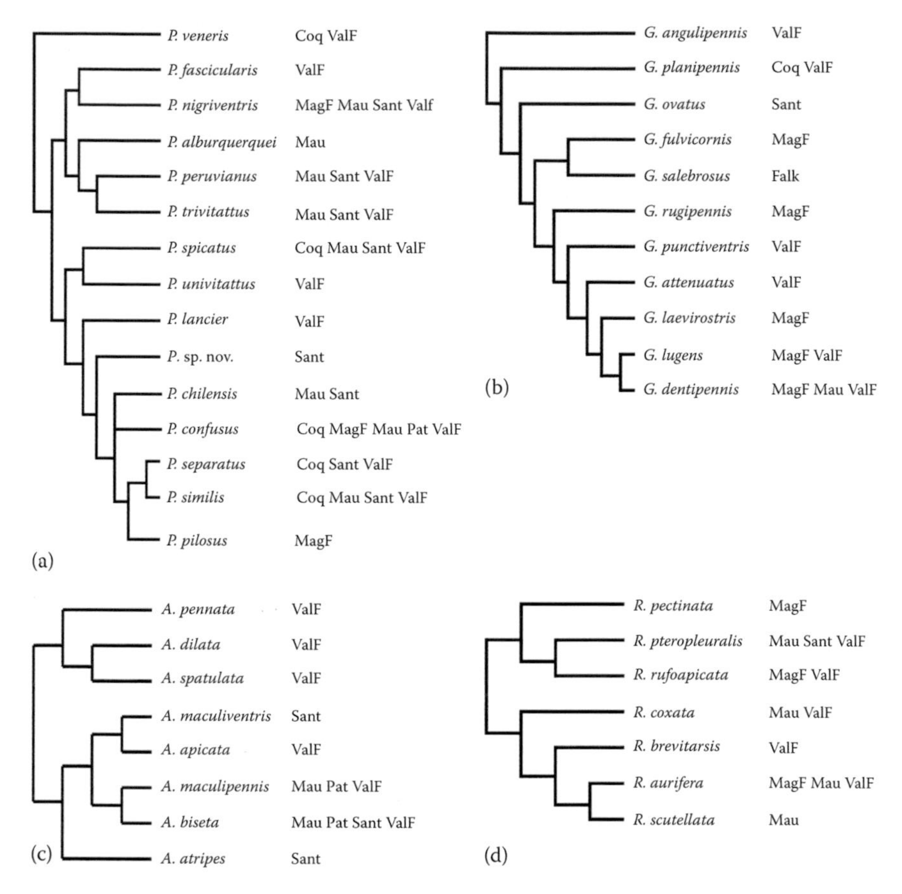

Figure 5.21 Taxon-area cladograms of taxa of the Subantarctic and Central Chilean subregions analyzed by Soares and de Carvalho (2005). (Modified from Soares, E.D.G. and de Carvalho, C.J.B., Biogeography of *Palpibracus* (Diptera: Muscidae): An integrative study using panbiogeography, parsimony analysis of endemicity and component analysis, In: Llorente, J. and Morrone, J.J. [Eds.], *Regionalización biogeográfica en Iberoamérica y tópicos afines,* Las Prensas de Ciencias, UNAM, Mexico City, pp. 485–494, 2005. With permission.) (a) *Palpibracus,* (b) *Germainiellus,* (c) *Apsil* and (d) *Reynoldsia.* Coq = Coquimbo, Falk = Falkland Islands, MagF = Magellanic Forest, Mau = Maule, Pat = Patagonia, Sant = Santiago and ValF = Valdivian Forest.

which showed a major area of endemism between 33°S and 42°S, subdivided in two areas. Component analysis led to three cladograms, each one showing different relationships among the Maule, Valdivian Forest and Magellanic Forest provinces. Santiago province was the sister-group to this clade, Coquimbo was related to this group and Patagonia was basal to them.

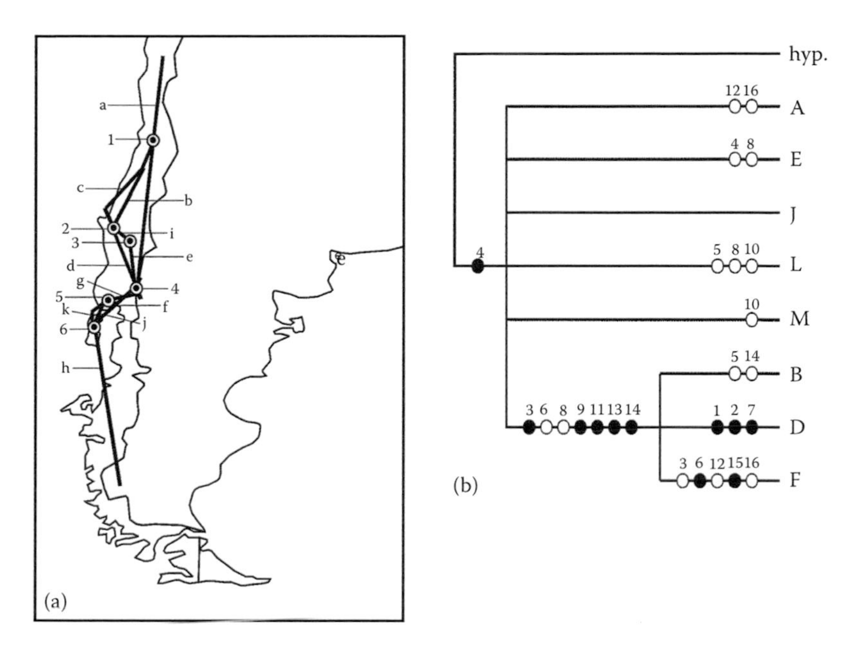

Figure 5.22 Results of the Soares and de Carvalho (2005) analysis of the Subantarctic and Central Chilean subregions (Modified from Soares, E.D.G. and de Carvalho, C.J.B., Biogeography of *Palpibracus* [Diptera: Muscidae]: An integrative study using panbiogeography, parsimony analysis of endemicity and component analysis, in Llorente, J. and Morrone, J.J. [Eds.], *Regionalización Biogeográfica en Iberoamérica y Tópicos Afines,* Las Prensas de Ciencias, UNAM, Mexico City, Mexico, pp. 485–494, 2005. With permission.) (a) Generalized tracks and nodes and (b) consensus cladogram of the 10 cladograms obtained in the parsimony analysis of endemicity.

Main conclusions: The species of *Palpibracus* showed a high degree of sympatry, their diversification probably took place in the southern endemism area indicated by the parsimony analysis of endemicity, which corresponded to the northern part of the Valdivian Forest. The phylogenetic analysis of *Palpibracus* and the generalized tracks indicated that most of the species diversified in the Valdivian Forest and the Magellanic Forest; the distribution of other species was mainly north of Chiloé island, in the Valdivian Forest and also in the Maule and Santiago provinces; and the distribution of the remaining species was more northerly, mostly north of 42°S. These results suggested a biotic connection between Central Chile and the Subantarctic subregion. Component analysis showed that Central Chile was paraphyletic and that the southern component, Santiago, was closer to Subantarctic elements than to Coquimbo. Santiago and the provinces of the Subantarctic subregion shared many taxa whereas Coquimbo had mainly endemic taxa. The separation of the Magellanic Forest, Maule and Valdivian Forest was a more recent event and their biotas were very similar. The relationship found in component analysis disagreed with Posadas and Morrone (2001), who stated that the Subantarctic subregion was paraphyletic, joining Maule, Valdivian Forest and Central Chile. This study also found the paraphyly of the Subantarctic subregion in the parsimony analysis of endemicity and the analysis of individual tracks and phylogeny of *Palpibracus* together. The northern part of the Valdivian Forest presented two different patterns, suggesting a vicariant event around 42°S. Menu-Marque et al. (2000) also found a generalized track joining the Valdivian Forest, Maule and Santiago, and another grouping the Valdivian Forest, Magellanic Forest and Magellanic Moorland.

The geological history of the area can explain the different patterns of relationship found. Events of marine introgression changed the continent conformation between 26 and 20 m.y.a. in the areas today comprising Chiloé Island, Concepción and Valparaíso, which were islands isolated from the continent (Donato et al., 2003). Three successive marine introgressions happened between 15 and 11 m.y.a., separating the Andean and Neotropical regions. Fossil records suggested that 1520 m.y.a. (Miocene) was the minimum age for Muscidae; *Palpibracus* could have diversified under the influence of these vicariance events associated with several concurrent glacial periods. These events changed the distributional boundaries of the biota of this region several times. *Palpibracus* diversification around 41°S–42°S was congruent with the marine introgressions of 26–20 m.y.a., that separated Chiloé Island from the southern Valdivian Forest and to the Maule (Donato et al., 2003). The last glacial age covered the region south of 42°S with ice from 2 m.y.a. to 15,000 years ago, thereby imposing northern distributional limits for the biota. The few records of *Palpibracus* south of 42°S suggested that the Quaternary glaciation events could have influenced its distributional patterns. Hence, species found in Patagonia and southern Chile probably reached these areas by dispersal. These patterns can be associated with the gradual migration of the biota that occurred following the retreat of the ice sheets and with the re-establishment of temperate forests in the region.

Magellanic Moorland province

Austral Littoral region (in part)—Reiche, 1905: map (regionalization).
Magellanian-Falklandian province (in part)—Skottsberg, 1905: 416 (regionalization).
South Chilean-Fuegian province (in part)—Skottsberg, 1905: 416 (regionalization).

Magellanic region (in part)—Goetsch, 1931: 2 (regionalization); O'Brien, 1971: 204 (regionalization).
Fuegian district (in part)—Osgood, 1943: 27 (regionalization).
Magellanic Moorland area—Godley, 1960: 467 (regionalization); Morrone et al., 1994: 110 (cladistic biogeography).
Magellanic Moorland zone—Kuschel, 1960: 544 (regionalization).
Austral Pacific region—Peña, 1966b: 219 (regionalization).
Southern Pacific region—Peña, 1966a: 16 (regionalization).
Austral Pacific zone—Cekalovic, 1974: 301 (regionalization); Artigas, 1975: map (regionalization).
Patagonian Ice zone—Cekalovic, 1974: 303 (regionalization).
Magellanic area (in part)—Artigas, 1975: map (regionalization).
Austroandean province (in part)—Rivas-Martínez and Navarro, 1994: map (regionalization).
Fuegian province (in part)—Rivas-Martínez and Navarro, 1994: map (regionalization).
Pacific region (in part)—Roig-Juñent, 1994b: 182 (cladistic biogeography).
Subpolar *Nothofagus* Forests ecoregion (in part)—Dinerstein et al., 1995: 96 (ecoregionalization); Brion and Ezcurra, 2017: 1 (ecoregionalization).
Magellanic Moorland province—Morrone, 1999: 14 (regionalization), 2000b: 7 (regionalization), 2001b: 126 (regionalization); Posadas and Morrone, 2001: 267 (cladistic biogeography), 2003: 72 (cladistic biogeography); Morrone, 2004a: 158 (regionalization); Soares and de Carvalho, 2005: 485 (evolutionary biogeography); Morrone, 2006: 485 (regionalization); Spinelli et al., 2006: 302 (parsimony analysis of endemicity); Escalante et al., 2009: 379 (endemicity analysis); Morrone, 2010: 38 (regionalization); Posadas, 2012: 2 (faunistics); Alfaro et al., 2013: 244 (faunistics); Jerez and Muñoz-Escobar, 2015: 201 (faunistics); Morrone, 2015b: 220 (regionalization); Arana et al., 2017: 421 (shapefiles); Romano et al., 2017: 445 (track analysis).
Magellanic Moorland district—Posadas et al., 2001: 1328 (conservation biogeography).
Boreal Austromagellanian province—Rivas-Martínez et al., 2011: 27 (regionalization).

Definition

The Magellanic Forest province to the east and the Pacific Ocean to the west limit Southern Chile and Argentina (Godley, 1960; Kuschel, 1960; Morrone, 2000b, 2001b, 2006, 2015b; Jerez and Muñoz-Escobar, 2015). It runs from Golfo de Penas (48°S) to Cape Horn. An annual rainfall of 3500–8000 mm, low temperature and strong winds characterize the area (Kuschel, 1960; Cekalovic, 1974).

Endemic and characteristic taxa

Morrone (2000b, 2015b) provided a list of endemic and characteristic taxa. Some examples included *Noterapion fuegianum* (Coleoptera: Brentidae); *Ceroglossus suturalis, Feroniomorpha lucida, Lissopterus, Metius malachiticus, Mimodromius nigrotestaceus* and *Nothocascellius hyadesii* (Figure 5.23a; Coleoptera: Carabidae); *Antarctobius rugirostris, A. yefacel, Anthonomus ornatus, Berberidicola exaratus* and *Telurus* (Coleoptera: Curculionidae); *Parochlus pilosus* (Diptera: Chironomidae); *Diamphidicus chilensis* (Diptera: Scatopsid ae); *Gigantodax brophyi* (Figure 5.23b) and *G. rufidulus* (Diptera: Simuliidae); and *Abrothrix hershkovitzi* and *A. markhami* (Rodentia: Cricetidae).

Vegetation

Prostrate dwarf shrubs, cushion plants (Figure 5.24), grass-like plants and bryophytes, with scattered clumps of evergreen forests in sheltered areas characterize the Moorlands (Figure 5.25; Kuschel, 1960; Cabrera and Willink, 1973; Domínguez Díaz et al., 2015; Brion and Ezcurra, 2017). Bogs comprise species of characteristic Austral genera, for example, *Astelia, Bolax, Caltha, Donatia, Drapetes, Gaimardia, Lepidothamnus* and *Phyllacne* (Brion and Ezcurra, 2017). Luebert and Pliscoff (2006) characterized the coastal antiboreal moorland as dominated by *Astelia pumila* (Asteliaceae) and *Donatia fascicularis* (Stylidaceae); other species are *Acaena pumila, Carex microglochin, Carpha alpina, Drosera uniflora, Gaimardia australis, Myrteola*

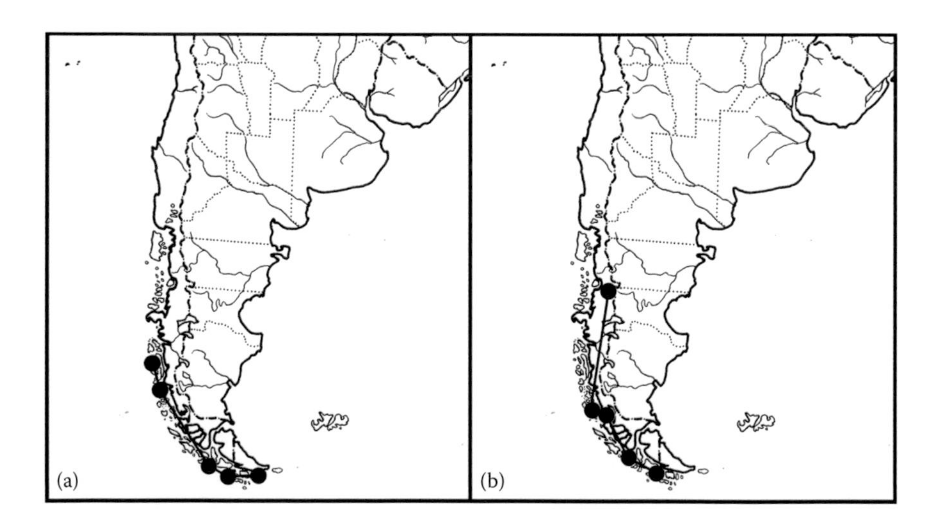

Figure 5.23 Maps with individual tracks in the Magellanic Moorland province. (a) *Nothocascellius hyadesii* (Coleoptera: Carabidae) and (b) *Gigantodax brophyi* (Diptera: Simuliidae).

Figure 5.24 *Azorella filamentosa* (Apiaceae), cushion plant characteristic of the Magellanic Moorland province, southern Chile. (Courtesy of Marcelo Arana.)

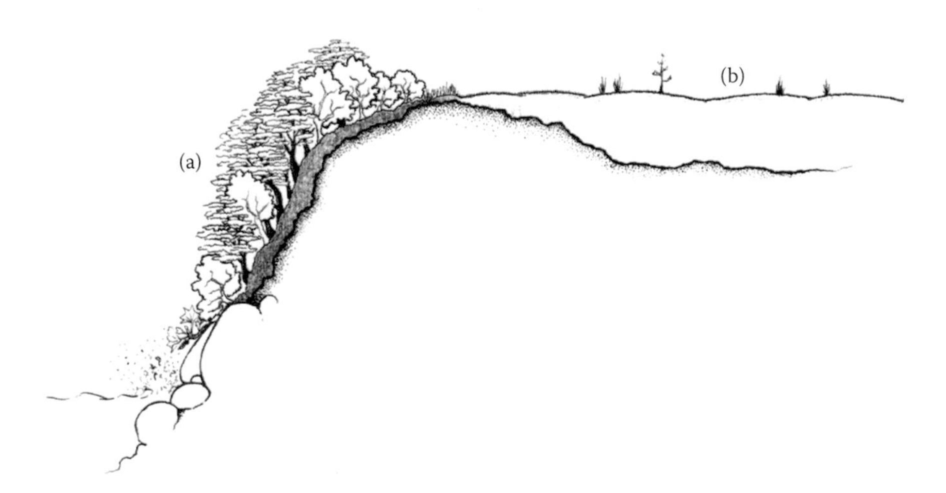

Figure 5.25 Vegetation in the Magellanic Moorland province. (Modified from Luebert, F. and Pliscoff, P., *Sinopsis Bioclimática y Vegetacional de Chile*. Editorial Universitaria, Santiago de Chile, 2006. With permission.) (a) Coastal moorland with *Astelia pumila* and *Donatia fascicularis* and (b) interior moorland.

nummularia, Nothofagus betuloides, Pilgerodendron uviferum, Pinguicula antarctica, Schoenus antarcticus, Senecio trifurcatus, Tapeinia pumila, Tepualia stipularis and *Tetroncium magellanicum.*

Biotic relationships

A track analysis based on plants, insects, crustaceans and mollusks (Morrone, 1992) showed that the southern part of Tierra del Fuego (Magellanic Forest and Magellanic Moorland provinces) was a node, with generalized tracks connecting it with the Falklands and Campbell (New Zealand) islands. Two cladistic biogeographical analyses based on species of Coleoptera (Morrone, 1993b; Morrone et al., 1994) showed that the Magellanic Mooorland province was related to the Magellanic Forest, Valdivian Forest and Falkland Islands provinces. Another cladistic biogeographical analysis, based on species of Diptera (Soares and de Carvalho, 2005), found that the Magellanic Moorland province was the sister area to the Central Chilean and Subantarctic subregions.

Falkland Islands province

Falkland Islands province—Engler 1882: 346 (regionalization); Morrone, 1999: 14 (regionalization), 2000b: 7 (regionalization), 2001b: 127 (regionalization); Posadas and Morrone, 2001: 267 (cladistic biogeography), 2003: 72 (cladistic biogeography); Morrone, 2004a: 158 (regionalization); Soares and de Carvalho, 2005: 485 (evolutionary biogeography); Morrone, 2006: 485 (regionalization); Escalante et al., 2009: 379 (endemicity analysis); Morrone, 2010: 38 (regionalization), 2011: 2085 (evolutionary biogeography); Posadas, 2008: 5 (faunistics); Morrone, 2015b: 221 (regionalization); Arana et al., 2017: 421 (shapefiles).

Magellanian-Falklandian province (in part)—Skottsberg, 1905: 416 (regionalization).

Insular province—Cabrera, 1951: 62 (regionalization), 1953: 107 (regionalization), 1971: 40 (regionalization); Cabrera and Willink, 1973: 102 (regionalization); Cabrera, 1976: 78 (regionalization); Apodaca et al., 2015a: 87 (regionalization).

Falkland Islands area—Morrone et al., 1994: 110 (cladistic biogeography).

Pacific region (in part)—Roig-Juñent, 1994b: 182 (cladistic biogeography).

Patagonian Grasslands ecoregion (in part)—Dinerstein et al., 1995: 102 (ecoregionalization); Dellafiore, 2017b: 1 (ecoregionalization).

Cortaderal Malvinian district—Roig, 1998: 142 (regionalization).

Fuegian-Malvinian district (in part)—Roig, 1998: 138 (regionalization).

South Atlantic Islands ecoregion—Burkart et al., 1999: 37 (ecoregionalization).

Falkland Islands district—Posadas et al., 2001: 1328 (conservation biogeography).
Insular Falkland province—Rivas-Martínez et al., 2011: 27 (regionalization).
Patagonian Steppe ecoregion (in part)—Dellafiore, 2017a: 1 (ecoregionalization).

Definition

Argentinean archipelago of the Falklands or Malvinas (Figure 5.26) and South Georgia island, in the south Atlantic Ocean (Ringuelet, 1955b; Cabrera and Willink, 1973; Morrone, 2000b, 2001b, 2006, 2011, 2015b). The archipelago comprises two main islands, East Falkland or Gran Malvina (5000 km^2) and West Falkland or Soledad (3500 km^2), and more than 230 smaller islands varying from 220 km^2 to a few m^2 (Moore, 1968; Robinson, 1984). The general relief is hilly, especially in West Falkland; some mountains reach 705 m (Mount Usborne). Climate is cool temperate oceanic, rainfall is low (mean annual precipitation at Port Stanley of 640 mm), with West Falkland tending to be drier (Broughton and McAdam, 2005).

The Falkland Islands were originally part of Africa. Prior to fragmentation of Gondwanaland, the Falkland Platform was adjacent to the east coast of South Africa (Marshall, 1994). The islands subsequently rotated 180° relative to South Africa while drifting to their current placement near southern South America.

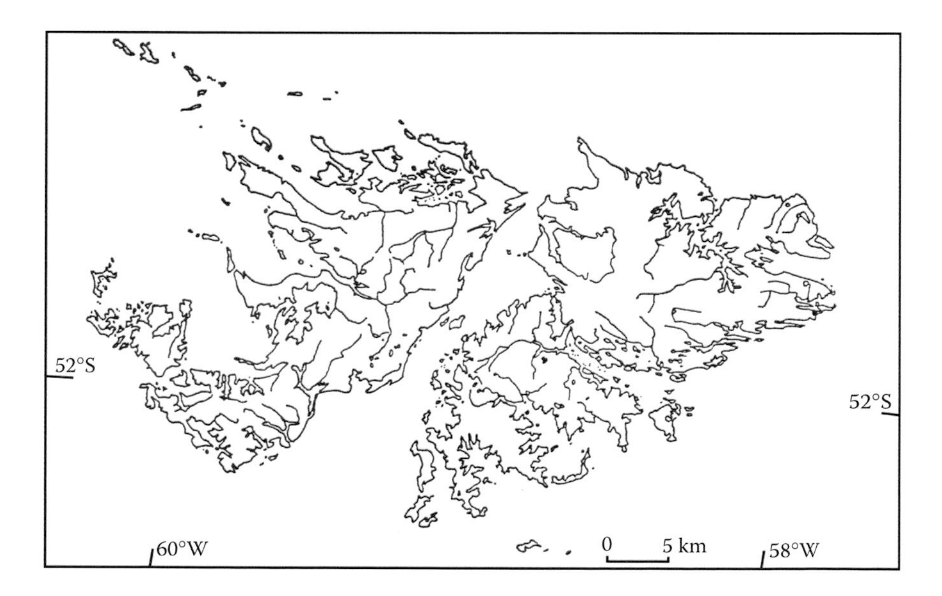

Figure 5.26 Map of the Falkland Islands.

Endemic and characteristic taxa

Morrone (2000b, 2015b) provided a list of endemic and characteristic taxa. Some examples included *Chevreulia lycopodioides, Erigeron incertus, Gamochaeta antarctica, Leucheria suaveolens, Nassauvia gaudichaudii, N. serpens, Senecio littoralis* and *S. vaginatus* (Asteraceae); *Calceolaria fothergillii* (Calceolariaceae); *Plantago moorei* (Plantaginaceae); *Notiomaso shackletonii* and *N. striatus* (Araneae: Lyniphiidae; Lavery, 2017); *Falklandia rumbolli* (Araneae: Orsolobidae; Lavery, 2017); *Tetragnatha insulata* (Araneae: Tetragnathidae; Lavery, 2017); *Chalcosphaerium enderleini* and *C. solox* (Coleoptera: Byrrhidae); *Pseudomigadops falklandicus, P. handkei* and *P. fuscus* (Coleoptera: Carabidae); *Crotonia macfadyeni* (Acari: Crotoniidae); *Antarctobius abditus, A. bidentatus, A. malvinensis, Caneorhinus biangulatus* (Figure 5.27a), *Cylydrorhinus lemniscatus* (Figure 5.27b), *Falklandius goliath, Germainiellus salebrosus, Malvinius compressiventris* (Figure 5.27c), *M. nordenskioeldi* (Figure 5.27d), *Lanteriella microphtalma* (Figure 5.28), *Morronia brevirostris, Puranius championi* and *P. scaber* (Coleoptera: Curculionidae); *Parudenus falklandicus* (Orthoptera: Raphidophoridae); *Chloephaga picta leucoptera* and *Tachyeres brachypterus* (Anseriformes: Anatidae); *Troglodytes aedon cobbi* (Passeriformes: Troglodytidae); *Turdus falcklandii* (Passeriformes: Turdidae); and *Dusicyon australis* (Carnivora: Canidae).

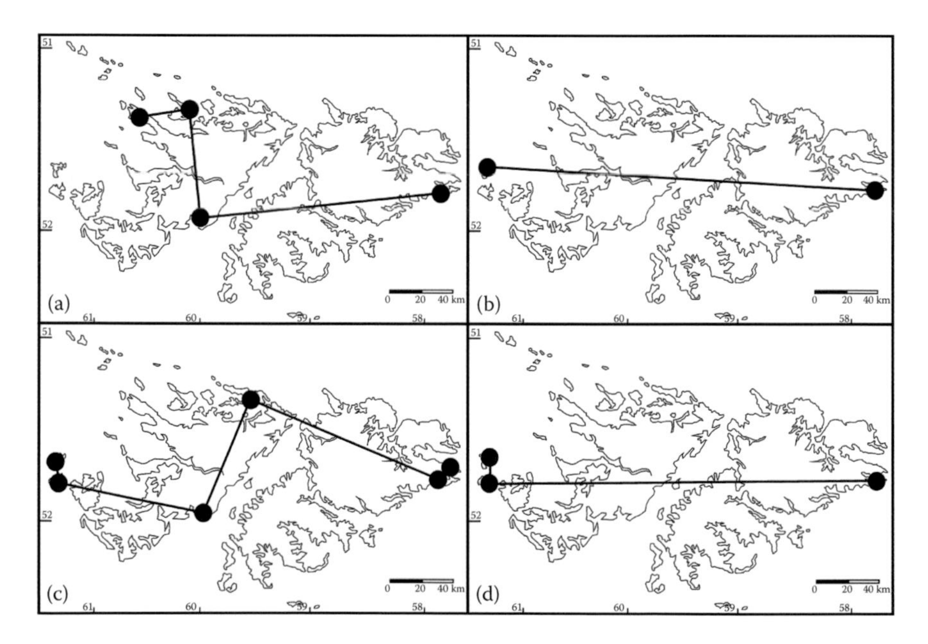

Figure 5.27 Maps with individual tracks of weevil species (Coleoptera: Curculionidae) endemic to the Falkland Islands. (a) *Caneorhinus biangulatus,* (b) *Cylydrorhinus lemniscatus,* (c) *Malvinius compressiventris* and (d) *M. nordenskioeldi.*

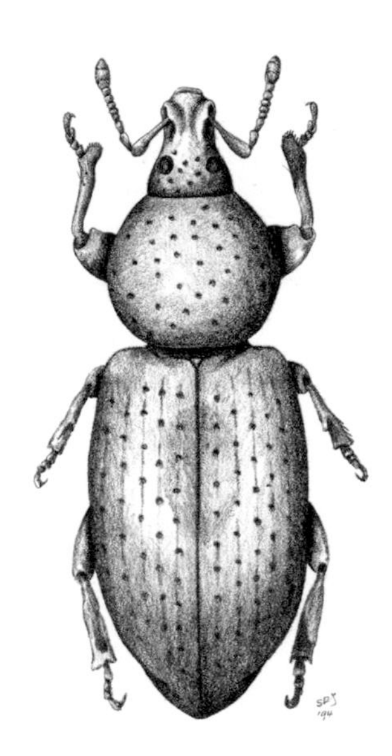

Figure 5.28 *Lanteriella microphthalma,* species of Listroderini (Coleoptera: Curculionidae) endemic to the Falkland Islands. (Courtesy of Sergio Roig-Juñent.)

Broughton and McAdam (2005) concluded that 13 out of the 171 species of vascular plants were endemic to the islands.

Vegetation

The Falkland Islands province consists of grasslands, steppes, fern beds and tundra (Cabrera and Willink, 1973; Dinerstein et al., 1995; Roig, 1998; Broughton and McAdam, 2005). Dominant plant species include *Abrotanella emarginata, Acaena adscendens, A. magellanica, A. microcephala, Astellia pumila, Azorella lycopodioides, A. selago, Baccharis magellanica, Blechnum tabulare, Bolax gummifera, Caltha appendiculata, Carex trifida, Cortaderia pilosa, Deschampsia flexuosa, Empetrum rubrum, Festuca erecta, Gaimardia australis, Juncus sheuzerioides, Pernettya pumila, Poa annua, P. antarctica, P. flabellata, Pratia repens, Senecio candicans, S. littoralis, S. vaginata, Uncinia smithii* and *Veronica elliptica* (Skottsberg, 1922; Moore, 1968; Cabrera, 1971, 1976). Moore (1968) identified the following vegetation types: (1) maritime tussock formation, in coastal areas, usually below 200 m and dominated by *Poa flabellata* (Poaceae); (2) oceanic heath: the most common vegetation covering the islands, with one association dominated by *Cortaderia* (Poaceae) and the other dominated by dwarf shrubs, with *Empetrum rubrum* and *Pernettya pumila* (Ericaceae); (3) feldmark: above 600 m, where cushion plants trend to predominate, with *Azorella selago, A. lycopodioides* and *Bolax gummifera* (Apiaceae), *Colobanthus subulatus* (Caryophyllaceae) and *Abrotanella emarginata* (Asteraceae); (4) fern and bog: with three associations, dominated by *Rostkovia* (Juncaceae), *Astelia* (Asteliaceae) and *Juncus scheuzerioides* (Juncaceae), respectively; (5) bush: dominated by *Chiliotrichum* (Asteraceae) and *Hebe* (Plantaginaceae); (6) littoral vegetation: dominated by *Senecio candidans* (Asteraceae) and *Ammophila arenaria* (Poaceae); and (7) freshwater vegetation: dominated by *Eleocharis melanostachys* (Cyperaceae) and *Myriophyllum elatinoides* (Haloragaceae). Paleobotanical data indicated that a podocarp-type forest was present in the islands in pre-Neogene times (Birnie and Roberts, 1986). Fire and grazing modified the vegetation of the Falkland Islands significantly (Buckland and Hammond, 1997).

Biotic relationships

Ringuelet (1955b) postulated the inclusion of the Falkland Islands in the Subantarctic subregion, based on an extensive analysis of several animal species, for example, fish, mammals, birds, insects, arachnids, crustaceans, annelids and mollusks. Moore (1968) surveyed the vascular plants and found a close relationship with southern South America, particularly Tierra del Fuego and the southern Andes. He also noted some connections with the Northern Hemisphere via the Andes, and with Australia and New Zealand via the Subantarctic islands.

A track analysis based on plants, insects, crustaceans and mollusks (Morrone, 1992) showed that the Falkland Islands were a node with generalized tracks connecting them with South Georgia, Tristan da Cunha-Gough, Crozet and Tierra del Fuego islands (Figure 5.29). Three cladistic biogeographical, based on species of Coleoptera (Morrone, 1993a, 2011; Morrone et al., 1994), showed that the Falkland Islands province was related to the Magellanic Forest, Valdivian Forest and Magellanic Moorland provinces. Another cladistic biogeographical analysis, based on species of Diptera (Soares and de Carvalho, 2005), found a closer relationship of the Falkland Islands province with the Maule, Valdivian Forest and Magellanic Forest provinces, but with the Magellanic Moorland province more distantly related to it.

Regionalization

This province comprises two districts: Falkland Islands and South Georgia Island (Roig, 1998; Morrone, 2011, 2015b).

Falkland Islands district

Falkland Islands district—Morrone, 2011: 2086 (evolutionary biogeography), 2015b: 221 (regionalization).

Definition: The district corresponds to the two main islands and surrounding smaller islands.

Endemic taxa: Basically those of the province.

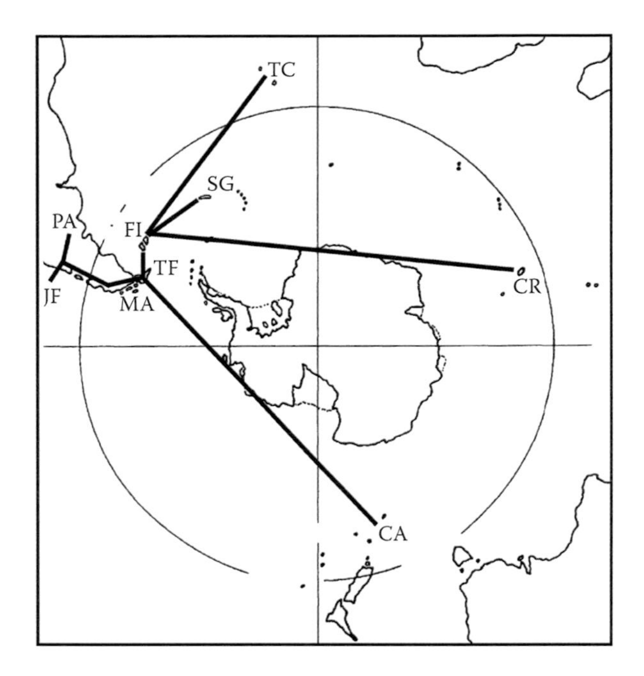

Figure 5.29 Generalized tracks and nodes based on an analysis of plants, insects, crustaceans and mollusks showing that the Falkland Islands are a node, with generalized tracks connecting them with South Georgia, Tristan da Cunha-Gough, Crozet and Tierra del Fuego islands. (Modified from Morrone, J.J., *Acta Entomol. Chil.*, 17, 157–174, 1992. With permission.) (CA) Campbell and other New Zealand Subantarctic Islands, (CR) Crozet, Marion and Prince Edward Islands, (FI) Falkland Islands, (JF) Juan Fernández Islands, (MA) Magellan area, (PA) Patagonia, (SG) South Georgia, (TC) Tristan da Cunha-Gough Islands and (TF) Tierra del Fuego.

South Georgia Island district

Georgias district—Roig, 1998: 142 (regionalization).

South Georgia Island district—Morrone, 2011: 2086 (evolutionary biogeography), 2015b: 221 (regionalization).

Insular Atlantical [sic] Antarctic province—Rivas-Martínez et al., 2011: 27 (regionalization).

Definition: The district corresponds to the island of South Georgia, situated in the south Atlantic, east of the Falklands (Gressitt, 1970). Rivas-Martínez et al. (2011) placed several islands peripheral to Antarctica, including South Georgia, in the Circumantarctic subkingdom.

Endemic taxa: *Ainudrilus dartnalli* (Oligochaeta: Tubificidae; Erséus and Grimm, 2002) and *Notiomaso grytivenkis* (Araneae: Linyphiidae; Lavery, 2017).

Cenocrons

Geological evidence suggests that during several glaciations of the Neogene and Quaternary times, the Falkland Islands might have connected with the continent (Posadas and Morrone, 2004). This evidence allows postulation of alternative vicariance and geodispersal events related to sea-level changes induced by glacial-eustatic agents (Morrone and Posadas, 2005). McDowall (2005) concluded that the stronger affinities of the biota of the Falkland Islands were with southern South America (especially Tierra del Fuego), and interpreted them by former dispersal. He also conjectured that the original biota of the Falklands, present when the proto-Falklands detached from South Africa and then made contact with the Patagonian shelf, was completely replaced because of climatic fluctuations and the displacement by propagule pressure imposed by the arrival of Patagonian taxa. Morrone and Posadas (2005) considered that the islands were land-connected to southern South America during the past 130 m.y.a. at least during five events when sea level was 100 m lower than the present, providing several opportunities for dispersal.

Papadopoulou et al. (2009) analyzed the age and persistence of the biota of the Falklands, based on the phylogenetic analysis of 55 species of beetles belonging to 13 families. They estimated the divergence times between congeneric sister species of Carabidae and Curculionidae. Based on their results, Papadopoulou et al. (2009) suggested a great age for the Falkland lineages, that likely preceded the Pleistocene glaciations, and considered that dispersal or temporary land bridges at times of low eustatic sea level, in pre-Pleistocene time and possibly earlier (>10 m.y.a), may have established the connections with southern South America.

Case study: Integrative biogeographic analysis of the weevils of the Falkland Islands

Title: "Island evolutionary biogeography: Analysis of the weevils (Coleoptera: Curculionidae) of the Falkland Islands (Islas Malvinas)." (Morrone, J. J., *Journal of Biogeography*, 38: 2078–2090, 2011.)

Goal: To undertake an evolutionary biogeographical analysis of the weevils (Coleoptera: Curculionidae) of the Falkland Islands, integrating distributional, phylogenetic and molecular studies.

Location: Falkland Islands, southern South America.

Methods: The analysis followed five steps, each corresponding to a specific question, method and technique. The study used track analysis and methods for identifying areas of endemism to identify biotas, which are the basic units of evolutionary biogeography. The study used cladistic biogeography based on phylogenetic data to test the historical relationships among these biotas. Based on the results of these analyses, the study achieved a biogeographical regionalization. The study incorporated intraspecific phylogeography and molecular clocks were to help identify the cenocrons that became integrated in the biotia. Finally, the study integrated the geological and biological knowledge available to construct a geobiotic scenario.

Results: Based on the individual tracks of 23 weevil species, the study identified six generalized tracks (Figure 5.30):

1. *Maule–Valdivian forests*: Supported by *Chileudius varians, Hybreoleptops tuberculifer, Megalometis spinifer* (Figure 5.31), *Strangaliodes mutuarius, Falklandius peckorum* and *Puranius fasciculiger.*
2. *Magellanic forest*: Supported by *Caneorhinus lineatus, C. tessellatus, Antarctobius lacunosus, Falklandiopsis magellanica* and *Telurus dissimilis.*
3. *Magellanic moorland*: Supported by *Antarctobius rugirostris* and *A. yefacel.*

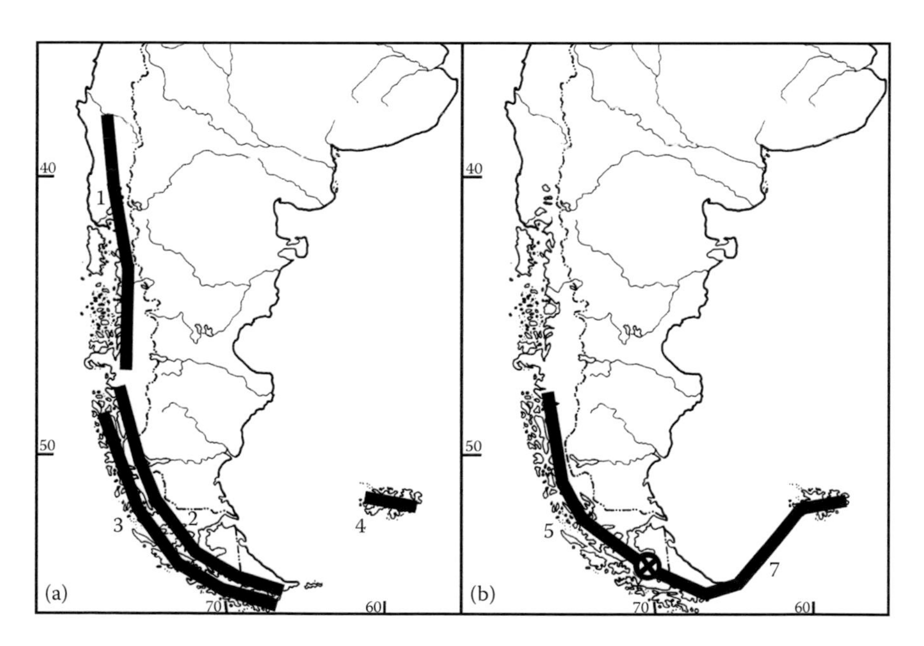

Figure 5.30 Generalized tracks and node based on species of Subantarctic Curculionidae. (Modified from Morrone, J.J., *J. Biogeogr.*, 38, 2078–2090, 2011. With permission.) (a) Four generalized tracks and (b) two generalized tracks and one node. (1) Maule–Valdivian Forests generalized track, (2) Magellanic Forest generalized track, (3) Magellanic Moorland generalized track, (4) Falkland Islands generalized track, (5) Magellanic Forest–Magellanic Moorland generalized track, (6) node and (7) Magellanic Forest–Falkland Islands generalized track.

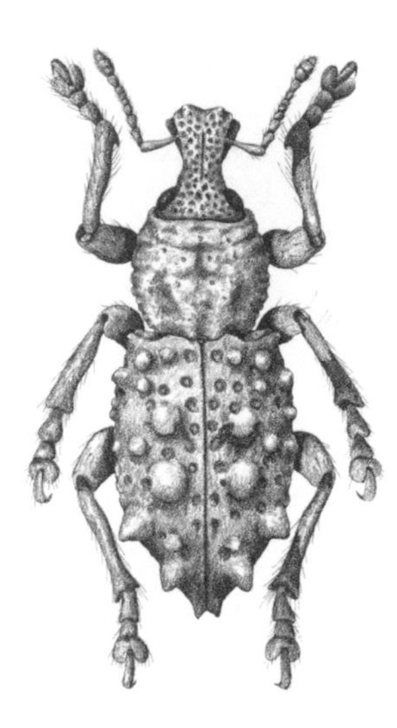

Figure 5.31 *Megalometis spinifer,* weevil species endemic to the Maule and Valdivian Forests (Coleoptera: Curculionidae). (Courtesy of Sergio Roig-Juñent.)

4. *Falkland Islands*: Supported by *Caneorhinus biangulatus, Cylydrorhinus caudiculatus, Cylydrorhinus lemniscatus, Morronia brevirostris, Malvinius compressiventris, M. nordenskioeldi, Antarctobius abditus, A. bidentatus, A. falklandicus, A. malvinensis, A. vulsus, Falklandius goliath, F. kuscheli, F. turbificatus, Lanteriella microphtalma, Puranius championi, P. exsculpticollis* and *P. scaber.*
5. *Magellanic forest–magellanic moorland*: Supported by *Antarctobius germaini, A. hyadesii* and *Telurus caudiculatus.*
6. *Magellanic forest–falkland Islands*: Supported by *Falklandiellus suffodens* and *Falklandius antarcticus.*

The study identified a node in the Magellanic Forest, based on the overlap of the Magellanic Forest–Magellanic Moorland and the Magellanic Forest–Falkland Islands generalized tracks.

In the second step, the study obtained six taxon-area cladograms and seven paralogy-free subtrees derived from them for the taxa for which phylogenetic hypotheses were available, for example, *Cylydrorhinus* generic group, *Strangaliodes* generic group, *Antarctobius, Falklandius* generic group, *Germainiellus* and *Puranius.* The parsimony analysis of the data matrix led to a single general area cladogram, which shows the following sequence of vicariance events: (Magellanic Moorland [Maule–Valdivian Forests [Magellanic Forest, Falkland Islands]]).

The study classified the Falkland Islands in the Austral kingdom, Andean region and Subantarctic subregion. Within the latter, the study treated them as a biogeographical province with two districts: Falkland Islands and South Georgia Island.

Based on the choronogram of Papadopoulou et al. (2009), the study concluded that the split of Falkland species from their sister species predated the Oligocene. In some cases, the study calculated minimum ages of ca. 4 m.y.a. for some lineages. Based on the distribution of the sister taxa, the study inferred a single Subantarctic cenocron. A possible exception is *Pentarthrum carmichaeli*, which may have dispersed recently to the islands in association with *Nothofagus* driftwood. Molecular clock estimates (Papadopoulou et al., 2009) suggested that separation of species endemic to the Falklands from their sister species pre-dated the Oligocene. The study found minimum ages of the Falkland weevil lineages were greater than previously suggested; for example, the study estimated the basal internal split of the genus *Malvinius*, endemic to the Falklands, to be at least 4.17 ± 1.9 m.y.a. This Subantarctic cenocron represented an extension of the southern South American biota; its geodispersal to the Falklands might have occurred during the Early Oligocene (Morrone and Posadas, 2005).

The sequence of events in the general area cladogram and the geological, tectonic and paleontological information available allowed construction of a geobiotic scenario. The scenario consists of six steps (Figure 5.32): (1) development of the Subantarctic biota in southern South America, (2) arrival of the Falkland Islands crustal block from South Africa in the Early Cretaceous, (3) geodispersal of the Subantarctic cenocron from southern South America to the Falklands during the Early Oligocene, (4) vicariance of the Magellanic Moorland, (5) vicariance of the Maule–Valdivian Forests and (6) final vicariance between the Magellanic Forest and the Falkland Islands.

Main conclusions: The biotic components identified support the Subantarctic connection of the Falkland weevils as previously noted for other taxa, for example, insects, spiders, mites, several other invertebrates and birds. Although the study noted older African biotic connections occasionally (e.g., for amphipods [Barnard and Barnard, 1983], isopods [Green, 1974] and oligochaetes [Brinkhurst and Jamieson, 1971]), the study dismissed because they were poorly defined or in need of phylogenetic corroboration (McDowall, 2005). The Falkland biota seemed to be related to that of the rest of the Subantarctic subregion, in southern South America. Falkland weevils seemed to belong to a single Subantarctic cenocron, dating to at least the Early Oligocene when geodispersal from southern South America may have occurred. *Pentarthrum carmichaeli* was the only possible exception; it shared widespread distribution on other Subantarctic islands with *Kenodactylus audouini* Guérin-Ménéville (Coleoptera: Carabidae; Roig-Juñent and Domínguez, 2001), and

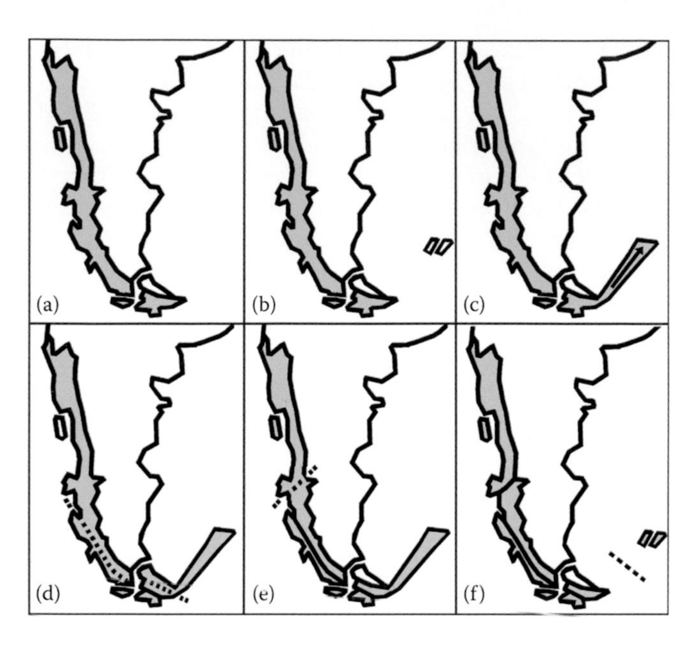

Figure 5.32 Geobiotic scenario explaining the biotic evolution of the Falkland biota. (Modified from Morrone, J.J., *J. Biogeogr.*, 38, 2078–2090, 2011. With permission.) (a) Development of the original Subantarctic biota, (b) arrival of the Falkland Islands crustal block, (c) geodispersal of taxa from southern South America to the Falklands, (d) vicariance of the Magellanic Moorland, (e) vicariance of the Maule–Valdivian forests and (f) vicariance between the Magellanic forest and the Falkland Islands.

may belong to a more recent cenocron. The Subantarctic cenocron may have replaced completely (or almost completely) an older African cenocron when the proto-Falklands made contact with the Patagonian continental shelf (McDowall, 2005).

Analyses from other taxa also suggested the existence of different cenocrons in the Falklands; palaeogeographical investigations established that at 200 m.y.a. the shelf region that makes up the Falkland Islands was part of eastern South Africa. The region moved from this position with the break-up of Gondwanaland (Marshall, 1994). The Falkland Islands crustal block moved westwards with the South American plate, to which it has been linked since before the break-up of Africa and South America around 130 m.y.a. None of the taxa analyzed to date appear to reflect these ancient geological affinities with Africa, whereas phylogenetic and track analyses strongly corroborated relationships with South American taxa. Sea-level changes during the Tertiary and specifically the most recent glacial cycles might have resulted in temporary land connections between the Falklands and southern South America. This connection would require a drop of sea level of only 100–150 m. Periglacial conditions might have been severe enough to cause the local extinction of most terrestrial species, and studies postulated that Falklands insects might have colonized the islands during the last 15–10 kyr (Buckland and Hammond, 1997). The analysis of Papadopoulou et al. (2009), however, showed that the separation of the Falkland endemics and their sister species pre-dated the Pleistocene. Furthermore, microfossils showed that the Falklands harboured podocarp-type forests with some *Nothofagus* and tree ferns in the Early Oligocene (Birnie and Roberts, 1986), and listroderine weevils were known from Pliocene forests in Antarctica (Ashworth and Kuschel, 2003). A geobiotic scenario implying vicariance events related to sea-level variations, induced by glacial-eustatic agents, could explain the distributional patterns analyzed in this study. Species endemic to the Falklands did not constitute monophyletic clades but had their sister groups in continental areas of southern South America (Posadas and Morrone, 2004). Data on the age of the Falkland weevil taxa suggested that Pleistocene glaciations did not extirpate the existing biota and that ancient lineages persisted in situ, possibly for >10 m.y.a. (Papadopoulou et al., 2009). There was fossil evidence showing that the Falkland forests originally inhabited by these weevils dated back at least to the Early Oligocene (Birnie and Roberts, 1986).

Juan Fernández province

Juan Fernández region—Engler 1882: 346 (regionalization).
Islands region (in part)—Goetsch, 1931: 2 (regionalization).
Juan Fernández province—Cabrera and Willink, 1973: 103 (regionalization); Morrone, 2015b: 222 (regionalization).
Fernandezian region (in part)—Takhtajan, 1986: 252 (regionalization); Cox, 2001: 519 (regionalization).
Juan Fernández Islands province—Rivas-Martínez and Navarro, 1994: map (regionalization); Morrone, 1999: 14 (regionalization), 2000b: 8 (regionalization), 2001b: 121 (regionalization); Posadas and Morrone, 2003: 72 (cladistic biogeography); Morrone, 2004a: 158 (regionalization); Soares and de Carvalho, 2005): 485 (evolutionary biogeography); Morrone, 2006: 485 (regionalization); Escalante et al., 2009: 379 (endemicity analysis).
Insular Juan Fernández province—Rivas-Martínez et al., 2011: 27 (regionalization).
Juan Fernández Islands Temperate Forests ecoregion—Bernardello and Stuessy, 2017: 1 (ecoregionalization).

Figure 5.33 Panoramic view of the Juan Fernández archipelago. (Courtesy of Andrés Moreira-Muñoz.)

Definition

The Juan Fernández province consists of the Chilean islands of Masatierra or Robinson Crusoe, Masafuera or Alejandro Selkirk, and Santa Clara (Figure 5.33), which are situated in the Pacific Ocean, 600 km west of Valparaíso, approximately at 33°S (Skottsberg, 1920; Kuschel, 1961; Cabrera and Willink, 1973; Stuessy et al., 1984; Sanderson et al., 1987; Morrone, 2000b, 2001b, 2006, 2015b; Moreira-Muñoz and Elórtegui Francioli, 2014; Soto et al., 2017). Masatierra is closest to the continent and Masafuera is situated 150 km further west (Stuessy et al., 1990). The islands are of volcanic origin. Masatierra is approximately 4 m.y., whereas Masafuera is 1–2 m.y. (Stuessy et al., 1984). Climate is warm, with rain throughout the year, but with more precipitation in winter than in summer (Kuschel, 1961).

Endemic and characteristic taxa

Morrone (2000b, 2015b) provided a list of endemic and characteristic taxa. Some examples included *Breutelia masafuerae* and *Philonis glabrata* (Bartramiaceae); *Cyptodon crassinervis* (Cryphaeaceae); *Fissidens crassicuspes* and *F. fernandezianum* (Fissidentaceae); *Ptychomitrium fernandezianum* (Ptychomitriaceae); *Dicksonia berteriana* (Figure 5.34a; Dicksoniaceae); *Juania* (Arecaceae); *Centaurodendron, Rhetinodendron, Robinsonia* (Figure 5.34b) *and Yunquea* (Asteraceae); *Azara serrata* (Flacourtiaceae); Lactoridaceae;

Figure 5.34 Two endemic plant taxa of the Juan Fernández province. (a) *Dicksonia berteriana* (Dicksoniaceae) and (b) *Robinsonia* (Asteraceae). (Courtesy of Andrés Moreira-Muñoz.)

Agrostis masafuerana, Chusquea fernandeziana, Megalachne berteroana, M. masafuerana and *Podophorus bromoides* (Poaceae); *Cuminia* (Lamiaceae); *Rhaphitamnus venustus* (Verbenaceae); *Anolethrus*, Juanorhinini, *Pachystylus* and *Strongylopterus ovatus* (Coleoptera: Curculionidae); *Conchopterella kuscheli* and *C. maculata* (Neuroptera: Hemerobiidae); *Physothrips skottsbergi* (Thysanoptera: Thripidae); *Sephanoides fernandensis* (Apodiformes: Trochilidae); *Aphrastura masafuerae* (Passeriformes: Furnariidae); and *Anairetes fernandezianus* (Passeriformes: Tyrannidae). The islands harbor a high level of endemism (Stuessy et al., 1984, 1990; Crawford et al., 1992;

Baeza et al., 2002; Ruiz et al., 2004; Moreira-Muñoz, 2011; Moreira-Muñoz and Elórtegui Francioli, 2014; Urbina-Casanova et al., 2015; Bernardello and Stuessy, 2017). The most speciose plant family is Asteraceae, with four endemic genera: *Centaurodendron, Dendroseris, Robinsonia* and *Yunquea* (Bernardello and Stuessy, 2017).

Vegetation

The Juan Fernández vegetation consists of forests, shrublands and grasslands (Cabrera and Willink, 1973). Forests are dense, particularly around the higher hills of Masatierra (Kuschel, 1961). On Masatierra island, there are six vegetation zones: grasslands (0–100 m altitude), introduced shrubs (100–300 m), tall forests (300–500 m), lower montane forests (500–700 m), tree fern forests (700–750 m) and high brushwood on exposed cliffs (500–850 m); whereas on Masafuera island, zones are grassland slopes (0–400 m), deep ravines (0–500 m), lower montane forests (400–600 m), upper montane forests (600–950 m), high brushwood (950–1100 m) and an *alpine* zone (1100–1300 m; Bernardello and Stuessy, 2017). Dominant plant species include *Azara fernandeziana, Coprosma hookeri, Diksonia fernandeziana, Drimys confertifolia, Dysopsis hirsuta, Empetrum rubrum, Escallonia callcottiae, Fagara mayu, Juania australis, Myrceugenia schulzei, Pernettya rigida, Rhaphithamnus venustus, Robinsonia gayana, R. gracilis* and *Ugni selkirkii* (Cabrera and Willink, 1973; Stuessy et al., 1984).

Biotic relationships

The biota of the Juan Fernández Islands shows a close relationship with the biota of the remaining Subantarctic provinces, although Pacific and Neotropical elements are also present (Kuschel, 1961; Soto et al., 2017). Track analysis based on plants, insects, crustaceans and mollusks supported this (Morrone, 1992), showing that a generalized track with the Magellanic Moorland and Falkland Islands provinces connects the Juan Fernández Islands. Moreira-Muñoz (2011) also noted the floristic similarity between these islands and the Magellanic area. Kuschel (1961) noted a biotic relationship with the Desventuradas islands (San Félix and San Ambrosio islands) and considered that their floras were slightly similar, but the insect fauna evidenced a close relationship.

Cenocrons

The biota of the Juan Fernández Islands originated in the continental portion of the Subantarctic subregion (Kuschel, 1961; Stuessy et al., 1984; Ruiz et al., 2004; Soto et al., 2017). Skottsberg (1920) distinguished the following elements: Antarcto-Tertiary, Neotropical-Andean, Arcto-Tertiary, Paleotropical and Austral of wide-ranging seaside species. Bernardello et al. (2006) suggested that in addition to Subantarctic elements, the flora of the Juan Fernández Islands also show pantropical, Australian, New Zealand and Pacific colonizers.

Stuessy et al. (1990) examined the general patterns of speciation of the flora of the islands and found that anagenesis account for 71% of the endemic species, anacladogenesis for 24% and cladogenesis for 5%, concluding that rapid morphological evolution accompanied by small genetic change was consistent with the two former speciation patterns. Takayama et al. (2014) analyzed the species of *Robinsonia* (Asteraceae) and found five species originated by cladogenesis on the older Masatierra island, whereas a single species endemic to the younger Masafuera island was an anagenetic derivative of a species from the former island. Ruiz et al. (2004) analyzed four plant taxa of the Juan Fernández Islands and established their phylogentic relationships and genetic divergence with their continental sister taxa (Figure 5.35). Estimates of times of divergence between the Juan Fernández endemics and their sister taxa were between 188,500 and 2,580,000 years. The divergence between species endemic to the Masafuera Island and their continental relatives was lower than the

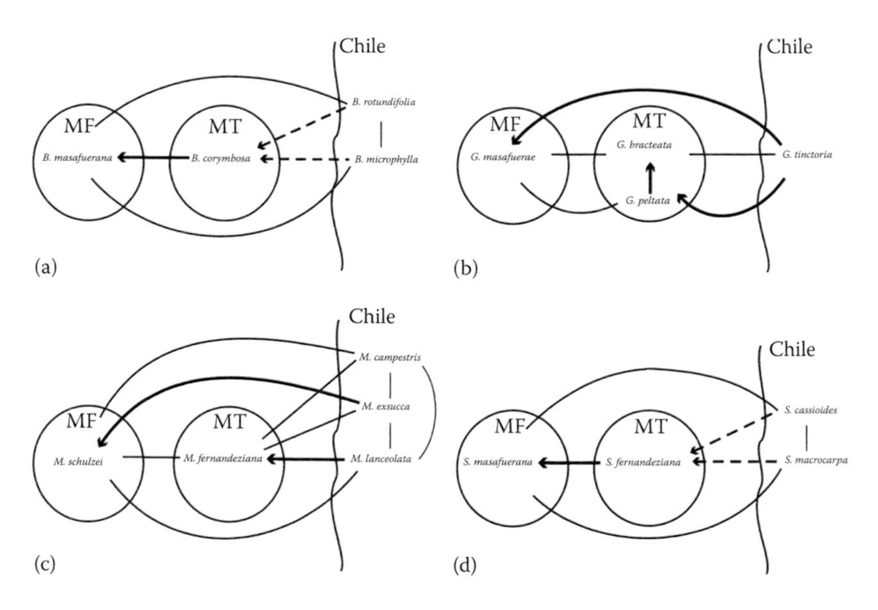

Figure 5.35 Diagrammatic representation of plant lineages from the Juan Fernández Islands and continental Chile. (Modified from Ruiz, E. et al., *Taxon*, 53, 321–332, 2004.) (a) *Berberis* (Berberidaceae), (b) *Gunnera* (Gunneraceae), (c) *Myrceugenia* (Myrtaceae) and (d) *Sophora* (Fabaceae). (MF) Masafuera or Alejandro Selkirk Island and (MT) Masatierra or Robinson Crusoe Island.

divergence between the species of Masatierra Island and their continental sister taxa. Genetic divergence was also greater between continental species and Masatierra species than between species pairs in Masatierra-Masafuera islands.

Soto et al. (2017) analyzed the spider species of the genus *Philisca* (Anyphaenidae), which included six species endemic to the Juan Fernández Islands and six species from the Valdivian Forest province. The species of the Juan Fernández Islands formed a monophyletic group nested within the continental species, the study estimated their split from their sister taxon at 2.21 m.y.a., which postdates the geological origin of the islands (4 m.y.a.). The study concluded that a single dispersal event followed by rapid diversification could explain the biogeographical history of this taxon.

Case study: Modes of speciation of the plant genus Robinsonia *in the Juan Fernández Islands*

Title: "Relationships and genetic consequences of contrasting modes of speciation among endemic species of *Robinsonia* (A steraceae, Senecioneae) of the Juan Fernández Archipelago, Chile, based on AFLPs and SSRs." (Takayama, K., P. López-Sepúlveda, J. Greimler, D. J. Crawford, P. Peñailillo, M. Baeza, E. Ruiz, G. Kohl, K. Tremetsberger, A. Gatica, L. Letelier, P. Novoa, J. Novak and T. F. Stuessy, *New Phytologist*, 205: 415–428, 2014.)

Goal: To estimate the levels of genetic variation within and among populations and species of *Robinsonia* (Asteraceae), endemic to the Juan Fernández Islands, to determine their relationships and to contrast their modes of speciation.

Location: Juan Fernández Islands, Chile.

Methods: The study analyzed population genetic structure by amplified fragment length polymorphism (AFLP) and microsatellite (simple sequence repeat, SSR) markers from 286 and 320 individuals, respectively, in 28 populations from the Juan Fernández Archipelago (Figure 5.36). The study constructed a neighbor-net phenogram (Bryant and Moulton, 2004) using the software SplitsTree4 ver. 4.10 (Huson and Bryant, 2006) based on a Nei–Li distance matrix calculated from the AFLP matrix.

Results: The neighbor-joining phenogram based on AFLP data using all individuals of *Robinsonia* showed all individuals clearly sorted into respective species, and the latter were distinct. The two sections, sect. *Eleutherolepis* (including *R. gracilis, R. evenia* and *R. masafuerae*) and sect. *Robinsonia* (including *R. gayana, R. thurifera* and *R. saxatilis*), were also distinct from each other. The study showed the close relationship of the populations of *R. masafuerae* on Masafuera to the

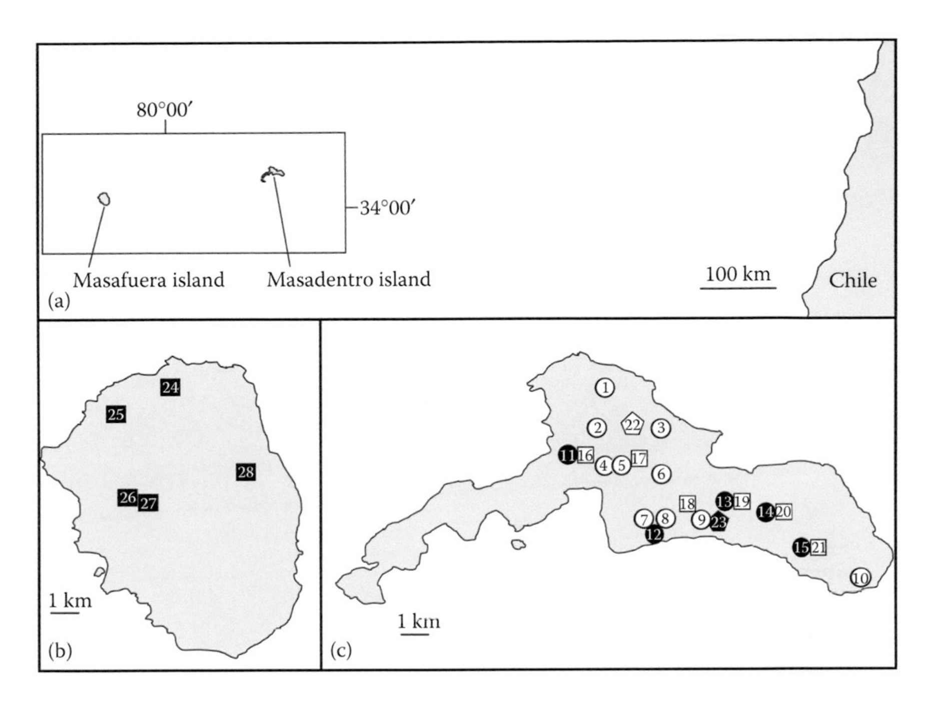

Figure 5.36 Map with the location of the Juan Fernández Islands and populations of *Robinsonia* (Asteraceae) sampled. (Modified from Takayama, K. et al., *New Phytol.*, 205, 415–428, 2014. With permission.) (a) Juan Fernández archipelago, (b) Masafuera Island and (c) Masadentro Island. Open squares, *R. evenia;* open squares, *R. gayana;* closed circles, *R. gracilis;* closed pentagon, *R. saxatilis;* open pentagon, *R. thurifera* and closed squares, *R. masafuerae.*

populations of *R. evenia* on Masatierra. In the neighbor-joining tree using microsatellite data (Figure 5.37), each species was well isolated, and the study showed a close relationship of the populations of *R. masafuerae* to the populations of *R. evenia.*

Main conclusions: The available data are consistent with the hypothesis that species of *Robinsonia* on Masatierra form an adaptively radiating complex of cladogenetic origin, and that *R. masafuerae* on Masafuera evolved anagenetically. Use of the hierarchical phylogenetic structure and known geological ages of the islands suggested a scenario for these origins. The first splitting event after immigration to Masatierra, some 3.5 m.y.a., led to the separation of *R. berteroi* (subgen. *Rhetinodendron*), which was the most divergent of all taxa within the complex. The other clade (subgen. *Robinsonia*) split into three main lineages, leading to the three presently recognized sections (*Eleutherolepis, Robinsonia* and *Symphyochaeta*). Differentiation occurred within them, with

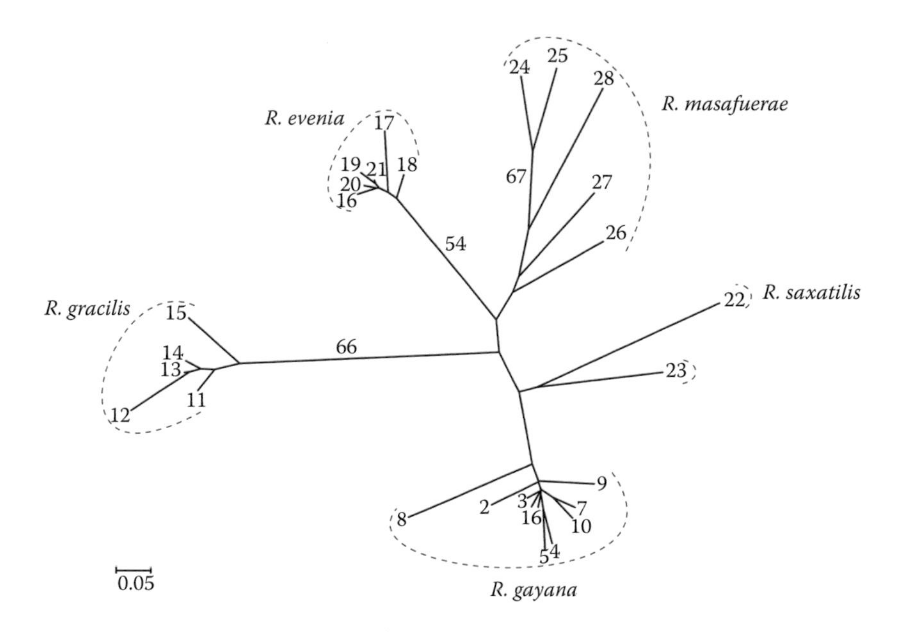

Figure 5.37 Neighbor-joining of the 28 populations of *Robinsonia* subgen. *Robinsonia* based on distance of microsatellite data. (Modified from Takayama, K. et al., *New Phytol.*, 205, 415–428, 2014. With permission.)

R. gayana and *R. thurifera* speciating within sect. *Robinsonia*, and *R. gracilis* and *R. evenia* separating within sect. *Eleutherolepis*. The study hypothesized that propagules of *R. evenia* dispersed to the younger island after its formation 1–2 m.y.a., and this immigrant population led to the anagenetically derived *R. masafuerae*.

chapter six

The Central Chilean subregion

The Central Chilean subregion corresponds to Central Chile between 26°S and 37°S (Morrone, 1996a, 2000a, 2001b, 2006, 2015b; Morrone et al., 1997). Its limits are to the south with the Subantarctic subregion and to the east and north with the South American transition zone. Based on its rich biota, the Central Chilean subregion is one of the world's biodiversity hotspots (Myers et al., 2000). Studies have treated it as a transitional zone between the Atacaman desert to the north and the Valdivian forest to the south (Locklin, 2017).

In this chapter, I characterize the Central Chilean subregion, provide its endemic and characteristic taxa and discuss its relationships. Studies divide it into two provinces, Coquimban and Santiagan, and six districts. I characterize the provinces and districts, describe their vegetation and discuss their relationships. Additionally, I discuss one case study.

Central Chilean subregion

Central Chilean province—Hauman, 1931: 62 (regionalization); Cabrera and Willink, 1973: 92 (regionalization); Willink, 1988: 206 (regionalization); Morrone, 1994a: 190 (parsimony analysis of endemicity), 1996a: 107 (regionalization); Posadas et al., 1997: 2 (parsimony analysis of endemicity).

Chilean district—Cabrera and Yepes, 1940: 16 (regionalization).

Chilean province—Mello-Leitão, 1943: 130 (regionalization); Ringuelet, 1975: 107 (regionalization); Ezcurra et al., 2014: 28 (biotic evolution).

Chile province (in part)—Fittkau, 1969: 642 (regionalization).

Southern Andes area (in part)—Sick, 1969: 465 (regionalization).

Chilean subcenter (in part)—Müller, 1973: 101 (regionalization).

Central Andean zone (in part)—Artigas, 1975: map (regionalization).

Chilean Sclerophyll Forest province—Udvardy, 1975: 41 (regionalization).

Chilean subregion—Paulson, 1979: 170 (regionalization); Flint, 1989: 1 (regionalization); Rivas-Martínez and Navarro, 1994: map (regionalization).

Central Chilean region—Rivas-Martínez and Tovar, 1983: 516 (regionalization).

Central Chile area—Coscarón and Coscarón-Arias, 1995: 726 (areas of endemism); Morrone et al., 1997: 25 (cladistic biogeography); Roig-Juñent et al., 2006: 408 (cladistic biogeography).
Chilean Matorral ecoregion—Dinerstein et al., 1995: 103 (ecoregionalization); Locklin, 2017: 1 (ecoregionalization).
Central Chilean subregion—Morrone, 1999: 13 (regionalization); Ocampo and Morrone, 1999: 21 (faunistics); Morrone, 2000a: 99 (regionalization), 2001b: 114 (regionalization); Posadas and Morrone, 2001: 267 (cladistic biogeography); Donato et al., 2003: 340 (dispersal-vicariance analysis); Posadas and Morrone, 2003: 72 (cladistic biogeography); Morrone, 2004a: 158 (regionalization); Soares and de Carvalho, 2005: 485 (evolutionary biogeography); Morrone, 2006: 484 (regionalization); Pérez and Posadas, 2006: 1785 (systematic revision); Spinelli et al., 2006: 302 (parsimony analysis of endemicity); Casagranda et al., 2009: 19 (endemicity analysis); Escalante et al., 2009: 379 (endemicity analysis); Morrone, 2010: 38 (regionalization); Urtubey et al., 2010: 506 (track analysis and cladistic biogeography); Kutschker and Morrone, 2012: 541 (track analysis); Vivallo, 2013: 529 (systematic revision); Carrara and Flores, 2015: 47 (faunistics); Goin et al., 2015: 133 (biotic evolution); Leivas et al., 2015: 116 (faunistics); Morrone, 2015b: 209 (regionalization); Monckton, 2016: 125 (systematic revision and map); Urra, 2017a: 30 (faunistics), 2017b: 30 (faunistics).
Central region—Roig-Juñent and Domínguez, 2001: 557 (faunistics).
Middle Chilean-Patagonian region (in part)—Rivas-Martínez et al., 2011: 27 (regionalization).
Central Chile ecoregion—Moreira-Muñoz, 2014: 221 (endemism).

Endemic and characteristic taxa

Morrone (2000a, 2006) provided a list of endemic and characteristic taxa. Some examples included *Triptilion spinosum* (Figure 6.1a; Asteraceae); *Gordius chilensis* (Gordioidea: Gordiidae); *Missulena tussulena, Plesiolena bonneti* and *P. jorgelina* (Araneae: Actinopodidae); *Apodrassodes quilpuensis* (Araneae: Gnaphosidae); *Platnickia elegans* (Figure 6.1b; Araneae: Zodariidae; Grismado and Platnick, 2008); *Gigantochernes franzi* (Pseudoscorpiones: Chernetidae); *Sedna* (Solifugae: Ammotrechidae); *Aegla papudo* (Figure 6.1c; Decapoda: Aeglidae); *Parastacus pugnax* (Decapoda: Parastacidae); *Mendizabalia* (Coleoptera: Buprestidae); *Eurymetopum maculatum* (Coleoptera: Cleridae); *Listroderes curvipes, L. nodifer* and *L. robustus* species groups (Coleoptera: Curculionidae); *Polyncus scaber* (Coleoptera: Trogidae); *Araucnephia, Araucnephioides* and *Gigantodax minor* (Diptera: Simuliidae); *Neuquenaphis* (Heteroptera: Aphidae); *Bradynobaenus wagenknechti* (Hymenoptera: Bradynobaenidae);

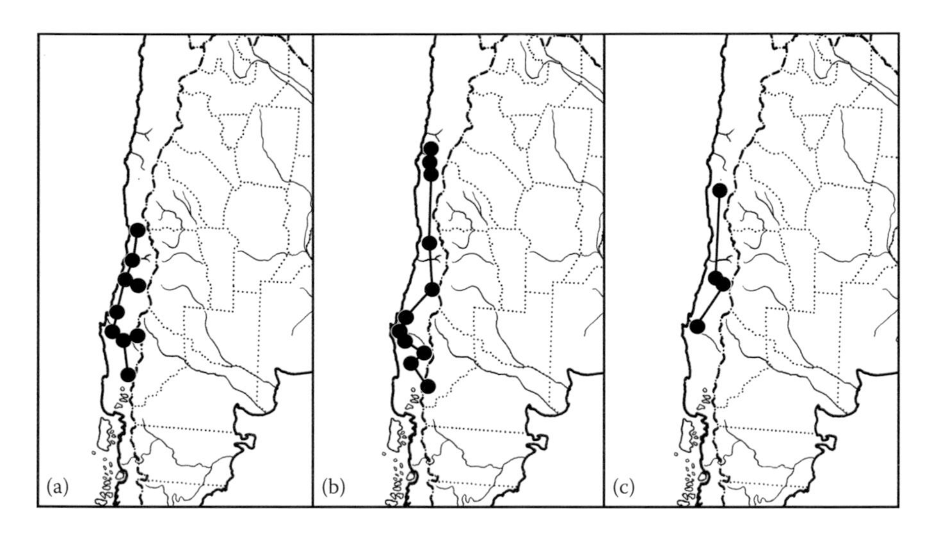

Figure 6.1 Maps with individual tracks in the Central Chilean subregion. (a) *Triptilion spinosum* (Asteraceae), (b) *Platnickia elegans* (Araneae: Zodariidae) and (c) *Aegla papudo* (Decapoda: Aeglidae).

Palaephatus albicerus (Lepidoptera: Palaephatidae); *Rheopetalia apicalis* (Odonata: Austropetaliidae); *Phenes raptor centralis* (Odonata: Petaluridae); *Moluchacris* (Orthoptera: Tristiridae); *Asthenes humicola* (Passeriformes: Furnariidae); *Abrocoma bennetti* (Rodentia: Abrocomidae); and *Octodon degus, O. lunatus* and *Spalacopus cyanus* (Rodentia: Octodontidae). Roig-Juñent and Domínguez (2001) considered that this area harbors 38.9% of endemic species of Carabidae (Coleoptera).

Biotic relationships

A track analysis based on weevil (Coleoptera: Curculionidae) species (Morrone, 1996b) showed that the Central Chilean subregion was related to the Maule province of the Subantarctic subregion. Another track analysis based on weevil species (Morrone, 1994a), a cladistic biogeographical analysis based on plant and animal taxa (Morrone et al., 1997) and fossil evidence (Troncoso and Romero, 1998) indicated that this subregion was closely related to the Subantarctic subregion.

Regionalization

The Central Chilean subregion comprises two provinces (Figure 6.2): Coquimban and Santiagan.

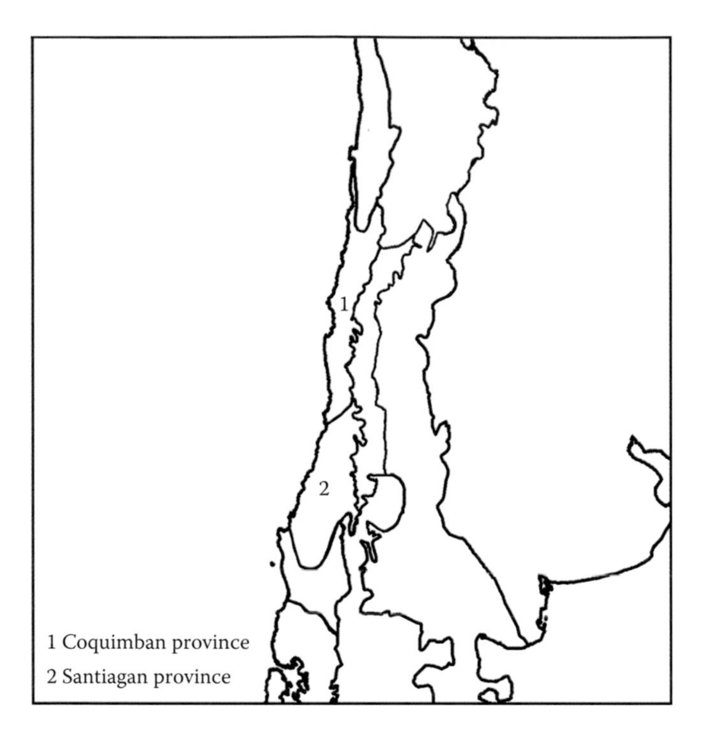

Figure 6.2 Map of the provinces of the Central Chilean subregion.

Case study: Cladistic biogeographical analysis of Central Chile

Title: "A cladistic biogeographic analysis of Central Chile." (Morrone, J. J., L. Katinas and J. V. Crisci, *Journal of Comparative Biology,* 2: 25–41, 1997.)

Goal: To undertake a cladistic biogeographic analysis based on plant and animal taxa and provide a natural delimitation of Central Chile and its areas of endemism.

Location: Chile.

Methods: The study analyzed 24 genera of Asteraceae, Buprestidae and Curculionidae (Coleoptera), and Gnaphosidae (Araneae). The study accomplished the delimitation of Central Chile by mapping species ranges, where overlap in ranges of two or more species determines an area of endemism (Müller, 1973). The cladistic biogeographical analysis was based on 10 taxon cladograms, using four methods and the appropriate software: component analysis (Nelson and Platnick, 1981) with Component 1.5 (Page, 1989); Brooks parsimony analysis (Brooks, 1985, 1990) with Hennig86 (Farris, 1988); three-area statements (Nelson and Ladiges, 1991a) with TAS (Nelson and

Ladiges, 1991b) and Hennig86; and paralogy-free subtree analysis (Nelson and Ladiges, 1996) with TASS (Nelson and Ladiges, 1995). The study quantified the amounts of difference between any resulting general cladograms obtained by calculating the items of error (Nelson and Platnick, 1981) with Component 1.5. Additionally, the study performed a parsimony analysis of endemicity (PAE), dividing Central Chile into 13 units corresponding to one-degree latitude with Hennig86.

Results: The study mapped distributions of species analyzed and their overlap indicated four major areas within Central Chile: Coquimbo, Santiago, Curicó and Ñuble. Because of the relationships of Central Chile with the Subantarctic subregion, the study figured a portion of the latter and included it in the analysis.

The cladistic biogeographical analysis using the different methods produced five different general area cladograms, and the items of error led the study to choose one of them (Figure 6.3) where the pattern of relationships was as follows: ([Coquimbo, Santiago], [Curicó, [Ñuble, Subantarctic]]). The parsimony analysis of endemicity resulted in one cladogram (Figure 6.3) with three groups: a northern group (areas A–F), an intermediate group (areas G–J) and a southern group (areas K–M). The groups corresponded mainly to the areas of the general area cladogram: B = Coquimbo, D + E = Santiago, G = Curicó, H = Ñuble and I–M = Subantarctic.

Main conclusions: Central Chile as delimited traditionally did not constitute a natural biogeographical area because Curicó and Ñuble were more closely related to a part of the Subantarctic subregion rather than to Coquimbo and Santiago. The parsimony analysis of endemicity, a completely different approach, agreed with this interpretation. According to these results, it seemed that Curicó and Ñuble represented an area of overlap of elements from the Central Chilean and Subantarctic subregions, whereas the area comprising Coquimbo plus Santiago would represent the region with the more typical Central Chilean biota.

Coquimban province

Savanna zone (in part)—Mann, 1960: 27 (regionalization).
Scrubland zone—Mann, 1960: 23 (regionalization).
Coquimban district—Cabrera and Willink, 1973: 91 (regionalization).
Coquimbo zone—Artigas, 1975: map (regionalization).
Desertic Mediterranean Chilean province—Rivas-Martínez and Navarro, 1994: map (regionalization).
Desertic Mesochilean province—Rivas-Martínez and Navarro, 1994: map (regionalization).

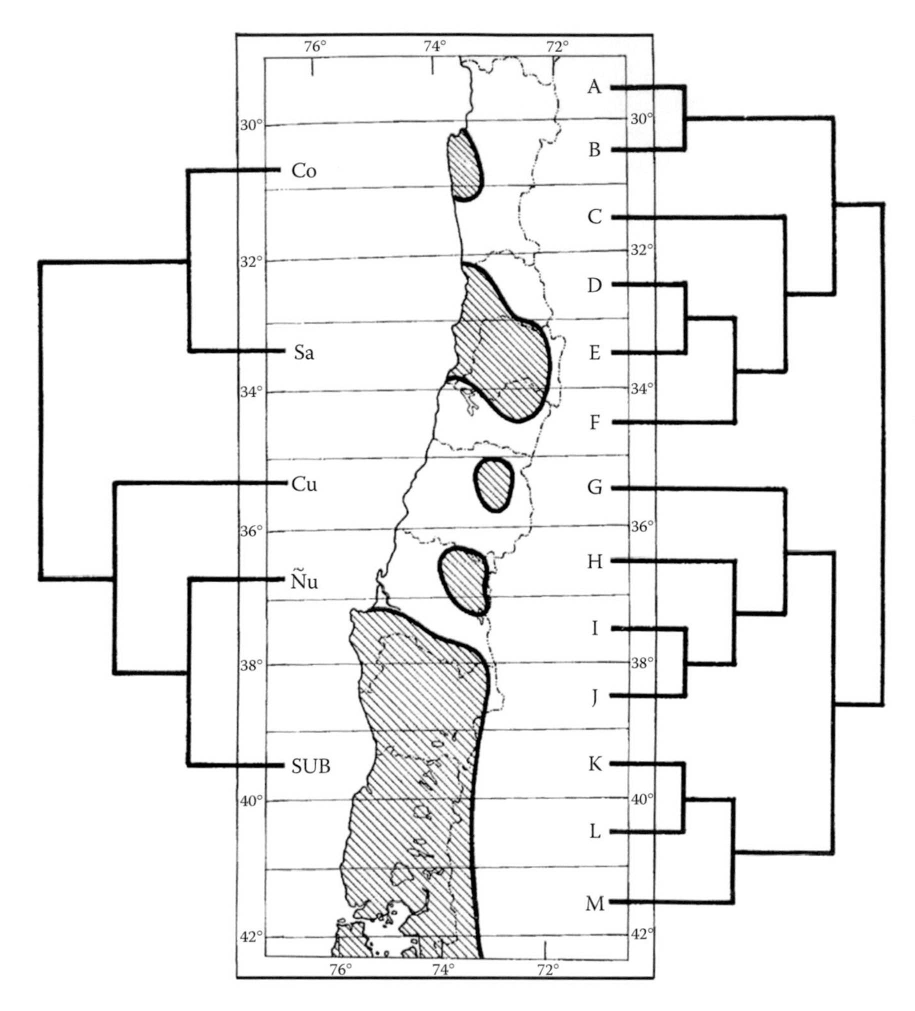

Figure 6.3 Area cladograms in the Central Chilean subregion. (Modified from Morrone, J.J. et al., *J. Comp. Biol.*, 2, 25–41, 1997. With permission.) Left, general area cladogram; right, area cladogram obtained from the parsimony analysis of endemicity.

Coquimbo province—Morrone, 1999: 13 (regionalization), 2001b: 115 (regionalization), 2004a: 158 (regionalization); Soares and de Carvalho, 2005: 485 (evolutionary biogeography); Morrone, 2006: 484 (regionalization), 2010: 38 (regionalization); Flores and Pizarro-Araya, 2012: 7 (systematic revision); Alfaro et al., 2013: 236 (faunistics), 2014: 387 (faunistics); Ferretti, 2015: 13 (dispersal-vicariance analysis); Leivas et al., 2015: 116 (faunistics).

Coquimban province—Morrone, 2000a: 99 (regionalization), 2015b: 212 (regionalization); Martínez et al., 2017: 479 (track analysis and regionalization); Monckton, 2016: 125 (systematic revision and map).
Coquimbo area—Roig-Juñent et al., 2006: 408 (cladistic biogeography).
Coquimbo region—Urbina-Casanova et al., 2015: 273 (floristics).

Definition

North Central Chile, between 26°S and 33°S (Morrone, 2000a, 2001b, 2015b).

Endemic and characteristic taxa

Morrone (2000a, 2015b) provided a list of endemic and characteristic taxa. Some examples include *Chuquiraga ulicina* subsp. *acicularis*, *Triptilion cordifolium* and *T. globosum* (Figure 6.4a; Asteraceae); *Echemoides chilensis*, *E. illapel* and *E. tofo* (Araneae: Gnaphosidae); *Migas vellardi* (Araneae: Migidae); *Bradynobaenus gayi* (Hymenoptera: Bradynobaenidae); *Conognatha obenbergeri* and *Lasionota rouleti roitmani* (Coleoptera: Buprestidae); *Cnemalobus coquimbanus*, *C. cyaneus*, *C. nuria*, *C. pegnai* and *Notoperyphus cekalovici* (Coleoptera: Carabidae); *Listroderes angusticeps*, *L. howdenae* and *L. robustus* (Figure 6.4b; Coleoptera: Curculionidae); *Praocis sanguinolenta*, *P. subsulcata*,

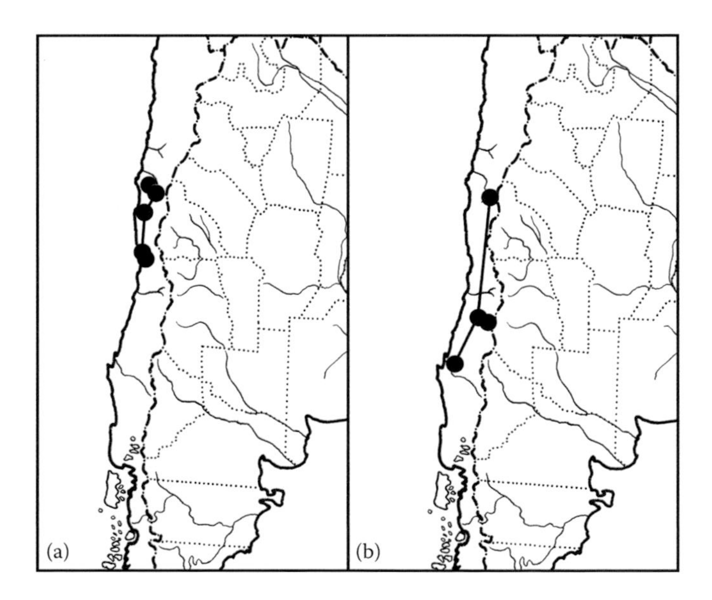

Figure 6.4 Maps with individual tracks in the Coquimban province. (a) *Triptilion globosum* (Asteraceae) and (b) *Listroderes robustus* (Coleoptera: Curculionidae).

P. marginata, P. spinolai, P. parva and *P. tibialis* (Coleoptera: Tenebrionidae); *Araucnephioides schlingeri* (Diptera: Simuliidae); and *Chinchilla lanigera* (Rodentia: Chinchillidae).

Vegetation

The Coquimban province consists of xeric vegetation, including desert scrubland and thorn forest (Figures 6.5 and 6.6; Luebert and Pliscoff, 2006). Dominant plant species include *Argylia* spp., *Balbisia peduncularis, Calandrinia* spp., *Copiapoa marginata, Echinopsis littoralis, Eulychnia acida, E. breviflora, E. tenuis, Heliotropium stenophyllum, Lithraea caustica, Oxalis gigantea, Puya chilensis* and *Skytanthus acutus* (Cabrera and Willink, 1973; Rivas-Martínez et al., 2011).

Figure 6.5 Xeric vegetation typical of the Coquimban province, central Chile. (a) Shrubs and herbs and (b) cacti. (Courtesy of Sergio Roig-Juñent.)

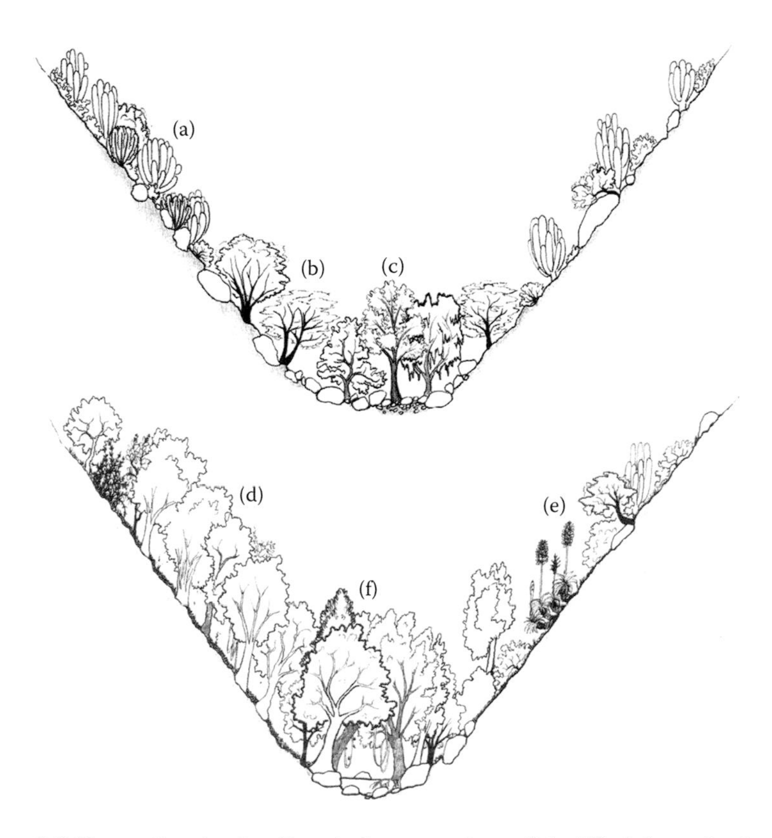

Figure 6.6 Vegetation in the Coquimban province. (Modified from Luebert, F. and Pliscoff, P., *Sinopsis Bioclimática y Vegetacional de Chile*. Editorial Universitaria, Santiago de Chile, 2006.) (a)–(c) Desert scrubland, (d)–(f) xeric forest with *Acacia caven*. (a) Desert scrub, (b) sclerophyllous forest, (c) ravine forest with *Salix humboldtiana*, (d) sclerophyllous forest with *Cryptocarpa alba* and *Peumus boldus*, (e) shrubs with *Puya berteroniana* and *Colliguaja odorifera* and (f) ravine forest with *Beilschmiedia miersii* and *Crinodendron patagua*.

Biotic relationships

O'Brien (1971) considered that the entomofauna of the Coquimban province was related to the entomofauna of the Atacama province. A parsimony analysis of endemicity and a cladistic biogeographical analysis based on plant and animal species (Morrone et al., 1997) showed that it is the sister area of the Santiagan province. Another cladistic biogeographical analysis, based on species of Diptera (Soares and de Carvalho, 2005), showed the Coquimban province as sister area to a clade comprising the Santiagan, Maule, Valdivian Forest, Magellanic Forest and Falkland Islands provinces.

Regionalization

Morrone (2015b) treated three areas identified by Peña (1966a, b) as the Central Andean Cordillera, Coquimban Desert and Intermediate Desert districts.

Central Andean Cordillera district

Cordilleras from Southern Coquimbo to Chillán region—Reiche, 1905: map (regionalization).

Central Andean Cordillera region—Peña, 1966a: 10 (regionalization), 1966b: 216 (regionalization).

Central Andean zone (in part)—Artigas, 1975: map (regionalization).

Central Andean Mountain range—Flores and Vidal, 2000: 193 (systematic revision).

Central Andean Cordillera district—Morrone, 2015b: 212 (regionalization); Monckton, 2016: 125 (systematic revision and map).

Definition: The Central Andean Cordillera district extends from 26°S to 33°S in the Andean portion of the province; its vegetation is characteristic, with dense cloud cover, strong rainfall and heavy snowfall during winter (Peña, 1966a). Cecilia Ezcurra (pers. comm.) suggests more careful analysis of the boundaries between this district and the Cuyan High Andean province.

Endemic taxa: *Acalodegma vidali* (Coleoptera: Cerambycidae) and *Geonemides ater* (Coleoptera: Curculionidae).

Vegetation: Within this district, Luebert and Pliscoff (2006) identified different types of scrublands. Between 1500 and 2500 m, *Adesmia longipes* (Fabaceae) and *Senecio bipontini* (Asteraceae) dominate the low scrubland vegetation; other species are *Acaena macrocephala, Adesmia retusa, Empetrum rubrum, Festuca thermarum, Hypochaeris arenaria, Luzula racemosa, Nassauvia revoluta, Perezia pedicularifolia, Poa alopecurus, Senecio subdiscoideus* and *Valeriana fonckii*. At 3500–4500 m, *Adesmia subterranea* and *A. echinus* (Fabaceae) dominate the low scrubland; other species are *Anarthrophyllum gayanum, Azorella cryptantha, A. madreporica, Calceolaria pinifolia, Chaetanthera minuta, Cistante picta, Doniophyton weddellii, Oreopolus macarnthus, Viola chrysantha* and *V. frigida*. Additionally, there are grasslands with abundant *Chaetanthera sphaerodalis* (Asteraceae); other species are *Adesmia capitellata, Chaetanthera pulvinata, Lenzia chamaepitys, Nototriche auricoma, N. holosericea, Senecio socompae, S. sundtii* and *Stipa frigida*.

Coquimban Desert district

Coquimbo to Maule Littoral region—Reiche, 1905: map (regionalization).
Coquimban Desert region—Peña, 1966a: 9 (regionalization), 1966b: 215 (regionalization).
Coquimban Desert subregion—O'Brien, 1971: 199 (regionalization).
Coquimbo Desert area—Roig-Juñent, 1994b: 180 (cladistic biogeography).
Coquimban Desert area—Roig-Juñent and Flores, 2001: 257 (cladistic biogeography).
Coquimban Desert district—Morrone, 2015b: 212 (regionalization); Monckton, 2016: 125 (systematic revision and map).

Definition: The Coquimban Desert district extends from 29°S to 33°S; during part of the year marine fog covers it, which increases its humidity (Peña, 1966a).

Endemic taxa: *Chilicola mantagua* (Hymenoptera: Colletidae; Monckton, 2016); *Cnemalobus convexus* and *Cicindelidia trifasciata australis* (Coleoptera: Carabidae); *Geosphaeropterus meridionalis* (Coleoptera: Curculionidae; Elgueta, 2013); *Auladera crenicosta, Praocis hirtella* and *P. tibialis* (Coleoptera: Tenebrionidae); and *Elasmoderus lutescens* (Orthoptera: Tristiridae; Roig-Juñent, 1994a, 1994b; Flores and Pizarro-Araya, 2012).

Vegetation: Within this district, Luebert and Pliscoff (2006) identified different types of scrublands. In littoral areas of Atacama and Coquimbo, from sea level to 300 m, *Heliotropium stenophyllum* (Boraginaceae) and *Oxalis gigantea* (Oxalidaceae) dominate the open scrubland; other species are *Adesmia tenella, Bahia ambrosoides, Cryptantha glomerulata, Chuquiraga ulicina, Cistanthe coquimbensis, Encelia canescens, Erodium cicutarium, Haplopappus cerberoanus, Lobelia polyphylla, Nolana crassulifolia* and *Schismus arabicus*. In interior arid zones, from sea level to 1200 m, dominant species in the scrubland are *Heliotropium stenophyllum* (Boraginaceae) and *Flourensia thurifera* (Asteraceae); other species are *Adesmia bedwellii, Bromus berterianus, Carica chilensis, Cordia decandra, Bahia ambrosioides, Edorium cicutarius, Eulychnia acida, Opuntia berterii, Porlieria chilensis, Proustia cuneifolia* and *Senna cumingi. Flourensia thurifera* (Asteraceae) and *Colliguaya odirifera* (Euphorbiaceae) dominate other scrubland; other species are *Adesmia confusa, Bridgesia incisaefolia, Cordia decandra, Ephedra chilensis, Gutierrezia resinosa, Pasithaea coerulea, Proustia cinerea* and *Vulpia minurus*. In the central coastal area, from seal level to 500 m, *Bahia ambrosioides* (Asteraceae) and *Puya chilensis* (Bromeliacae) dominated scrubland; other species were *Baccharis macraei, Bromus berterianus, Eulychnia acida, Fuchsia lycioides, Oxalis gigantea, Plantago hispidula, Solanum pinnatum* and *Vulpia myurus*.

Intermediate Desert district

Intermediate Desert region—Peña, 1966a: 8 (regionalization).
Copiapó Steppe zone—Artigas, 1975: map (regionalization).
Intermediate Desert area—Roig-Juñent and Flores, 2001: 257 (cladistic biogeography).
Intermediate Desert district—Morrone, 2015b: 212 (regionalization); Monckton, 2016: 125 (systematic revision and map).

Definition: The Intermediate Desert district occupies a vast area between 25°S to 29°S, having canyons with abundant shrubs, cacti, grasses and annual herbs (Peña, 1966a).

Endemic taxa: *Chilicola erithropoda, C. guanicoe, C. neffi* and *C. packeri* (Colletidae; Monckton, 2016); *Geosphaeropterus chango* and *G pegnai* (Coleoptera: Curculionidae; Elgueta, 2013); and *Praocis medevedevi* and *P. marginata* (Coleoptera: Tenebrionidae; Flores and Pizarro-Araya, 2012).

Vegetation: Within this district, Luebert and Pliscoff (2006) identified different types of scrublands. Between 200 and 1500 m, dominant species of the scrubland are *Skytanthus acutus* (Apocynaceae) and *Atriplex deserticola* (Amarantaceae); other species are *Alona rostrata, Argylia radiata, Aristolochia chilensis, Caesalpinia angulata, Calandrinia calycina, Chorizanthe commisuralis, Cryptantha parviflora, Euphorbia copiapina, Fagonia chilensis, Nolana baccata, Tetragonia copiapina* and *Viola polypoda*. Between 700 and 2500 m, dominant species are *Nolana leptophylla* (Solanaceae) and *Cistanthe salsoloides* (Montiaceae); other species are *Adesmia atacamensis, Atriplex deserticola, Dinemandra ericoides, Heliotropium chenopodiaceum, Lycium minutifolium, Malesherbia deserticola, Nolana flaccida* and *Reyesia parviflora*.

Cenocrons

Some studies hypothesized that a northward shift of the cooler-wetter southern climates and vegetational belts during the Quaternary pluvial (= glacial) periods explained the presence of relict forests in the Coquimban province. The conditions these forests provided apparently allowed the dispersal of animal taxa from the north, as the representatives of the family Cricetidae (Rodentia; Caviedes and Iriarte, 1989).

Santiagan province

Santiagan district—Osgood, 1943: 27 (regionalization).
Savanna zone (in part)—Mann, 1960: 27 (regionalization).
Santiagan region—O'Brien, 1971: 202 (regionalization).
Santiagan area—Artigas, 1975: map (regionalization).

Central Chilean province—Rivas-Martínez and Navarro, 1994: map (regionalization); Rivas-Martínez and Navarro, 1994: map (regionalization).

Chilean Winter-Rain Forests ecoregion—Dinerstein et al., 1995: 96 (ecoregionalization).

Santiagan province—Morrone, 1999: 13 (regionalization), 2000a: 100 (regionalization), 2015b: 212 (regionalization); Monckton, 2016: 125 (systematic revision and map); Martínez et al., 2017: 479 (track analysis and regionalization).

Santiago province—Morrone, 2001b: 116 (regionalization), 2004a: 158 (regionalization); Soares and de Carvalho, 2005: 485 (evolutionary biogeography); Morrone, 2006: 484 (regionalization); Pérez and Posadas, 2006: 1784 (faunistics); Casagranda et al., 2009: 21 (endemicity analysis); Chani-Posse, 2010: 45 (faunistics); Morrone, 2010: 38 (regionalization); Flores and Pizarro-Araya, 2012: 7 (systematic revision); Alfaro et al., 2013: 243 (faunistics), 2014: 387 (faunistics); Ferretti, 2015: 13 (dispersal-vicariance analysis); Urra, 2017a: 30 (faunistics), 2017b: 30 (faunistics).

Definition

The Santiagan province consists of South Central Chile, between 33°S and 38°S south latitude (Morrone, 2000a, 2001b, 2015b). This province comprises three different physiographic units: the Coastal Range in the west, the Andean Range in the east and the Central Depression between them (Donoso, 1996).

Endemic and characteristic taxa

Morrone (2000a, 2015b) provided a list of endemic and characteristic taxa. Some examples included *Triptilion berteroi* (Asteraceae); *Crinodendron patagua* (Elaeocarpaceae); *Fuchsia* sect. *Kierschlegeria* (Onagraceae); *Acanthogonatus huaquen* and *A. quilocura* (Araneae: Nemesiidae); *Echemoides cekalovici* (Araneae: Gnaphosidae); *Aegla laevis talcahuano* (Figure 6.7a; Decapoda: Aeglidae); *Phyllopetalia altarensis*, *P. pudu* and *P. apicalis* (Odonata: Austropetaliidae); *Molucachris cinerascens, M. migripes* and *Peplacris* (Orthoptera: Tristiridae); *Atacamita biimpresa* and *Mendizabalia germaini cyranovoridis* (Coleoptera: Buprestidae); *Eurymetopum brevevittatum* and *E. inerme* (Coleoptera: Cleridae); *Hybreoleptops roseus* (Coleoptera: Curculionidae); *Pogonomyrmex bispinosum* (Hymenoptera: Formicidae); *Nanniresthenia penai* (Heteroptera: Miridae); *Tesserodoniella elguetai* (Coleoptera: Scarabaeidae); *Auladera andicola, A. atronitens, A. rugicollis, Callyntra cantillana, Nycterinus rugiceps australis* and *Praocis rufipes*

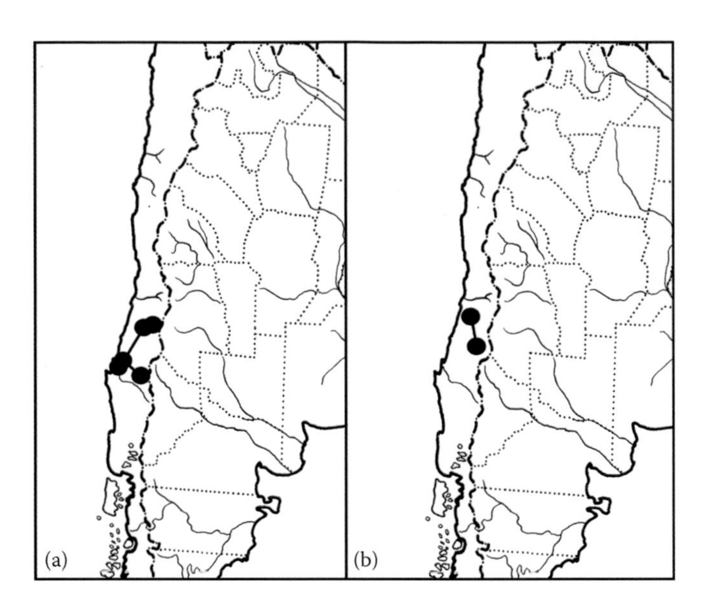

Figure 6.7 Maps with individual tracks in the Santiagan province. (a) *Aegla laevis talcahuano* (Decapoda: Aeglidae) and (b) *Cyanoliseus patagonus byroni* (Psittaciformes: Psittacidae).

(Coleoptera: Tenebrionidae); *Gigantodax luispenai* (Diptera: Simuliidae); *Euneomys noei* (Rodentia: Cricetidae); *Cyanoliseus patagonus byroni* (Figure 6.7b; Psittaciformes: Psittacidae); and *Pteroptochos castaneus* (Passeriformes: Rhinocryptidae).

Vegetation

The Santiagan province consists of xeric vegetation, including thorn savannas and shrublands (Figure 6.8), with small Mediterranean forests with *Nothofagus* at higher altitudes and elevations (Dinerstein et al., 1995; Donoso, 1996). The vegetation is well-adapted to summer dryness and winter precipitations (Donoso, 1996).

Biotic relationships

A parsimony analysis of endemicity and a cladistic biogeographical analysis based on plant and animal species (Morrone et al., 1997) showed that it was the sister area of the Coquimban province. Another cladistic biogeographical analysis, based on species of Diptera (Soares and de Carvalho, 2005), found that the Santiagan province was more closely related to the Valdivian Forest, Maule, Magellanic Forest and Falkland Islands provinces rather than to the Coquimban province.

Figure 6.8 Vegetation typical of the Santiagan province. (a) Shrubs and cacti and (b) shrubs. (Courtesy of Andrés Moreira-Muñoz.)

Regionalization

Morrone (2015b) treated three areas identified by Peña (1966a,b) as the Central Coastal Cordillera, Central Valley and Southern Andean Cordillera districts.

Central Coastal Cordillera district

Central Coastal Cordillera region—Peña, 1966a: 11 (regionalization), 1966b: 217 (regionalization).

Litoral Cauquenes zone—Artigas, 1975: map (regionalization).

Central Coastal mountain range—Flores and Vidal, 2000: 193 (systematic revision).

Central Coastal Cordillera area—Roig-Juñent and Flores, 2001: 258 (cladistic biogeography).

Central Coastal Cordillera district—Morrone, 2015b: 213 (regionalization); Monckton, 2016: 125 (systematic revision and map).

Definition: The Central Coastal Cordillera district encompasses the Coastal Range of the province from 32°S to 36°S (Peña, 1966a).

Endemic taxa: *Chilicola curvapeligrosa* (Hymenoptera: Colletidae; Monckton, 2016), and *Pristidactylus alvaroi* and *P. patagonica* (Squamata: Liolaemidae).

Vegetation: Within this district, Luebert and Pliscoff (2006) identified different vegetation types. In the coast of Valparaíso and northern Libertador Bernardo O'Higgins, between sea level and 500 m, *Acacia caven* (Fabaceae) and *Maytenus boaria* (Celastraceae) dominate the thorn forest; other species are *Baccharis linearis, Berberis chilensis, Bromus berterianus, Gochnatia foliolosa, Maytenus boaria, Mehlenbeckia hastulata, Retanilla trinervia* and *Vulpia myuros.* In northern Valparaíso and southern Coquimbo, *Peumus boldus* (Monimiaceae) and *Schinus latifolius* (Anacardiaceae) dominate the Mediterranean sclerophyllous arborescent shrubland; other species are *Anisomeria litoralis, Azara celastrina, Bahia ambrosioides, Eupatorium glechonophyllum, Fuchsia lycioides, Lobelia polyphylla* and *Podanthus mitiqui.* In the western slopes of the Cordillera de la Costa in Libertador Bernardo O'Higgins and Maule, between sea level and 800 m, *Lithraea caustica* (Anarcardiaceae) and *Azara integrifolia* (Salicaceae) dominate the Mediterranean sclerophyllous forest; other species are *Adiantum chilense, Alstroemeria revoluta, Blechnum hastatus, Bomarea salsilla, Chusquea cumingii, Escallonia revoluta, Lomatia hirsuta, Proustia pyrifolia, Ribes punctatum, Sophora macrocarpa* and *Triptilion spinosum.*

Central Valley district

Central Valley region—Peña, 1966a: 11 (regionalization), 1966b: 216 (regionalization).

Central Valley zone—Artigas, 1975: map (regionalization).

Interior Cauquenes zone—Artigas, 1975: map (regionalization).

Central Valley area—Roig-Juñent, 1994b: 180 (cladistic biogeography).

Andean Cordillera and Central Valley area (in part)—Díaz Gómez et al., 2009: 4 (dispersal-vicariance anal).

Cauquenes zone—Artigas, 1975: map (regionalization).

Central Valley district—Morrone, 2015b: 213 (regionalization); Monckton, 2016: 125 (systematic revision and map).

Definition: The Central Valley district occupies the central portion of the province in the Central Depression; cultivated lands have almost completely replaced its original thorn forest vegetation (Peña, 1966a).

Endemic taxa: *Chrysobothris bothrideres* (Coleoptera: Buprestidae), *Eurymetopum bispinosum* (Coleoptera: Cleridae) and *Myrmecodema nycterinoides* (Coleoptera: Tenebrionidae; Roig-Juñent, 1994b).

Vegetation: Within this district, Luebert and Pliscoff (2006) identified different vegetation types. In the Libertador Bernardo O'Higgins, Metropolitana de Santiago and Valparaíso regions, between 200 and 800 m, *Acacia caven* (Fabaceae) and *Prosopis chilensis* (Fabaceae) dominate the Mediterranean thorn forest; other species are *Baccharis linearis, B. paniculata, Cestrum parqui, Echinopsis chiloensis, Maytenus boaria, Porlieria chilensis, Proustia cuneifolia, Quillaja saponaria* and *Solanum ligustrinum.* In the Libertador Bernardo O'Higgins and Maule regions, between 100 and 900 m, *Acacia caven* (Fabaceae) and *Lithraea caustica* (Anacardiaceae) dominate the Mediterranean thorn forest; other species are *Agostis tenuis, Briza minor, Cestrum parqui, Medicago hispída, Muehlenbeckia hastulata, Peumus boldus, Podanthus mitiqui, Proustia cuneifolia, Trevoa quinquenervia* and *Vulpia myurus.*

Southern Andean Cordillera district

Southern Andean Cordillera region—Peña, 1966a: 11 (regionalization), 1966b: 217 (regionalization).

Central Andean zone (in part)—Artigas, 1975: map (regionalization).

Southern Andean mountain range—Flores and Vidal, 2000: 193 (systematic revision).

Andean Cordillera and Central Valley area (in part)—Díaz Gómez et al., 2009: 4 (dispersal-vicariance analysis).

Southern Andean Cordillera district—Morrone, 2015b: 213 (regionalization); Monckton, 2016: 125 (systematic revision and map).

Definition: The Southern Andean Cordillera district extends in the Andean Range from 33°S to 38°S; it is more humid than the Central Coastal Cordillera district (Peña, 1966a). Cecilia Ezcurra (pers. comm.) suggests a more careful analysis of the boundaries between this district and the Cuyan High Andean province.

Endemic taxa: *Chilicola katherinae* and *C. randolphi* (Hymenoptera: Colletidae; Monckton, 2016) and *Falsopraocis australis* (Coleoptera: Tenebrionidae).

Vegetation: In the western slopes of the Andes in the Maule, Libertador Bernardo O'Higgins, Metropolitana de Santiago, Valparaíso and southern Coquimbo regions, between 2000 and 2600 m, Luebert and Pliscoff (2006) identified the Mediterranean low scrubland as dominated by *Laretia acaulis* (Apiaceae) and *Berberis empetrifolia* (Berberidaceae); other species are *Adesmia aegiceras, Anarthrophyllum gayanum, Azorella madreporica, Bromus setifolius, Doniophyton weddellii,*

Hordeum comosum, Perezia carthamoides, Phleum alpinum, Senecio clarioneifolius and *S. multicaulis.* In higher places, above 3000 m, *Nastanthus spathulatus* (Calyceraceae) and *Menonvillea spathulata* (Brassicaceae) dominate the Andean Mediterranean grassland; other species are *Azorella trifurcata, Chaetanthera pulvinata, Junellia uniflora, Lenzia chamaepitys, Nototriche auricoma, Senecio pissisii* and *Stipa frigida.*

Cenocrons

Temperate forests similar to those of the Valdivian Forest province would have developed in the Santiagan province during the Pleistocene (Caviedes and Iriarte, 1989).

chapter seven

The Patagonian subregion

The Patagonian subregion corresponds to southern Argentina, from central Mendoza, widening to the south through Neuquén, Río Negro, Chubut and Santa Cruz, to northern Tierra del Fuego; and reaching Chile in some Andean areas of Maule, Biobío, Malleco, Aisén and Magallanes (Morrone, 2001b,c, 2006, 2015b). From an ecological viewpoint, one can consider the Patagonian subregion a cool semi-desert (Soriano, 1983).

In this chapter I characterize the Patagonian subregion and the single province recognized within it, provide their endemic and characteristic taxa and discuss their relationships. Within the Patagonian province I recognize five subprovinces and seven districts, detail their endemic taxa and characterize their vegetation. Finally, I discuss a case study referring to the areas of endemism of the province based on insect taxa.

Patagonian subregion

Patagonian formation—Lorentz, 1876: 92 (regionalization).
Argentinean Patagonian province—Engler, 1882: 346 (regionalization).
Patagonian region—Delachaux, 1920: 127 (regionalization); Rivas-Martínez and Tovar, 1983: 516.
Patagonian Steppes area—Hauman, 1920: 47 (regionalization); Holdgate, 1960: 560 (regionalization); Hueck, 1966: 3 (regionalization).
Patagonian Steppes province—Hauman, 1931: 62 (regionalization).
Patagonian Desert area—Parodi, 1934: 171 (regionalization).
Patagonian district—Cabrera and Yepes, 1940: 15 (regionalization).
Patagonian Savannas and Deserts area—Hueck, 1957: 40 (regionalization).
Patagonian Steppes zone—Kuschel, 1960: 546 (regionalization).
Patagonian dominion—Ringuelet, 1961: 160 (biotic evolution); Maury, 1979: 711 (faunistics).
Patagonian Steppe region—Peña, 1966a: 15 (regionalization); Roig-Juñent and Domínguez, 2001: 558 (faunistics).
Patagonia province—Fittkau, 1969: 642 (regionalization); Baranzelli et al., 2014: 752 (phylogeographic analysis).

East Patagonia area—Sick, 1969: 452 (regionalization).
Patagonian center—Müller, 1973: 151 (regionalization); Cracraft, 1985: 36 (regionalization).
Steppe zone—Cekalovic, 1974: 308 (regionalization).
Patagonia area—Coscarón and Coscarón-Arias, 1995: 726 (areas of endemism); Apodaca et al., 2015b: 5 (dispersal-vicariance analysis).
Patagonian area—Acosta and Maury, 1998a: 554 (faunistics), 1998b: 573 (faunistics).
Patagonian subregion—Morrone, 1999: 15 (regionalization); Ocampo and Morrone, 1999: 21 (faunistics); Morrone, 2001b: 129 (regionalization), 2001c: 2 (regionalization); Posadas and Morrone, 2001: 267 (cladistic biogeography); Morrone, 2004a: 158 (regionalization); Soares and de Carvalho, 2005: 485 (evolutionary biogeography); Flores and Chani-Posse, 2005: 575 (systematic revision); Morrone, 2006: 486 (regionalization); Casagranda et al., 2009: 19 (endemicity analysis); Escalante et al., 2009: 379 (endemicity analysis); Morrone, 2010: 38 (regionalization); Urtubey et al., 2010: 506 (track analysis and cladistic biogeography); Kutschker and Morrone, 2012: 541 (track analysis); Posadas, 2012: 2 (faunistics); Campos-Soldini et al., 2013: 21 (track analysis); Fergnani et al., 2013: 296 (cluster analysis); Vivallo, 2013: 529 (systematic revision); Omad, 2014: 565 (systematic revision); Carrara and Flores, 2015: 47 (faunistics); Morrone, 2015b: 222 (regionalization); Ruiz et al., 2016: 389 (track analysis); Arana et al., 2017: 421 (shapefiles); Romano et al., 2017: 445 (track analysis).
Patagonian Steppe subregion—Spinelli et al., 2006: 302 (parsimony analysis of endemicity).
Middle Chilean-Patagonian region (in part)—Rivas-Martínez et al., 2011: 27 (regionalization).
Patagónica province—Brignone et al., 2016: 327 (systematic revision).

Endemic and characteristic taxa

See section on the Patagonian province for detail.

Biotic relationships

See section on the Patagonian province for detail.

Regionalization

The endemicity analysis of Domínguez et al. (2006) recognized a single province (Figure 7.1; Morrone, 2015b).

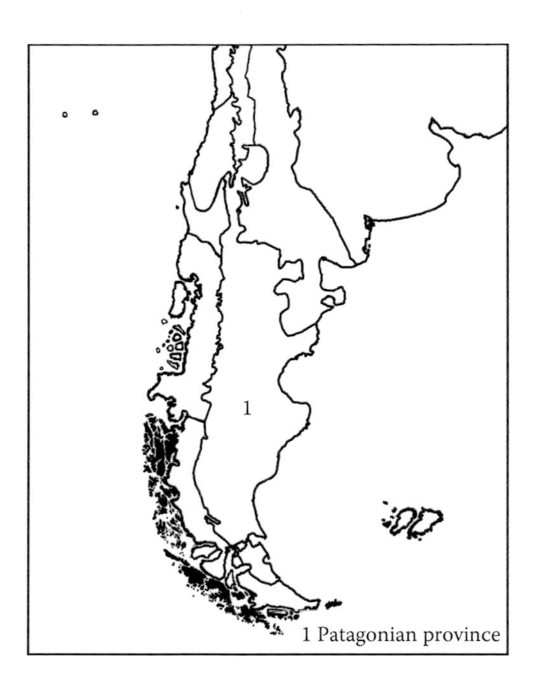

Figure 7.1 Map of the single province of the Patagonian subregion.

Patagonian province

Patagonian province—Mello-Leitão, 1939: 605 (regionalization), 1943: 130 (regionalization); Soriano, 1949: 198 (regionalization), 1950: 33 (regionalization); Cabrera, 1951: 54 (regionalization), 1953: 107 (regionalization), 1971: 33 (regionalization); Cabrera and Willink, 1973: 93 (regionalization); Ringuelet, 1975: 107 (regionalization); Udvardy, 1975: 41 (regionalization); Cabrera, 1976: 64 (regionalization); Willink, 1988: 206 (regionalization), 1991: 138 (regionalization); Morrone, 1994a: 190 (parsimony analysis of endemicity), 1996a: 108 (regionalization); Posadas et al., 1997: 2 (parsimony analysis of endemicity); Carpintero, 1998: 148 (faunistics); León et al., 1998: 127 (regionalization); Roig, 1998: 137 (regionalization); Sérsic et al., 2011: 477 (phylogeographic analysis); Carrara and Flores, 2013: 100 (endemicity analysis); Ezcurra et al., 2014: 28 (biotic evolution); Apodaca et al., 2015a: 93 (regionalization); Hechem et al., 2015: 6 (parsimony analysis of endemicity and track analysis); Morrone, 2015b: 223 (regionalization); Ruiz et al., 2016: 385 (track analysis); Arana et al., 2017: 421 (shapefiles); Martínez et al., 2017: 479 (track analysis and regionalization).

Patagonian district—Osgood, 1943: 27 (regionalization).

Steppe zone—Mann, 1960: 41 (regionalization).
Patagonian Steppe ecoregion—Dinerstein et al., 1995: 102 (ecoregionalization); Burkart et al., 1999: 33 (ecoregionalization); Dellafiore, 2017a: 1 (ecoregionalization).
Central Patagonia province—Morrone, 2001b: 130 (regionalization), 2004a: 158 (regionalization), 2006: 486 (regionalization), 2010: 38 (regionalization); Morrone et al., 2002: 4 (track analysis and regionalization); Posadas, 2012: 2 (faunistics); Campos-Soldini et al., 2013: 21 (track analysis); Ferretti et al., 2014b: 2 (parsimony analysis of endemicity); Flores and Cheli, 2014: 285 (faunistics); Jerez and Muñoz-Escobar, 2015: 207 (faunistics).
Central Patagonian province—Morrone, 2001c: 4 (regionalization); Flores, 2004: 593 (systematic revision).
Patagonian Steppe region—Roig-Juñent et al., 2001: 558 (faunistics).
Central Patagonia area—Roig-Juñent et al., 2006: 408 (cladistic biogeography); Díaz Gómez, 2009: 4 (dispersal-vicariance analysis).
Patagonian Steppe province—Roig et al., 2009: 164 (regionalization).

Definition

The Patagonian province covers southern Argentina, from central Mendoza to southern Santa Cruz, and southeastern Chile, including the northern half of the island of Tierra del Fuego (Cabrera, 1976; Roig, 1998; Morrone, 2001b,c, 2006, 2015b). The province is rather flat with some low-lying mountains, plateaus and plains (Figure 7.2; Dellafiore, 2017a). Climate is very dry and cold

Figure 7.2 Plains and plateaus characteristic of the Patagonian province, Santa Cruz river, Argentina. (Courtesy of Sergio Roig-Juñent.)

with snow in winter, frosts nearly year-round, 200–300 mm of rain per year and an average temperature below 8°C (Cabrera, 1976; Dellafiore, 2017a). Geotectonic units include the North Patagonian and Deseado nesocratons and six Middle to Upper Mesozoic sedimentary basins: Neuquén, Colorado, Ñirihuau, Valdés, San Jorge Gulf and Magellan (Soriano, 1983).

Endemic and characteristic taxa

Morrone (2001b,c, 2015b) provided a list of endemic and characteristic taxa. Some examples included *Azorella monantha* (Figure 7.3a; Apiaceae);

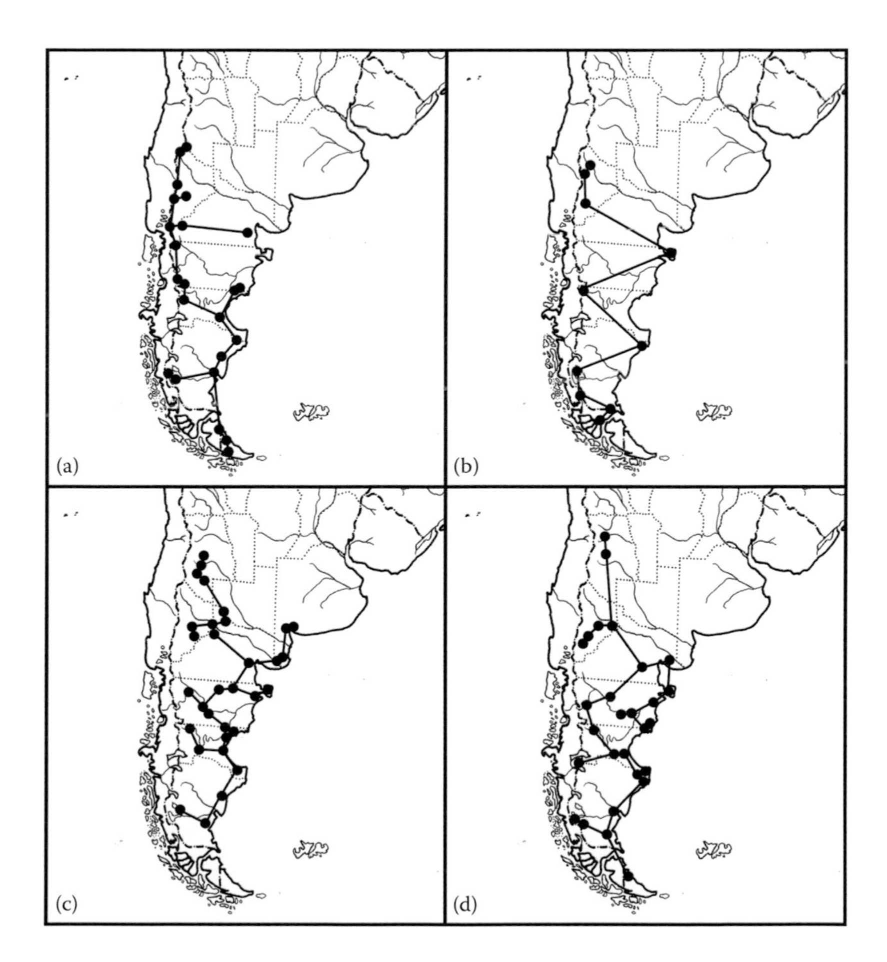

Figure 7.3 Maps with individual tracks in the Patagonian province. (a) *Azorella monantha* (Apiaceae), (b) *Acrostomus* (Coleoptera: Curculionidae), (c) *Epipedonota cristallisata* (Coleoptera: Tenebrionidae) and (d) *Mitragenius araneiformis* (Coleoptera: Tenebrionidae).

Chuquiraga aurea, Leucheria floribunda, L. gilliesii, L. rosea, L. runcinata, L. scrobiculata, Nassauvia uniflora and *N. ruizii* (Asteraceae); *Anarthrophyllum desideratum* (Fabaceae); *Urophonius granulatus* (Scorpiones: Bothriuridae); *Atrichopogus aridus, A. endemicus* and *A. inacayali* (Diptera: Ceratopogonidae); *Acrostomus* (Morrone, 1994c; Figure 7.3b), *Cylydrorhinus echinosoma, C. oblongus* and *Hyperoides balfourbrownei* (Coleoptera: Curculionidae); *Taurocerastes patagonicus* (Coleoptera: Geotrupidae); *Athlia giaii and Scybalophagus patagonicus* (Coleoptera: Scarabaeidae); *Epipedonota cristallisata* (Flores and Vidal, 2001; Figure 7.3c), *Mitragenius araneiformis* (Figure 7.3d; Coleoptera: Tenebrionidae; Flores, 2000); *Bufonacris bruchi* and *Eremopachys bergi* (Orthoptera: Tristiridae); *Ctenomys colburni, C. emilianus* and *C. haigi* (Rodentia: Ctenomyidae); *Lestodelphis halli* (Didelphimorphia: Didelphidae); *Liolaemus boulengeri* and *L. fitzingeri* (Squamata: Liolaemidae); *Tinamotis ingoulfi* (Tinamiformes: Tinamidae); *Podiceps gallardoi* (Podicipediformes: Podicipedidae); *Enicognathus ferrugineus* (Psittaciformes: Psittacidae); *Zaedyus pichyi* (Xenarthra: Dasypodidae); and *Lyncodon patagonicus* (Carnivora: Mustelidae).

Vegetation

The Patagonian province consists of shrub steppes, grasslands and deserts (Cabrera and Willink, 1973; Soriano, 1983; Roig, 1998). There are many xerophytic plants highly adapted for protection against drought (Figure 7.4).

Figure 7.4 Shrub steppes of the Patagonian province, Lake Nahuel Huapi, Río Negro, Argentina. (Courtesy of María Marta Cigliano.)

Dwarf and cushion shrubs are the most common vegetation types. Shrub species belong to the genera *Acantholippia, Benthamiella, Nassauvia* and *Verbena,* whereas cushion plants belong to the genera *Mulinum* and *Brachyclados* (Dellafiore, 2017a). Dominant plant species include *Agrostis pyrogea, Anarthrophyllum rigidum, Burkartia lanigera, Chuquiraga aurea, C. avellanedae, Colliguaya integerrima, Distichlis scoparia, D. spicata, Festuca argentina, F. gracillina, F. pallescens, Haplopappus pectinatus, Juncus leuserii, Junellia congesta, J. tridens, Mulinum spinosum, Nassauvia ameghinoi, N. axillaris, N. glomerulosa, Pantacantha* spp., *Poa atripidiformis, P. ligularis, Schinus marchandii, Senecio bracteolatus, S. filaginoides, S. patagonicus* and *Trevoa patagonica* (Cabrera, 1971, 1976; Cabrera and Willink, 1973; León et al., 1998; Hechem et al., 2015).

Biotic relationships

Cabrera (1976) considered that this province was related to the Puna and Cuyan High Andean provinces. Additionally, he considered the relationships with the adjacent Monte province irrelevant. Davis et al. (1997) considered that plant taxa have a strong Andean relationship.

Regionalization

Morrone et al. (2002) conducted a track analysis of Coleoptera species of the Central Patagonia province (*sensu* Morrone, 2001b). Based on the individual tracks of 93 beetle species belonging to the families Carabidae and Tenebrionidae, the study identified three generalized tracks (Figure 7.5a), which represent natural biogeographical units identified as districts (Figure 7.5b):

1. *Payunia district*: In the northwestern corner province, in southern Mendoza and northern Neuquén provinces, in Argentina. Identification supported by *Cnemalobus mendozensis, Nyctelia cicatricula, N. difficilis, N. garciae, N. gebieni, N. grandis, N. kulzeri, N. laevis, N. planata, N. producta, N. roigi, Omopheres ardoini, Psectrascelis atra, P. gigas, P. grandis, P. hirta, P. lucida* and *P. neuquenensis.*
2. *Central district*: Including the greatest part of the province, from southern Río Negro to central Santa Cruz province, in Argentina. Identification supported by *Antarctiola laevis, A. laevigata, Barypus longitarsis, Cnemalobus curtisii, Metius crassusculus, Mimodromius martinezi, Parhypates chilensis chubutensis, Trechisibus bruchi, T. cristinensis, T. topali, Epipedonota elegantula, E. willinki, Nyctelia blairi, N. caudata, N. consularis, N. darwini, N. discoidalis, N. fitzroyi, N. freyi, N. guerini, N. latiplicata, N. newporti, N. plicata, N. quadriplicata, N. sallaei, N. stephensi, N. undatipennis, N. vidalae, N. westwoodi, Patagonogenius acutangulus,*

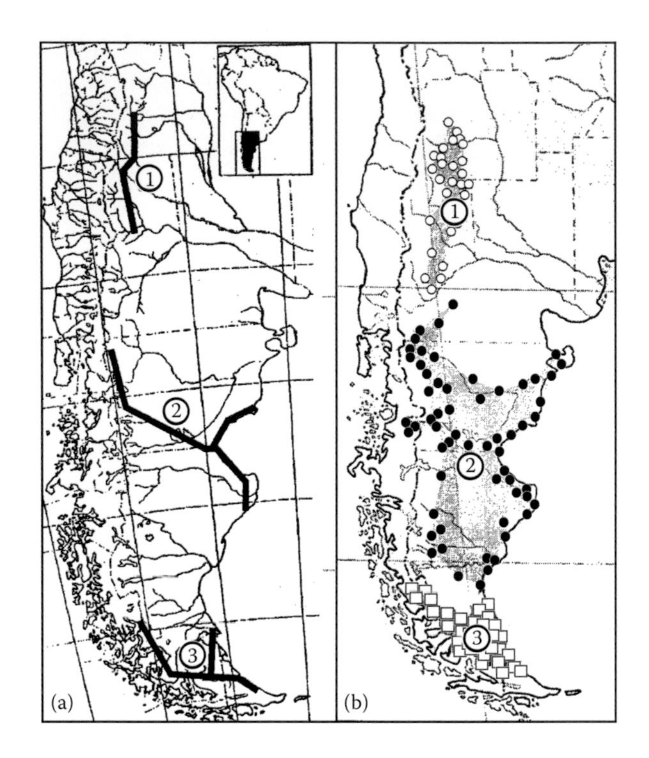

Figure 7.5 Generalized tracks and areas of endemism based on Coleoptera of the Patagonian province. (Modified from Morrone, J.J. et al., *Revista del Museo Argentino de Ciencias Naturales, nueva serie*, 4, 1–6, 2002. With permission.) (a) Generalized tracks and (b) districts, showing the localities of the species belonging to them. (1) Payunia generalized track and district, (2) Central Patagonian generalized track and district and (3) Fuegian generalized track and district.

P. quadricollis, *P. gentilii*, *Platesthes pilosa*, *Praocis sellata*, *P. sellata bergi*, *P. denseciliata*, *Psectrascelis convexipennis*, *P. latithorax*, *P. maximus*, *P. punctipennis*, *P. punctulata*, *P. sulcicollis* and *Scotobius contrerasi*.

3. *Fuegian district*: Extending from southern Santa Cruz province in Argentina to northern Tierra del Fuego, in Argentina and Chile.

The study concludes that these districts coincide with three geological basins: the Payunia district with the Neuquén basin, the Central district with the San Jorge basin and the Fuegian district with the Magellanes basin. Identification supported by *Feroniola bradytoides, Metius annulicornis, M. pogonoides, Notaphiellus cekalovici, Notholopha atrum, N. epistomale, Pseudomigadops nigrocoeruleus, Pycnochilla fallaciosa, Trechisibus magellanus, T. rectangulus, T. stricticollis, Epipedonota tricostata, Neopraocis reflexicollis, Nyctelia bremi, N. corrugata, N. fallax, N. granulata, N. multicristata, N. solieri, Platesthes depressa, P. silphoides, P. burmeisteri, P. similis, P. unicosta, P. nigra, P. granulipennis and Praocis striolicollis.*

Domínguez et al. (2006) conducted an analysis of endemicity of the Patagonian steppe based on several insect taxa, for example, Orthoptera of the family Tristiridae (genera *Eremopachys, Tristira, Circracris, Bufonacris* and *Tropisdosthetus*) and Coleoptera of the families Carabidae (*Barypus, Migadops, Pseudomigadops, Cnemalobus, Notaphus, Notholopha, Peryphus, Bembidarenas, Pycnochila, Anisostichus, Mimodromius, Chaudoirina, Carboniella, Feroniola, Trirammatus, Metius, Trechisibus* and *Tetragonoderes*), Curculionidae (*Acrostomus, Puranius, Listroderes, Falklandiopsis* and *Cylydrorhinus*), Tenebrionidae (*Mitragenius, Patagoniogenus, Epipedonota, Psectrascelis, Nyctelia, Omopheres, Neopraocis, Plathestes* and *Scotobius*), Geotrupidae (*Taurocerastes*), Trogidae (*Polynoncus*) and Scarabaeidae (*Aulacopalpus, Megathopa, Sybalophagus, Myloxena, Faargia* and *Athlia*). The study identified five major areas of endemism and seven subareas subordinated to them.

Morrone (2015b) recognized five subprovinces and seven districts for the Patagonian province based on the areas recognized previously by several studies (Soriano, 1950, 1956; Cabrera, 1971, 1976; Roig, 1998; Morrone, 1999; Morrone et al., 2002; Domínguez et al., 2006). This scheme followed closely the scheme by Domínguez et al. (2006).

A more recent track analysis (Hechem et al., 2015) was based on species of 52 plant genera belonging to 18 different families, for example, Asteraceae (26 species), Fabaceae (21 species), Solanaceae (13 species), Brassicaceae (11 species), Poaceae (9 species) and Verbenaceae (7 species). The analysis found two larger generalized tracks (Figure 7.6):

1. *Septentrional track* (*ABCDE*): Northern Payunia, Southern Payunia, Septentrional Subandean and Chubut districts, and Western Patagonia subprovince. Identification supported by *Chuquiraga avellanedae, Haplopappus diplopappus, Senecio ganganensis, Anarthrophyllum ornithopodum, Pappostipa ibarii* and *Junellia patagonica.*
2. *Austral track* (*FGH*): Santa Cruz and Meridional Subandean and the Fuegian subprovince. Identification supported by *Plantago correae, Nicoraepoa pugionifolia, Rytidosperma virescens* var. *parvispicum* and *Benthamiella sorianoi.*

The analysis of Hechem et al. (2015) also led to corroborate some areas supported by two or more generalized tracks: Western Patagonian subprovince (supported by *Senecio sandwithii* and *Heliotropium pinnatisectum*), Septentrional Subandean district (supported by *Menonvillea comberi* and *Boopis raffaellii*), Chubut district (supported by *Chuquiraga aurea, Nardophyllum patagonicum, Nassauvia juniperina, Senecio gilliesii* var. *dasycarpus, S. mustersii* var. *mustersii, Chilocardamum castellanosii, Xerodraba colobanthoides, X. glebaria, Pterocactus hickenii, Boopis chubutensis, Carex nelmesiana, Adesmia gramindea, A. neglecta, Astragalus colhuensis, Prosopis denudans* var. *stenocarpa, Frankenia patagonica, Sphaeralcea tehuelches, Fabiana nana, Jaborosa*

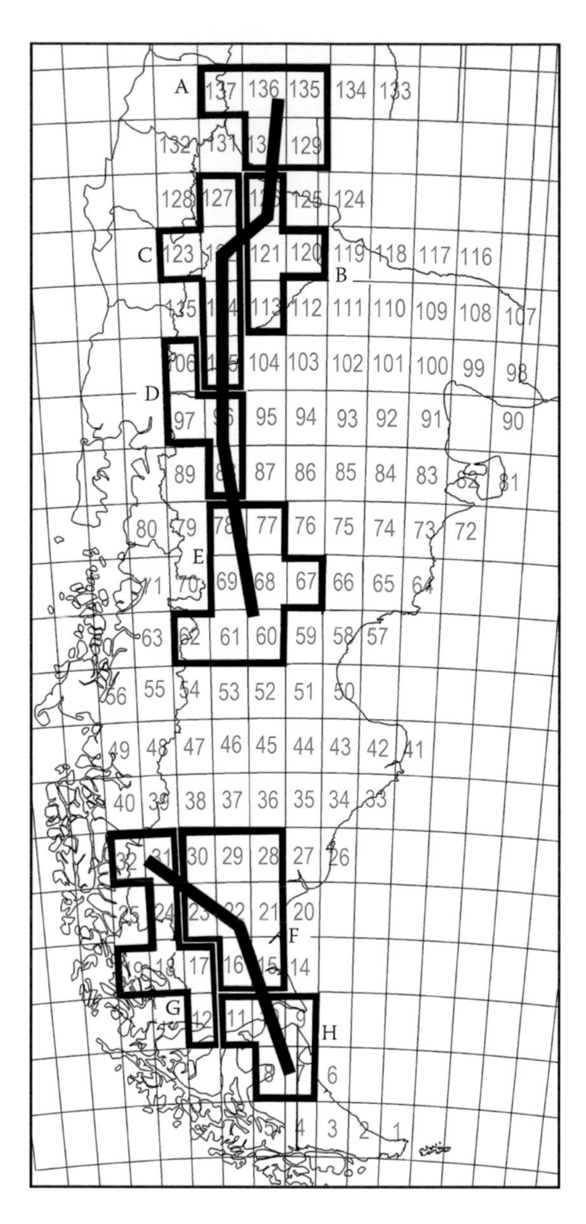

Figure 7.6 Generalized tracks based on plant taxa of the Patagonian province. (Modified from Hechem, V. et al., *Darwiniana*, 3, 5–20, 2015. With permission.) (A) Northern Payunia district, (B) Southern Payunia district, (C) Western Patagonian subprovince, (D) Septentrional Subandean district, (E) Chubut district, (F) Santa Cruz district, (G) Meridional Subandean district and (H) Fuegian subprovince.

chubutensis, Nicotiana ameghinoi, Mulguraea tetragonocalyx, Neosparton patagonicum and *Larrea ameghinoi*) and Santa Cruz district (supported by *Nassauvia sceptrum, Senecio desideratus, Sarcodraba karraikensis, Xerodraba lycopodioides, Adesmia silvestrii, Pappostipa ameghinoi var. ameghinoi, P. chubutensis, P. nana, Benthamiella pycnophylloides* and *B. skottsbergii*). This analysis failed to recover the Payunia, Fuegian and Subandean subprovinces as natural units, in contrast to the analysis by Domínguez et al. (2006) based on insect taxa. Future analyses combining plant and animal taxa should provide a more robust regionalization.

Central Patagonian subprovince

Central Patagonian district—Soriano, 1950: 33 (regionalization), 1956: 324 (regionalization); Cabrera, 1951: 56 (regionalization), 1971: 34 (regionalization); Cabrera and Willink, 1973: 94 (regionalization); Cabrera, 1976: 68 (regionalization); León et al., 1998: 129 (regionalization); Roig, 1998: 138 (regionalization); Flores, 2004: 605 (systematic revision); Morrone et al., 2002: 3 (track analysis and regionalization).

Central district—Soriano, 1983: 448 (regionalization); León et al., 1998: 129 (regionalization).

Septentrional Patagonian province—Rivas-Martínez and Navarro, 1994: map (regionalization).

Central Patagonia area—Roig-Juñent, 1994b: 183 (cladistic biogeography); Roig-Juñent and Flores 2001: 258 (cladistic biogeography); Flores and Roig-Juñent 2001: 315 (cladistic biogeography); Domínguez et al., 2006: 1534 (endemicity analysis).

Central Patagonian province—Morrone, 1999: 15 (regionalization).

North Patagonian province—Rivas-Martínez et al., 2011: 27 (regionalization).

Central Patagonian subprovince—Hechem et al., 2015: 6 (parsimony analysis of endemicity and track analysis); Morrone, 2015b: 224 (regionalization).

Definition: The Central Patagonian subprovince corresponds to the largest part of the Patagonian province, covering central Río Negro, Chubut and almost all of Santa Cruz (Soriano, 1950, 1956; Roig-Juñent, 1994b, Cabrera and Willink, 1973; León et al., 1998: 129). The subprovince encompasses the most arid portion of the province, with an annual precipitation less than 200 mm (León et al., 1998).

Endemic taxa: *Barypus chubutensis* and *B. longitarsis* (Coleoptera: Carabidae); *Taurocerastes patagonicus* (Scarabaeidae); *Epipedonota tricostata, Platesthes silphoides, P. burmeisteri, P. kuscheli* and *P. granulipennis* (Coleoptera: Tenebrionidae); and *Pompilocalus catriel* (Hymenoptera: Pompilidae; Domínguez et al., 2006).

Districts: The Central Patagonian subprovince comprises the Chubut, San Jorge Gulf and Santa Cruz districts.

Chubut district

Chubut subdistrict—Soriano, 1956: 324 (regionalization); León et al., 1998: 129 (regionalization).

Chubutian subdistrict—Cabrera, 1976: 68 (regionalization); Soriano, 1983: 448 (regionalization); León et al., 1998: 129 (regionalization); Roig, 1998: 138 (regionalization).

Chubutian area—Domínguez et al., 2006: 1534 (endemicity analysis).

Chubut district—Hechem et al., 2015: 6 (parsimony analysis of endemicity and track analysis); Morrone, 2015b: 224 (regionalization).

Definition: The Chubut district corresponds to the northern portion of the subprovince in Chubut, Neuquén and Río Negro, at the boundary of the Monte province (Soriano, 1983). The species *Chuquiraga avellanedae* (Asteraceae) is common in all the communities (León et al., 1998).

Endemic taxa: *Senecio gilliesii* var. *dasycarpus, S. mustersii* var. *mustersii* and *Nardophyllum patagonicum* (Asteraceae); *Chilocardamum castellanosii, Xerodraba colobanthoides* and *X. glebaria* (Brassicaceae); *Pterocactus hickenii* (Cactaceae); *Boopis chubutensis* (Calyceraceae); *Carex nelmesiana* (Cyperaceae); *Adesmia graminidea, A. neglecta, Astragalus colhuensis* and *Prosopis denudans* var. *stenocarpa* (Fabaceae); *Sphaeralcea tehuelches* (Malvaceae); and *Fabiana nana, Jaborosa chubutensis and Nicotiana ameghinoi* (Solanaceae; Hechem et al., 2015).

Vegetation: León et al. (1998) characterized the shrub steppe as dominated by *Nassauvia glomerulosa, N. ulicina* and *Chuquiraga aurea* (Asteraceae); other species are *Acaena caespitosa, Acantholippia seriphioides, Brachyclados caespitosus, Hoffmanseggia trifoliata, Lycium chilense, Pleurophora patagonica, Perezia lanigera, Schinus polygamus, Stipa ameghinoi, S. humilis* and *S. ibari.* In northeastern Chubut, in areas limiting to the Monte province, *Chuquiraga avellaneda* (Asteraceae) dominates the shrub steppe; other species are *Lycium aneghinoi, L. chilense, Prosopis denudans* and *Verbena ligustrina.* In central Chubut, *Colliguaya integerrima* (Euphorbiaceae) dominates the shrub steppe; other species are *Adesmia boroniodes, Anarthrophyllum desideratum, A. rigidum, Berberis heterophylla, Lycium chilense, Nardophyllum obtusifolium, Schinus polygamus* and *Verbena ligustrina.*

San Jorge Gulf district

San Jorge Gulf district—Soriano, 1950: 33 (regionalization); Cabrera, 1951: 57 (regionalization); Soriano, 1956: 324 (regionalization); Cabrera, 1971: 35 (regionalization); Cabrera and Willink, 1973: 95 (regionalization); Cabrera, 1976: 69 (regionalization); Soriano, 1983: 451 (regionalization); León et al., 1998: 131 (regionalization); Roig, 1998: 139 (regionalization); Hechem et al., 2015: 6 (parsimony analysis of endemicity and track analysis); Morrone, 2015b: 224 (regionalization).

San Jorge Gulf province—Morrone 1999: 15 (regionalization).
Coastal Patagonia area—Roig-Juñent et al., 2006: 408 (cladistic biogeography).
Definition: It extends approximately from Cabo Raso to Punta Casamayor, surrounding the San Jorge Gulf (Soriano, 1983).
Vegetation: The San Jorge Gulf district consists of shrub and grassy steppes. León et al. (1998) characterized the high shrub steppe as dominated by *Stipa humilis* and *S. speciosa* (Poaceae) and *Colliguaya integerrima* (Euphorbiaceae); other species are *Baccharis darwini, Grindelia chilensis, Mutisia retrorsa, Nassauvia ulicina, Perezia recurvatta* ssp. *becki* and *Phacellia magellanica. Trevoa patagonica* (Rhamnaceae) is characteristic of another shrub steppe. *Festuca argentina* and *F. pallescens* (Poaceae), *Nardophyllum obtusifolium* and *Senecio filaginoides* (Asteraceae), *Adesmia campestris* (Fabaceae) and *Mulinum spinosum* (Apiaceae) dominate the shrub and grassy steppe; other species are *Adesmia lotoides, Azorella* spp., *Nassauvia darwini, Mulinum halei* and *Perezia patagonica.*

Santa Cruz district

Santa Cruz subdistrict—Soriano, 1956: 324 (regionalization); León et al., 1998: 129 (regionalization).
Santacruzian subdistrict—Cabrera, 1976: 69 (regionalization); Soriano, 1983: 450 (regionalization); Roig, 1998: 138 (regionalization).
Meridional Patagonian province—Rivas-Martínez and Navarro, 1994: map (regionalization).
Santacrucian area—Domínguez et al., 2006: 1534 (endemicity analysis).
South Patagonian province—Rivas-Martínez et al., 2011: 27 (regionalization).
Santa Cruz district—Hechem et al., 2015: 6 (parsimony analysis of endemicity and track analysis); Morrone, 2015b: 224 (regionalization).
Definition: Santa Cruz district covers most of the subprovince between 47°S and 51°S, from the seashore to the Andes (Soriano, 1983). The species *Verbena tridens* (Verbenaceae) is common in all the communities (León et al., 1998).
Endemic taxa: *Nassauvia sceptrum* and *Senecio desideratus* (Asteraceae); *Sarcodraba karraikensis* and *Xerodraba lycopodioides* (Brassicaceae); *Adesmia silvestrii* (Fabaceae); *Pappostipa nana* (Poaceae); *Benthamiella skottsbergii* (Solanaceae); and *Epipedonota tricostata* (Coleoptera: Tenebrionidae; Domínguez et al., 2006; Hechem et al., 2015).
Vegetation: South and north of the Santa Cruz river basin, León et al. (1998) characterized the shrub steppe as dominated by *Verbena tridens* (Verbenaceae); other species are *Acaena poeppigiana, Azorella caespitosa, Festuca pyrogea, Nassauvia darwini, Stipa ibari* and *S. neaei.* West of the Cardiel and Strobel lakes, *Nardophyllum obtusifolium* and *N. bryoides* (Asteraceae) dominated the shrub steppe; the other species is *Festuca pallescens.*

Fuegian subprovince

Fuegian district—Cabrera, 1971: 36 (regionalization); Cabrera and Willink, 1973: 95 (regionalization); Cabrera, 1976: 70 (regionalization); Flores, 2004: 605 (systematic revision); Morrone et al., 2002: 4 (track analysis and regionalization).

Steppe zone (in part)—Cekalovic, 1974: 308 (regionalization).

Fuegian province (in part)—Rivas-Martínez and Navarro, 1994: map (regionalization).

Austral Patagonia area—Roig-Juñent, 1994b: 182 (cladistic biogeography); Flores and Roig-Juñent, 2001: 315 (cladistic biogeography); Roig-Juñent and Flores, 2001: 259 (cladistic biogeography); Domínguez et al., 2006: 1534 (endemicity analysis); Roig-Juñent et al., 2006: 408 (cladistic biogeography).

Patagonian Grasslands ecoregion (in part)—Dinerstein et al., 1995: 102 (ecoregionalization); Dellafiore, 2017b: 1 (ecoregionalization).

Magellanic district—León et al., 1998: 133 (regionalization).

Fuegian-Malvinian district (in part)—Roig, 1998: 138 (regionalization).

Humid Magellanic Steppe of *Festuca gracillima* and Murtillares district—Roig, 1998: 141 (regionalization).

Fuegian Patagonian province—Morrone, 1999: 16 (regionalization).

Fuegian subprovince—Hechem et al., 2015: 6 (parsimony analysis of endemicity and track analysis); Morrone, 2015b: 224 (regionalization).

Definition: The Fuegian province corresponds to the steppes of Tierra del Fuego (Cabrera, 1971).

Endemic taxa: *Barypus clivinoides* (Coleoptera: Carabidae); *Nyctelia granulata, Platesthes depressa, P. nigra, P. similis* and *P. unicosta* (Coleoptera: Tenebrionidae); and *Bufonacris terrestris* and *Tristira magellanica* (Orthoptera: Tristiridae; Roig-Juñent, 1994b; Domínguez et al., 2006).

Vegetation: León et al. (1998) characterized the xeric grass steppe as dominated by *Festuca gracillima* (Poaceae) and *Nardophyllum bryoides* (Asteraceae); other species are *Azorella fuegiana, Calceolaria uniflora, Carex andina, Rhytidosperma virescens* and *Viola maculata.* In more humid places, *Empetrum rubrum* (Empetraceae), *Chiliotrichum diffusum* (Asteraceae) and *Festuca gracillima* (Poaceae) dominate the grass steppe; other species are *Berberis empetrifolia, Deschampsia flexuosa, Gentianella magellanica, Geum magellanicum, Hierocloe pusilla, Perezic pilifera, Pernettya pumila, Poa rigidifolia, Primula magellanica* and *Ranunculus peduncularis.*

Payunia subprovince

Payunia district—Cabrera, 1971: 34 (regionalization); Cabrera and Willink, 1973: 94 (regionalization); Cabrera, 1976: 66 (regionalization); Soriano, 1983: 440 (regionalization); León et al., 1998: 134

(regionalization); Flores, 2004: 603 (systematic revision); Morrone et al., 2002: 2 (track analysis and regionalization).

Andean Mediterranean province—Rivas-Martínez and Navarro 1994: map (regionalization).

Payenia district—Roig, 1998: 138 (regionalization).

Payunia province—Morrone, 1999: 15 (regionalization).

Payunia area—Flores and Roig-Juñent, 2001: 315 (cladistic biogeography); Roig-Juñent and Flores, 2001: 258 (cladistic biogeography); Domínguez et al., 2006: 1534 (endemicity analysis); Roig-Juñent et al., 2006: 408 (cladistic biogeography); Díaz Gómez et al., 2009: 4 (dispersal-vicariance analysis).

Payunia subprovince—Hechem et al., 2015: 6 (parsimony analysis of endemicity and track analysis); Morrone, 2015b: 224 (regionalization).

Definition: The Payunia subprovince corresponds to the most septentrional part of the province, in southern Mendoza and northern Neuquén (Cabrera, 1971, 1976; Soriano, 1983; Morrone et al., 2002). The most characteristic Patagonian steppes are areas defined by exposure and elevation, intermingled with scrub vegetation, characteristic of the Monte province (Soriano, 1983).

Endemic taxa: *Berberis comberi* (Berberidaceae), *Condalia megacarpa* (Rhamnaceae), and *Prosopis castellanoi* and *Senna kurtzii* (Fabaceae; León et al., 1998).

Vegetation: León et al. (1998) characterized the shrub steppe above 1800 m as dominate by *Mulinum spinosum* and *Azorella* spp. (Apiaceae), *Adesmia* spp. (Fabaceae) and *Maihuenia* spp. (Cactaceae). Below 1400 m, *Astragalus pehuenches* and *Anarthrophyllum rigidum* (Fabaceae), *Ephedra ochreata* (Ephedraceae), *Stillingia patagonica* (Euphorbiaceae) and *Berberis grevilleana* (Breberidaceae) dominated the steppe.

Districts: The Payunia subprovince comprises the Northern Payunia and Southern Payunia districts.

Northern Payunia district

Northern Payunia area—Domínguez et al., 2006: 1534 (endemicity analysis).

Northern Payunia district—Hechem et al., 2015: 6 (parsimony analysis of endemicity and track analysis); Morrone, 2015b: 225 (regionalization).

Definition: The Northern Payunia district corresponds to southern Mendoza (Domínguez et al., 2006).

Endemic taxa: *Austrocactus bertinii* (Cactaceae); *Cnemalobus mendozensis* (Coleoptera: Carabidae); and *Nyctelia garciae* and *N. laevis* (Coleoptera: Tenebrionidae; Domínguez et al., 2006; Hechem et al., 2015).

Southern Payunia district

Southern Payunia area—Domínguez et al., 2006: 1534 (endemicity analysis).

Southern Payunia district—Hechem et al., 2015: 6 (parsimony analysis of endemicity and track analysis); Morrone, 2015b: 225 (regionalization).

Definition: The Southern Payunia district corresponds to northern Neuquén (Domínguez et al., 2006).

Endemic taxa: *Aylacophora deserticola* (Asteraceae), *Platesthes neuquensis* (Coleoptera: Tenebrionidae) and *Athlia parvicollis* (Coleoptera: Scarabaeidae; Domínguez et al., 2006; Hechem et al., 2015).

Subandean subprovince

Subandean district—Soriano, 1950: 33 (regionalization), 1956: 324 (regionalization); Cabrera, 1951: 55 (regionalization), 1971: 35 (regionalization); Cabrera and Willink, 1973: 95 (regionalization); Cabrera, 1976: 69 (regionalization); León et al., 1998: 132 (regionalization).

Aysén Cordillera region—Peña 1966a: 15 (regionalization).

Austral Altoandean district—Cabrera, 1971: 32 (regionalization), 1976: 57 (regionalization); Roig, 1998: 139 (regionalization).

Steppe zone (in part)—Cekalovic, 1974: 308 (regionalization).

Subandean Patagonian province—Morrone, 1999: 15 (regionalization), 2001c: 4 (regionalization).

Subandean Patagonia province—Morrone, 2001b: 132 (regionalization), 2004a: 158 (regionalization), 2006: 486 (regionalization), 2010: 38 (regionalization); Morrone et al., 2002: 4 (track analysis and regionalization); Ferretti et al., 2014b: 2 (parsimony analysis of endemicity); Flores and Cheli, 2014: 285 (faunistics); Omad, 2014: 565 (systematic revision).

Subandean subprovince—Hechem et al., 2015: 6 (parsimony analysis of endemicity and track analysis); Morrone, 2015b: 226 (regionalization).

Definition: The Subandean subprovince corresponds to a narrow portion of the province, along the Austral Cordillera, from Neuquén to Santa Cruz (Soriano, 1950, 1956; Cabrera, 1976; Morrone et al., 2002). The subprovince represents the boundary of the province with the Subantarctic subregion (León et al., 1998). Grasslands are characteristic (León et al., 1998).

Endemic taxa: *Anomophthalmus insolitus* (Figure 7.7) and *Cylydrorhinus costatus* (Coleoptera: Curculionidae).

Vegetation: The Subandean subprovince consists of grass steppe, dominated by Poaceae, with very few shrubs. León et al. (1998) characterized the steppe as dominated by *Festuca pallescens* (Poaceae) and *Lathyrus magellanicus* (Fabaceae); other species are *Deschampsia*

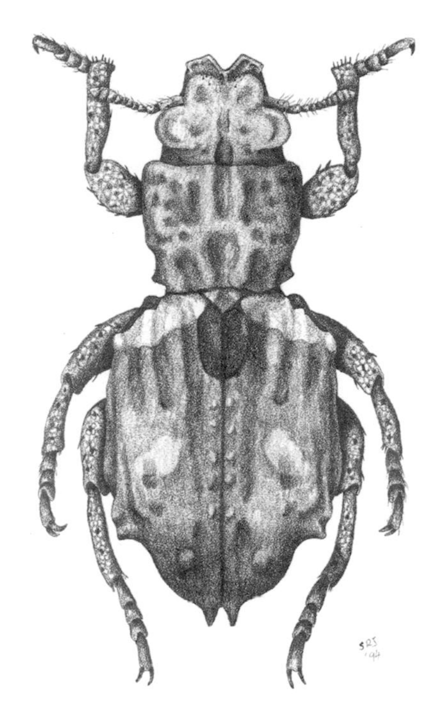

Figure 7.7 *Anomophthalmus insolitus* (Coleoptera: Curculionidae), weevil species endemic to the Subandean subprovince of the Patagonian province. (Courtesy of Sergio Roig-Juñent.)

elegantula, D. flexuosa, Elymus patagonicus, Festuca magellanica, F. pyrogea, Phleum commutatum and *Rhytidosperma virescens.*

Districts: The Subandean subprovince comprises the Austral High Andean, Meridional Subandean Patagonia and Septentrional Subandean Patagonia districts.

Austral High Andean district

Austral High Andean district—Cabrera, 1971: 30 (regionalization); Padró, 2017: 26 (parsimony analysis of endemicity and track analysis).

Definition: The High Andean district corresponds to the Cordilleran portion of the province of Neuquén (Padró, 2017).

Endemic taxa: *Blechnum microphyllum* (Blechnaceae); *Astragalus nivicola* and *A. patagonicus* (Fabaceae); *Calceolaria palenae* (Scrophulariaceae); *Discaria nana* (Rhamnaceae); *Gamocarpha alpina* (Calyceraceae);

Noccaea magellanica (Brassicaceae); *Polypogon monspeliensis* and *Rytidosperma picta* (Poaceae); and *Senecio pachyphyllos, S. portalesianus* and *S.subumbellatus* (Asteraceae; Padró, 2017)

Meridional Subandean Patagonia district

Meridional Subandean Patagonia area—Domínguez et al., 2006: 1534 (endemicity analysis).

Meridional Subandean Patagonia district—Hechem et al., 2015: 6 (parsimony analysis of endemicity and track analysis); Morrone, 2015b: 226 (regionalization).

Definition: The Meridional Subandean Patagonia district corresponds to the southern portion of the subprovince in Santa Cruz (Domínguez et al., 2006).

Endemic taxa: *Barypus painensis* (Coleoptera: Carabidae) and *Acrostomus magellanicus* (Coleoptera: Curculionidae; Domínguez et al., 2006).

Septentrional Subandean Patagonia district

Septentrional Subandean Patagonia area—Domínguez et al., 2006: 1534 (endemicity analysis).

Septentrional Subandean Patagonia district—Hechem et al., 2015: 6 (parsimony analysis of endemicity and track analysis); Morrone, 2015b: 226 (regionalization).

Definition: The Septentrional Subandean Patagonia district corresponds to the northern portion of the subprovince, in Neuquén and Río Negro (Domínguez et al., 2006).

Endemic taxa: *Boopis raffaelli* (Calyceraceae), *Menonvillea comberi* (Brassicaceae) and *Barypus minus* (Coleoptera: Carabidae; Domínguez et al., 2006; Hechem et al., 2015).

Western Patagonian subprovince

Western district—Soriano, 1950: 33 (regionalization), 1956: 324 (regionalization); Cabrera, 1976: 66 (regionalization); Soriano, 1983: 441 (regionalization); León et al., 1998: 128 (regionalization).

Western Patagonian district—Cabrera, 1951: 56 (regionalization), 1971: 34 (regionalization); Cabrera and Willink, 1973: 94 (regionalization).

Western Patagonia area—Roig-Juñent, 1994b: 183 (cladistic biogeography); Flores and Roig-Juñent, 2001: 315 (cladistic biogeography); Roig-Juñent and Flores, 2001 (cladistic biogeography): 258; Domínguez et al., 2006: 1534 (endemicity analysis); Roig-Juñent et al., 2006: 408 (cladistic biogeography); Díaz Gómez et al. 2009: 4 (dispersal-vicariance analysis).

Western Patagonian province—Morrone 1999: 15 (regionalization).

Western Patagonian subprovince—Hechem et al., 2015: 6 (parsimony analysis of endemicity and track analysis); Morrone, 2015b: 226 (regionalization).

Definition: The Western Patagonian subprovince corresponds to the westernmost part of the province, in a narrow strip from Neuquén to northwestern Santa Cruz (Cabrera, 1951; Soriano, 1983; León et al., 1998; Flores and Roig-Juñent, 2001). The boundaries of the Western Patagonian subprovince with the Subandean subprovince are unclear (León et al., 1998).

Endemic taxa: *Heliotropium pinnatisectum* (Boraginaceae); *Senecio sandwithii* (Asteraceae); *Barypus gentilii, B. neuquensis, Cnemalobus gentilii* and *C. neuquensis* (Coleoptera: Carabidae); and *Asidelia contracta, Nyctelia hayekae* and *N. wittmeri* (Coleoptera: Tenebrionidae; Roig-Juñent, 1994b; Domínguez et al., 2006; Hechem et al., 2015).

Vegetation: The Western Patagonian subprovince consists of shrub and grassy steppes, with most of the species belonging to the Poaceae. León et al. (1998) characterized the steppes as dominated by *Stipa speciosa, S. humilis* and *Poa lanuginosa* (Poaceae), *Adesmia campestris* (Fabaceae) and *Berberis heterophylla* (Berberidaceae); other species are *Bromus setifolius, Ephedra frustillata, Hordeum comosum, Lycium chilense, Mulinum spinosum, Schinus polygamus* and *Senecio filaginoides*. In the central part of the subprovince, other species include *Stillingia patagonica* (Euphorbiaceae), *Nassauvia axillaris* and *Nardophyllum parvifolium* (Asteraceae) and *Tetraglochin ameghinoi* (Rosaceae). In some places, there are dense populations of *Anarthrophyllum rigidum* (Fabaceae), *Verbena ligustrina* (Verbenaceae) and *Corynabutilon bicolor* (Malvaceae). In the higher portions, there are grassy steppes dominated by *Festuca pallescens, F. argentina, Poa ligularis* and *Stipa speciosa* (Poaceae).

Cenocrons

A review of several phylogeographical analyses of plant and vertebrate taxa (Sérsic et al., 2011) revealed some congruent patterns, suggesting Neogene and Quaternary geological events driving the biotic evolution of the Patagonian province. Sérsic et al. (2011) detected nine phylogeographic breaks, four of them shared by plants and vertebrates. Additionally, studies found several stable areas at both sides of the Colorado river, northern Río Negro, northern Chubut and Tierra del Fuego; as well as expansion routes from the east towards the southwest at different latitudes, from northern Neuquén towards the south, and from southern and northern Chubut with a south-north direction (Figure 7.8). Some of these expansion routes might be associated with secondary contact areas in the northern tip of the San Jorge Gulf, in the coastline of Río Negro province, in the central

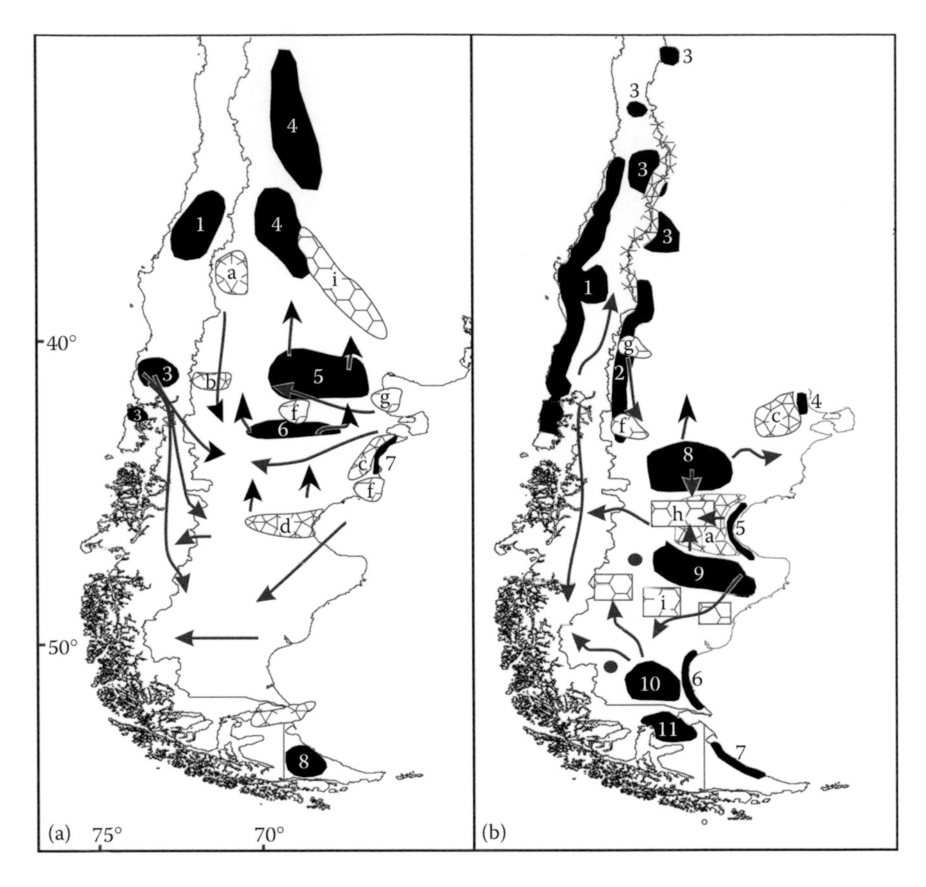

Figure 7.8 Stable areas and expansion routes detected in phylogeographic analyses in the Patagonian province. (a) Vertebrates and (b) plants. (Modified from Sérsic, A.N. et al., *Biol. J. Linnean Soc.*, 103, 475–494, 2011. With permission.) (1)–(11) Stable areas, (a)–(e) fragmentation and (f)–(i) areas of secondary contact.

area of the Somuncurá plateau and in the ecotone between the Patagonian and Monte provinces. Studies proposed areas that may have persisted as refugia along the coast of the Río Negro province, in San Jorge Gulf, central Chubut, southeastern and southern Santa Cruz and Tierra del Fuego.

Studies detected concordance in the distribution of different plant and animal lineages at a large spatial scale, although their temporal concordance is difficult to assess in the absence of data on divergence times (Sérsic et al., 2011). Hypotheses of different geological scenarios attempted to explain the observed patterns, including tectonic/orogenic events, volcanism and paleobasins in the Neogene and glacial cycles and coastline shifts in the Quaternary. The most recent breaks in the Patagonian province seem to be the result of younger (Quaternary) events which coincided with other studies that suggested a more recent origin of the steppe vegetation (e.g., Cosacov et al., 2013).

Case study: Areas of endemism in the Patagonian steppes based on insect taxa

Title: "Areas of endemism of the Patagonian steppe: An approach based on insect distributional patterns using endemicity analysis." (Domínguez, M. C., S. Roig-Juñent, J. J. Tassin, F. C. Ocampo and G. E. Flores, *Journal of Biogeography,* 33: 1527–1537, 2006.)

Goal: To delimit areas of endemism in the Patagonian steppe by means of an endemicity index, and to compare the resulting areas of endemism with those proposed by previous studies.

Location: Patagonian steppe, southern South America.

Methods: The study divided the area analyzed into grid-cells of 1° latitude by 1° longitude. The study gathered distributional data for 149 insect species from systematic revisions and from collections. The study constructed a matrix based on 1317 georeferenced data entries for the presence/absence of the different taxa; 19 of these grid-cells did not have Patagonian steppe vegetation but the study considered them in the analysis because widespread species that inhabit Patagonian steppe were recorded in them. The study used the method proposed by Szumik et al. (2002) and Szumik and Goloboff (2004) with software NDM/VNDM ver. 1.5 (Goloboff, 2004), by means of a score of endemicity given by the number of endemic species for each combination of grid-cells. To determine how many species appear as endemic, the study determined endemicity for each species following four possible criteria and the score of endemicity represented the sum of individual endemicity values for all species considered as endemic for a given set of grid-cells. The study performed different endemicity analyses through a heuristic search in which the program was set to find ensembles of grid-cells, adding/eliminating two grid-cells at a time. One can define these ensembles by either four or more endemic species or by three or more endemic species. The study rejected ensembles with more than 50% of species in common and retained those with the highest values.

Results: The analysis searching for sets with four or more endemic species resulted in six sets of grid-cells with the highest endemicity value. Analysing these sets with VNDM, it was possible to identify areas nested within other major areas, conflictive areas and disjunct areas. For this analysis, the study did not implement the option continuous distribution. The study found two main groups of sets: the first one between 37°S–43°S and 73°W–68°W, and the second between 49°S–56°S and 74°W–64°W. This first analysis found

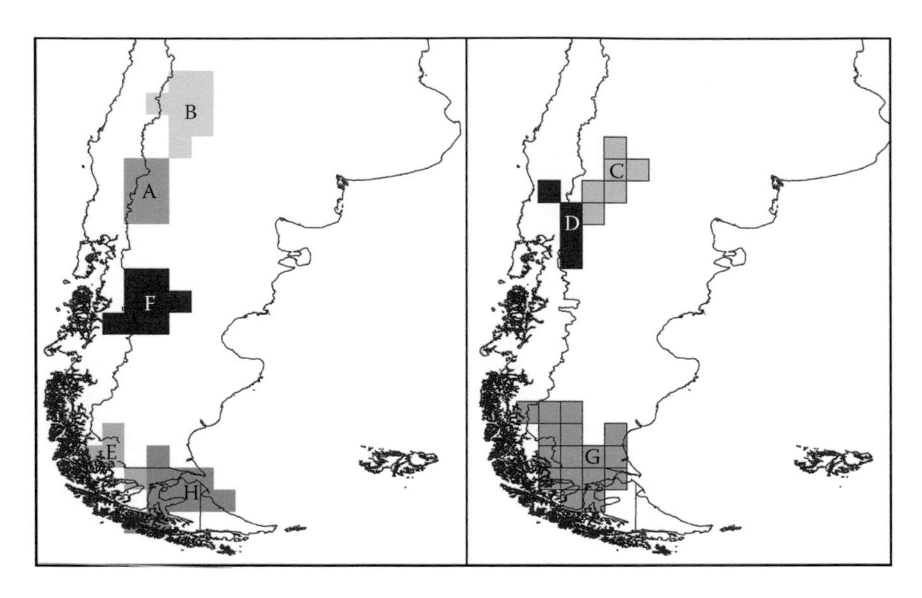

Figure 7.9 Areas of endemism in the Patagonian province found by Domínguez et al. (2006). (Modified from Domínguez, M.C. et al., *J. Biogeogr.*, 33, 1527–1537, 2006. With permission.) (A) Western Patagonia, (B) northern Payunia, (E) southern Subandean, (F) Chubutian, (H) Austral Patagonia, (C) southern Payunia, (D) northern Subandean and (G) Santacrucian.

three areas of endemism among the northern group of sets: the first defined by set 0 supported by 12 species corresponded to Western Patagonia (Figure 7.9); the second defined by set 1 supported by 5 species corresponded to the northern part of Subandean Patagonia; and the third defined by set 5 supported by 4 species corresponded to the southern part of Payunia. Among the southern groups of sets, the study recognized two areas of endemism, defined by sets 2 and 4: set 2 supported by five endemic species corresponded to the southern part of Subandean Patagonia and set 4 supported by seven endemic species corresponded to Austral Patagonia (Fuegian district).

The study performed a second analysis using the maximum polygon method to assume a continuous distribution for each species, considering each empty grid-cell within the maximum polygon of the distribution area of the species as an assumed/supposed presence. In this analysis, the study found the same areas of endemism as in the first analysis but with a higher endemicity value and larger total area. The study performed a third analysis considering continuous distributions of species and searching for areas of endemism defined by three or more species because the low number of species registered in many grid-cells may complicate the identification of areas of endemism in these grid-cells. The third analysis retained 12 sets of grid-cells, identifying the same previous areas of endemism and three new areas: set 5 supported by three species corresponded to the Chubutian district of Central Patagonia; set 6 also supported by three species was conflictive with set 3 (southern Payunia) because they shared one grid-cell but had no species in common; and set 6 corresponded to the northern part of Payunia. Sets 9, 10 and 11 constituted different alternative arrangements for the same area. Set 10 was conflictive with two other areas of endemism identified in the previous analyses, southern Subandean Patagonia and the Austral Patagonia, because they shared several grid-cells, but set 10 had no species in common with either one of them. Set 10 corresponded to the Santacrucian district of Central Patagonia.

The areas and subareas identified by Domínguez et al. (2006) are as follows:

1. *Western Patagonia*: Supported by *Barypus gentilii, B. neuquensis, B. schajovskoii, Cnemalobus gentilii, C. neuquensis, Patagonogenius elegans, P. atra, P. neuquensis, Nyctelia grandis, Polynoncus bullatus* and *Athlia giaii*.
2. *Payunia*: Including the Northern Payunia (Mendoza province; supported by *Cnemalobus mendozencis, Nyctelia garciae* and *N. laevis*) and Southern Payunia subareas (Neuquén province; supported by *Cnemalobus neuquensis, Psectrascelis hirta, Nyctelia gebieni* and *Athlia parvicollis*).
3. *Subandean Patagonia*: Including the Septentrional Subandean Patagonia (supported by *Barypus minus, Cnemalobus deplanatus, Mimodromius metallicus, Chaudoirina orfitlai* and *Tropisdosthetus bicarinatus*) and Meridional Subandean Patagonia subareas (supported by *Barypus painensis, Notaphiellus cekalovici, Tetragonoderes viridis cekalovici, Eremopachys bergi* and *Acrostomus malleganicus*).
4. *Central Patagonia*: Including the Chubutian (supported by *Barypus chubutensis, Circracris auris* and *Epipedonota subplana*) and Santacrucian subareas (supported by *Epipedonota tricostata, Nyctelia corrugatae* and *Taurocerastes patagonicus*).
5. *Austral Patagonia*: Supported by *Metius annulicornis, Metius fitavipleuris, Metius pogonoides, Tristira magellanica, Bufonacris terrestris, Neopraocis refitexocilis* and *Plathestes depressa*.

Main conclusions: The results of the analysis revealed several areas of endemism proposed by previous studies, such as Subandean Patagonia and Western Patagonia. In Central Patagonia the study recognized only two districts: Southern Payunia located in northern Patagonia, and Fuegian or Austral Patagonia in southern Patagonia.

Between them the study found no other area of endemism, probably because there was relatively little distributional data available for the taxa analyzed; for example, no information regarding the distribution of the 149 species recorded in this study was available for 21 grid-cells that belonged to the Patagonian steppe, 19 of which were from Central Patagonia. Furthermore, there were 41 grid-cells with only one species recorded, of which 16 belonged to Central Patagonia. Another reason why the study found no areas of endemism in the first two analyses in Central Patagonia could be the number of endemic species used to define these areas. Thus, the study performed a third analysis considering continuous distributions and setting a lower number of species (three instead of four) to define an area, allowing identification of the same five areas of the preceding analysis and three more: Northern Payunia, Chubutian and Santacrucian, which belonged to Central Patagonia. These eight areas represented almost all the natural areas proposed for the Patagonian steppe by other studies based on plant and insect species.

chapter eight

The South American transition zone

The South American transition zone comprises the highlands of the Andes between western Venezuela and Chile, the desert areas of coastal Peru and northern Chile, and central western Argentina (Morrone, 2004b, 2006, 2014a, 2017). It spans from sea level to 4500 m. The uplift of the Andes in the Neogene shaped it. This rise progresses from south to north and from west to east, with two major events: one in the Middle Miocene and another at the beginning of the Pliocene (Amarilla et al., 2015).

In this chapter, I characterize the South American transition zone briefly, detail its endemic and characteristic taxa and discuss its biotic relationships. Within the transition zone, I recognize seven provinces: Páramo, Desert, Puna, Atacama, Cuyan High Andean, Monte and Comechingones. I also characterize these provinces and their districts, discuss their endemic taxa and relationships and characterize their vegetation briefly. Finally, I provide three case studies.

South American transition zone

Peruvian subregion (in part)—Blyth, 1871: 428 (regionalization).
Argentinean subarea (in part)—Clarke, 1892: 381 (regionalization).
Tropical Andean subarea (in part)—Clarke, 1892: 381 (regionalization).
Patagonian subregion (in part)—Sclater and Sclater, 1899: 65 (regionalization); Kuschel, 1964: 447 (regionalization); Hershkovitz, 1969: 8 (regionalization); Kuschel, 1969: 712 (regionalization); Ojeda et al., 2002: 23 (biotic evolution).
Andean region (in part)—Shannon, 1927: 3 (regionalization); Good, 1947: 236 (regionalization); O'Brien, 1971: 198 (regionalization); Morain, 1984: 178 (textbook); Rivas-Martínez and Navarro, 1994: map (regionalization); Huber and Riina, 1997: 24 (glossary); Morrone, 2001a: 103 (regionalization), 2001b: 70 (regionalization), 2006: 483 (regionalization); Quijano-Abril et al., 2006: 1268 (track analysis); López et al., 2008: 1564 (parsimony analysis of endemicity and regionalization); Löwenberg-Neto and de Carvalho, 2009: 1751 (parsimony analysis of endemicity); Procheş and Ramdhani, 2012: 263 (cluster analysis and regionalization).

Andean dominion (in part)—Hauman, 1931: 60 (regionalization); Cabrera, 1951: 48 (regionalization); Orfila, 1941: 86 (regionalization); Cabrera, 1957: 335 (regionalization); Maury, 1979: 710 (faunistics); León et al., 1998: 127 (regionalization).

Andean-Patagonian subregion (in part)—Mello-Leitão, 1937: 232 (regionalization), 1943: 129 (regionalization); Ringuelet, 1961: 156 (biotic evol. and regionalization); Rapoport, 1968: 75 (biotic evol. and regionalization); Ringuelet, 1978: 255 (biotic evolution); Fittkau, 1969: 639 (regionalization); Paggi, 1990: 303 (regionalization).

Andean district (in part)—Cabrera and Yepes, 1940: 16 (regionalization).

Subandean district—Cabrera and Yepes, 1940: 15 (regionalization).

High Andean province—Cabrera, 1951: 49 (regionalization), 1953: 107 (regionalization), 1957: 337 (regionalization), 1958: 200 (regionalization); Cabrera and Willink, 1973: 84 (regionalization); Willink, 1991: 138 (regionalization); Huber and Riina, 1997: 270 (glossary); Ojeda et al., 2002: 24 (biotic evolution); Apodaca et al., 2015a: 82 (regionalization); Biganzoli and Zuloaga, 2015: 339 (floristics).

Andean province (in part)—Fittkau, 1969: 642 (regionalization).

Subandean province (in part)—Fittkau, 1969: 642 (regionalization).

Central Andes area—Sick, 1969: 463 (regionalization).

Andean-Patagonian dominion (in part)—Cabrera, 1971: 29 (regionalization), 1976: 50 (regionalization); Cabrera and Willink, 1973: 83 (regionalization); Willink, 1988: 206 (regionalization); Huber and Riina, 1997: 150 (glossary); Zuloaga et al., 1999: 18 (floristics); Ojeda et al., 2002: 24 (biotic evolution).

Austral subregion (in part)—Ringuelet, 1975: 107 (regionalization); Almirón et al., 1997: 23 (regionalization).

Andean subkingdom (in part)—Rivas-Martínez and Tovar, 1983: 516 (regionalization); Huber and Riina, 1997: 332 (glossary).

Argentine subregion (in part)—Smith, 1983: 462 (cluster analysis and regionalization).

Andean subregion (in part)—Morrone, 1994a: 190 (parsimony analysis of endemicity); Morrone, 1996a: 105 (regionalization); Posadas et al., 1997: 2 (parsimony analysis of endemicity).

Austroamerican subkingdom (in part)—Rivas-Martínez and Navarro, 1994: map (regionalization); Rivas-Martínez et al., 2011: 27 (regionalization).

Central Andes bioregion—Dinerstein et al., 1995: map 1 (ecoregionalization); Huber and Riina, 1997: 37 (glossary).

Neotemperate region (in part)—Amorim and Pires, 1996: 187 (regionalization).

Páramo-Punan subregion—Morrone, 1999: 11 (regionalization), 2001a: 106 (regionalization), 2001d: 1 (regionalization); Quijano-Abril et al., 2006: 1268 (track analysis).

South American transition zone—Morrone, 2004a: 158 (regionalization), 2004b: 42 (regionalization); Mihoč et al., 2006: 391 (parsimony analysis of endemicity and track analysis); Morrone, 2006: 482 (regionalization); Quijano-Abril et al., 2006: 1271 (track analysis); Alzate et al., 2008: 1252 (track analysis); Couri and de Carvalho, 2008: 2677 (biotic evolution); Roig-Juñent et al., 2008: 23 (biotic evolution); Roig et al., 2009: 164 (regionalization); de Carvalho and Couri, 2010: 295 (track analysis); Morrone, 2010: 37 (regionalization); Urtubey et al., 2010: 505 (track analysis and cladistic biogeography); Hechem et al., 2011: 46 (track analysis); Löwenberg-Neto et al., 2011: 1942 (macroecology); Luebert, 2011: 109 (biotic evolution); Coulleri and Ferrucci, 2012: 105 (track analysis); Ferretti et al., 2012: 1 (parsimony analysis of endemicity and track analysis); Mercado-Salas et al., 2012: 459 (track analysis); Alfaro et al., 2013: 243 (faunistics); Campos-Soldini et al., 2013: 16 (track analysis); Ferro, 2013: 323 (biotic evolution); Granara de Willink, 2014: 254 (faunistics); Huber, 2014: 134 (faunistics); Lamas et al., 2014: 955 (cladistic biogeography); Morrone, 2014a: 82 (regionalization), 2014b: 203 (cladistic biogeography); Ferretti, 2015: 3 (dispersal-vicariance analysis); Klassa and Santos, 2015: 520 (endemicity analysis); Amarilla et al., 2015: 1 (biotic evolution); Goin et al., 2015: 133 (biotic evolution); Leivas et al., 2015: 116 (faunistics); Morrone, 2015b: 210 (regionalization); Coelho et al., 2016: 28 (endemicity analysis); del Valle Elías and Aagesen, 2016: 161 (endemicity analysis); Ruiz et al., 2016: 385 (track analysis); Stonis et al., 2016: 561 (systematic revision); Arana et al., 2017: 421 (shapefiles); Escalante, 2017: 351 (parsimony analysis of endemicity); Martínez et al., 2017: 480 (track analysis and regionalization); Monckton, 2016: 125 (systematic revision and map); Morrone, 2017: 213 (regionalization).

Tropical South Andean superegion—Rivas-Martínez et al., 2011: 27 (regionalization).

Tropical South Andean region—Rivas-Martínez et al., 2011: 27 (regionalization).

Neotropical transition zone—Cione et al., 2015: 48 (biotic evolution).

Endemic and characteristic taxa

Morrone (2001d) provided a list of endemic and characteristic taxa. Some examples included *Ephedra multiflora* and *E. rupestris* (Ephedraceae); *Azorella compacta* (Apiaceae); *Adaetobdella cryptica* (Hirudinea: Glossiphoniidae); *Trachelopachys bicolor* (Araneae: Clubionidae); *Amblygnathus gilvipes peruanus* and *Notiobia aquilalorum* (Coleoptera: Carabidae); *Belostoma dallasi* (Heteroptera: Belostomatidae); *Crites*, *Incacris* and *Punacris* (Orthoptera: Tristiridae); *Cordibates*, *Melaphorus*, *Philorea* and *Pilobalia* (Coleoptera: Tenebrionidae); *Buteo poecilochrous* (Accipitriformes:

Accipitridae); *Phoenicopterus andinus* and *P. jamesi* (Phoenicopteriformes: Phoenicopteridae); *Diuca speculifera, Poospiza garleppi, Sicalis lutea* and *S. uropygialis* (Passeriformes: Thraupidae); *Picumnus dorbignyanus* and *Veniliornis frontalis* (Piciformes: Picidae); *Hippocamelus* (Cetartiodactyla: Cervidae); *Leopardus jacobita* and *Lynchailurus pajerus garleppi* (Carnivora: Felidae); *Sturnira bogotensis* (Chiroptera: Phyllostomidae); and *Chaetophractus nationi* (Xenarthra: Dasypodidae).

Biotic relationships

Urtubey et al. (2010) conducted a track analysis and a cladistic biogeographical analysis of the provinces of the South American transition zone and the subregions of the Neotropical and Andean regions based on Asteraceae. They obtained a general area cladogram (Figure 8.1) based on the area cladograms of *Barnadesia, Chuquiraga, Dasyphyllum, Hypochaeris* and the *Lucilia* generic group. This cladogram indicated a basic separation between the Atacaman, Monte and Cuyan High Andean provinces more closely related to the Andean region; whereas the Páramo, Desert and Puna provinces appeared to be closely related to the Neotropical region. These results corroborated the transitional character of the provinces assigned to the South American transition zone.

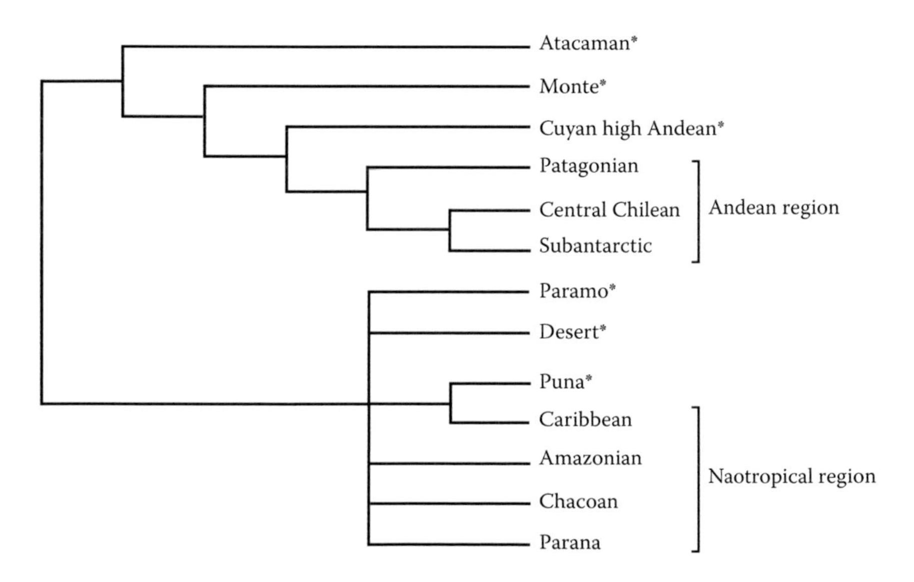

Figure 8.1 General area cladogram obtained by Urtubey et al. (2010) showing the relationships of the provinces of the South American transition zone and related areas. * denotes the provinces of the South American transition zone. (Modified from Urtubey, E. et al., *Taxon*, 59, 505–509, 2010. With permission.)

Regionalization

The South American transition zone comprises the Páramo, Desert, Puna, Atacama, Cuyan High Andean, Monte and Comechingones provinces (Figure 8.2).

Cenocrons

Roig-Juñent et al. (2008) distinguished four cenocrons based on insect taxa. The Pangeic cenocron consisted of old taxa present before the break-up of Pangaea, including two varieties: taxa found in eremic (xeric) environments and taxa from more humid habitats. The Gondwanic cenocron consisted of taxa present before the break-up of Gondwana, including five varieties: Gondwanic, Peripampasic Arc, Patagonian, autochthonous and

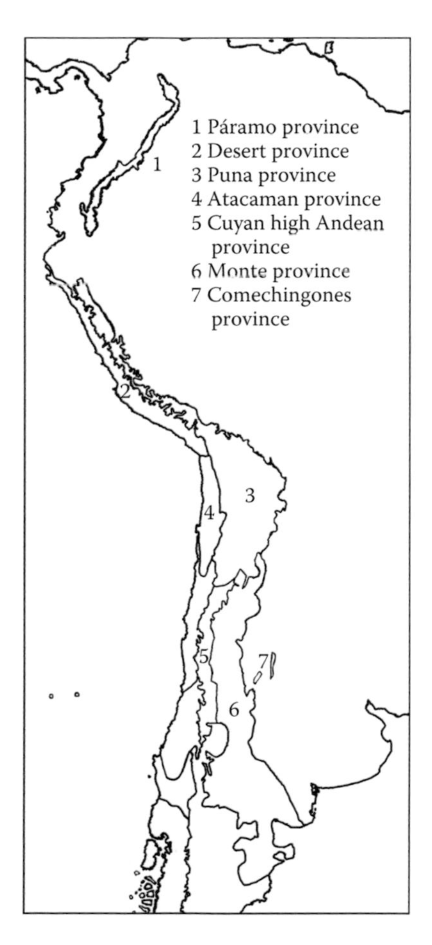

Figure 8.2 Map of the provinces of the South American transition zone.

Neotropical. The Holarctic cenocron consisted of taxa dispersed from the Nearctic region. The synanthropic cenocron included taxa dispersed to the areas within the last 500 years, associated with human kind.

Case study: Quaternary biogeography of the grass Munroa argentina

Title: "*Munroa argentina,* a grass of the South American transition zone, survived the Andean uplift, aridification and glaciations of the Quaternary." (Amarilla, L. D., A. M. Anton, J. O. Chiapella, M. M. Manifesto, D. F. Angulo and V. Sosa, *PLoS One,* 10: 1–21, 2015.)

Goal: To study the effect of the Andean uplift, climatic aridification since the Neogene and the Quaternary glaciation cycles in the evolution of the annual grass species *Munroa argentina.*

Location: Puna and Monte provinces, South American transition zone.

Methods: This study sampled a total of 152 accessions of *M. argentina* (Poaceae) from 20 localities covering its entire distribution range and its elevation gradient (1000–4200 m). The study isolated total genomic DNA from silica-gel-dried leaf tissue using the CTAB method (Doyle and Doyle, 1987). The study amplified the chloroplast and nuclear regions using a single amplification protocol following Peterson et al. (2010). The study assembled and edited sequences using Sequencher v4.1 (Gene Codes Corporation, Ann Arbor, Michigan). The study assayed amplified fragment length polymorphisms (AFLP) following the protocol of Lachmuth et al. (2010). The study constructed the parsimony network of haplotypes using TCS ver.1.2.1 (Clement et al., 2000) with the algorithm of Templeton et al. (1992). To estimate genetic differentiation among haplotypes, the study performed molecular dating under a Bayesian approach with BEAST ver. 2.1.3 (Bouckaert et al., 2014). The study estimated the divergence time of the haplotypes using the age obtained from the previous dating for Scleropogoninae and *Swallenia alexandrae.* The study ran analyses using a molecular clock model with uncorrelated rates, assuming a lognormal distribution of rates. For chloroplast data, the study calculated parameters of population diversity for each phylogroup derived from the haplotype network and phylogenetic analyses using Arlequin 3.1 (Excoffier et al., 2005).

The study estimated the ecological niche based on the available distribution records of *M. argentina.* A series of 19 variables summarizing aspects of climate represented environmental scenarios in the present and in the past (data obtained from WorldClim 1.4 [Hijmans et al., 2005]). For the Last Glacial Maximum analysis, the study used general circulation model simulations from two models: the Community Climate System Model (Collins et al., 2004) and the Model for Interdisciplinary Research on Climate (Hasumi and Emori, 2004).

Results: The study identified 41 haplotypes for the 152 individuals sampled and distributed in the Puna (Argentina and Bolivia), Prepuna (Argentina) and Monte (Argentina). The topology of the recovered phylogram was congruent with the result of the statistical parsimony network and identical to the chronogram obtained with BEAST (Figure 8.3). The study identified three major clades: the first formed by the Puna phylogroup and the other two by the Prepuna and Monte phylogroups. According to the molecular dating, the diversification of the phylogroups began in the Middle Pliocene–Late Pleistocene, approximately 3.4 m.y.a., separating the Puna phylogroup from the ancestor of the Prepuna-Monte phylogroups. The Prepuna

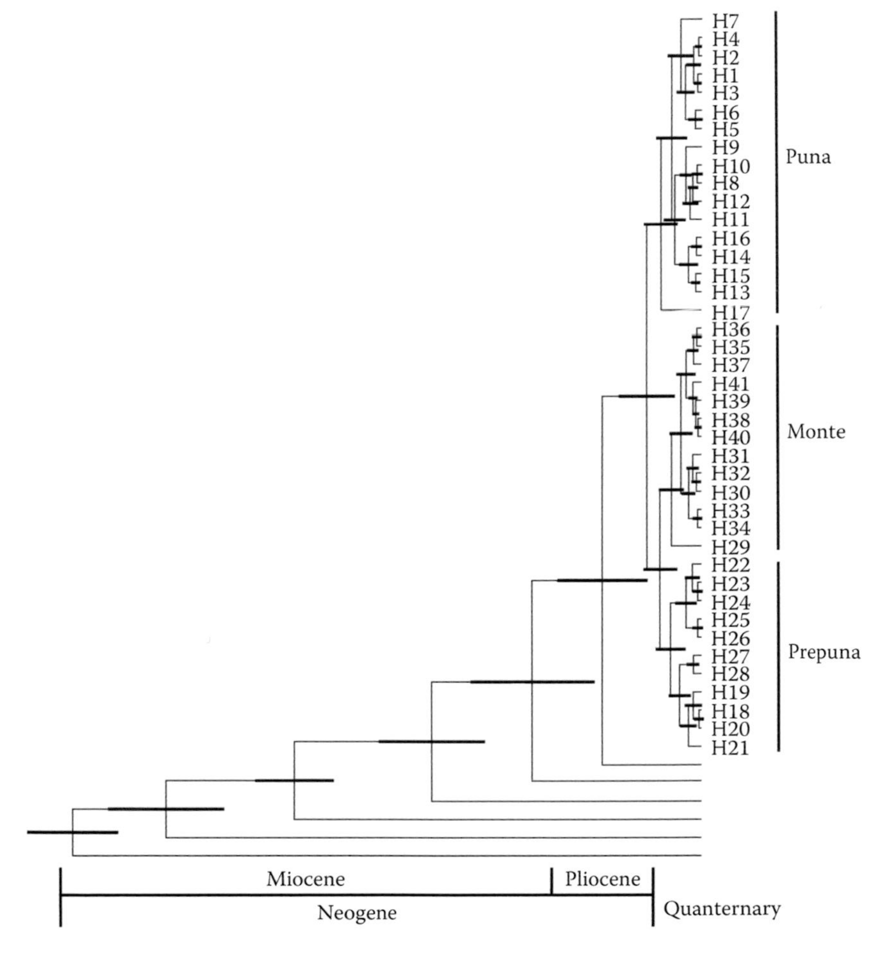

Figure 8.3 Chronogram of *Munroa argentina* haplotypes and other Chloridoideae obtained by Amarilla et al. (2015) based on the consensus tree from the Bayesian dating analysis. (Modified from Amarilla, L.D. et al., *PLoS One*, 10, 1–21, 2015.)

phylogroup diverged from the Monte phylogroup approximately 2.2 m.y.a. during the Pleistocene. The diversification of most of the haplotypes began 1.6 m.y.a. in the Mid-Pleistocene. The study identified two major genetic barriers: one separating Puna and Prepuna from Monte phylogroups, and the other separating Puna from Prepuna and Monte phylogroups. The Puna phylogroup may have increased over the past 90,000 years, the Prepuna phylogroup had a stable population size over the last 60,000 years and the Monte phylogroup may have increased over the past 85,000 years.

Niche distribution modeling for current climate conditions over-predicts the geographic distribution in the northern and southeastern extremes of *M. argentina* where studies have never recorded it. The potential distribution during the Last Interglacial Period was continuous and covered a more extensive area of suitable habitats from northern Prepuna province to the Patagonian subregion. The Last Glacial Maximum model indicated that suitable habitats were more extensive than they currently are (and less extensive than they were during the Interglacial) and over-predicts the geographic distribution further to the southeast.

Main conclusions: Geological, climate and genetic data suggested that the South American transition zone arose from the Middle Miocene to the Upper Pliocene and indicated the persistence of a semiarid climate between 8 and 3 m.y.a. that reached its highest aridity level about 6 m.y.a. (Hartley and Chong, 2002). The split between *M. argentina* and the ancestor shared with *M. andina* occurred 4.8 m.y.a. from the Late Miocene to the Late Pliocene when climate conditions were suitable. The split among phylogroups occurred in the Late Pliocene and mainly in the Pleistocene. These results suggested that, although the early stages of the evolutionary history of *M. argentina* linked to the aridification processes that gave rise to the South American transition zone and to the final uplift of the Southern Andes (5 m.y.a.), the following stages and conformation of its range linked instead to Quaternary glaciations. Climate oscillations in the Pleistocene were probably among the causes of differentiation among phylogroups of *M. argentina* through the fragmentation of an ancient, more widespread distribution. The results of this analysis revealed that since approximately 4 m.y.a., *Munroa argentina* has been able to persist despite orogenic and climate changes, and that its apparently once-continuous range has undergone fragmentation, with phylogroups with low historical gene flow among them and a strong genetic structure.

Páramo province

Incasic province (in part)—Mello-Leitão, 1943: 130 (regionalization); Fittkau, 1969: 642 (regionalization).

Páramo province—Cabrera, 1957: 335 (regionalization); Cabrera and Willink, 1973: 66 (regionalization); Morrone, 1994a: 190 (parsimony analysis of endemicity); Rivas-Martínez and Navarro, 1994: map (regionalization); Morrone, 1996a: 108 (regionalization); Posadas et al., 1997: 2 (parsimony analysis of endemicity); Katinas et al., 1999: 112 (track analysis); Morrone, 1999: 13 (regionalization), 2014a: 83 (regionalization); del Río et al., 2015: 1294 (track analysis and cladistic biogeography); Morrone, 2015b: 210 (regionalization); Stonis et al., 2016: 561 (systematic revision); Martínez et al., 2017: 479 (track analysis and regionalization); Morrone, 2017: 217 (regionalization).

Northern Andes area (in part)—Sick, 1969: 461 (regionalization); Porzecanski and Cracraft, 2005: 266 (parsimony analysis of endemicity).

North Andean center—Müller, 1973: 45 (regionalization); Cracraft, 1985: 62 (areas of endemism).

Bogotá subcenter—Müller, 1973: 46 (regionalization).

Colombian Montane province—Udvardy, 1975: 42 (regionalization); Huber and Riina, 1997: 130 (glossary).

Páramo region—Rivas-Martínez and Tovar, 1983: 516 (regionalization); Huber and Riina, 1997: 284 (glossary).

North Andean area—Coscarón and Coscarón-Arias, 1995: 726 (areas of endemism).

Cordillera Central Páramo ecoregion—Dinerstein et al., 1995: 102 (ecoregionalization); Huber and Riina, 1997: 248 (glossary).

Cordillera de Mérida Páramo ecoregion—Dinerstein et al., 1995: 102 (ecoregionalization); Huber and Riina, 1997: 248 (glossary).

Northern Andean Páramo ecoregion—Dinerstein et al., 1995: 102 (ecoregionalization); Huber and Riina, 1997: 249 (glossary).

Páramo ecoregion—Huber and Riina, 1997: 154 (glossary).

North Andean Páramo province—Morrone, 2001a: 107 (regionalization), 2001d: 3 (regionalization), 2004b: 45 (regionalization), 2006: 483 (regionalization); Alzate et al., 2008: 1252 (track analysis); Urtubey et al., 2010: 506 (track analysis and cladistic biogeography); Hechem et al., 2011: 46 (track analysis); Mercado-Salas et al., 2012: 459 (track analysis); Kutschker and Morrone, 2012: 543 (track analysis).

Norandean Páramo province—Quijano-Abril et al., 2006: 1270 (track analysis).

Northern Andean area (in part)—Apodaca et al., 2015b: 5 (dispersal-vicariance analysis).

Definition

The Páramo province comprises the high mountain peaks of the Andean cordillera of Venezuela, Colombia and Ecuador, from the upper forest line at 3000–3500 m upwards and below the permanent snowline at ca. 5000 m (Cabrera and Willink, 1973; Müller, 1973; Ringuelet, 1975; Rivas-Martínez and Tovar, 1983; Luteyn, 1992; Posadas et al., 1997; Gradstein, 1998; Rangel, 2000a,b; Morrone, 2001d, 2006, 2014a; Londoño et al., 2014). Glaciations strongly influence this high-elevation biome because the landscape is irregular, from very rough to flat and stretches from 3000 m to the perennial snowline a 4800–5000 m altitude (Luteyn, 1992). Studies debate whether this biome is natural or man-made. Luteyn (1992) considered that it is clear that páramos of the present are more extensive than in earlier times, beginning at much lower elevations because of the anthropogenic destruction of the forest. This province is part of the tropical Andean biodiversity hotspot characterized by its high endemism, which reaches 60% of the species (Hughes and Eastwood, 2006).

Endemic and characteristic taxa

Morrone (2014a) provided a list of endemic and characteristic taxa. Some examples included *Dicksonia stuebelli* (Dicksoniaceae); *Bomarea angustipetala, B. bredemeyerana* and *B. holtoni* (Alstroemeriaceae); *Blakiella, Espeletia* complex, *Floscaldasia, Ruilopezia atropurpurea* and *Westoniella* (Asteraceae); *Passiflora truxillensis* (Passifloraceae); *Aragoa* (Scrophulariaceae); *Bogotacris, Chibchacris* and *Timotes* (Orthoptera: Acrididae); Strengerianini (Figure 8.4a; Decapoda: Pseudothelphusidae); *Howdeniola sulcipennis, Minetes* and *Phyllothrox aristidis* (Coleoptera: Curculionidae); *Gigantodax cervicornis* (Figure 8.4b) and *G. siberianus* (Diptera: Simuliidae); *Atelopus tamaense* (Anura: Bufonidae); *Cavia porcellus anolaimae* (Rodentia: Caviidae); *Mazama rufina* (Cetartiodactyla: Cervidae); *Crypturellus kerriae* and *C. saltuarius* (Tinamiformes: Tinamidae); *Momotus momota olivaresii* (Coraciiformes: Momotidae); and *Muscisaxicola maculirostros niceforei* (Passeriformes: Tyrannidae).

Vegetation

Páramo province consists of different types of vegetation, including moorlands, xerophytic scrublands, grasslands and peat bogs (Figure 8.5; Cleef, 1978, 1981; van der Hammen and Cleef, 1986; Rangel et al., 1997; Gradstein, 1998). There are steppes of *Festuca* and *Deyeuxia* (Poaceae) with the typical *frailejones* (Asteraceae: Espeletiinae) and forests on the higher areas (Cabrera and Willink, 1973; Cuatrecasas, 1986; Monasterio, 1986; Sturm, 1990). Dominant plant species belong to the genera *Calamagrostis, Chusquea, Deyeuxia, Diplostephium, Cynoxys, Espeletia, Espeletiopsis,*

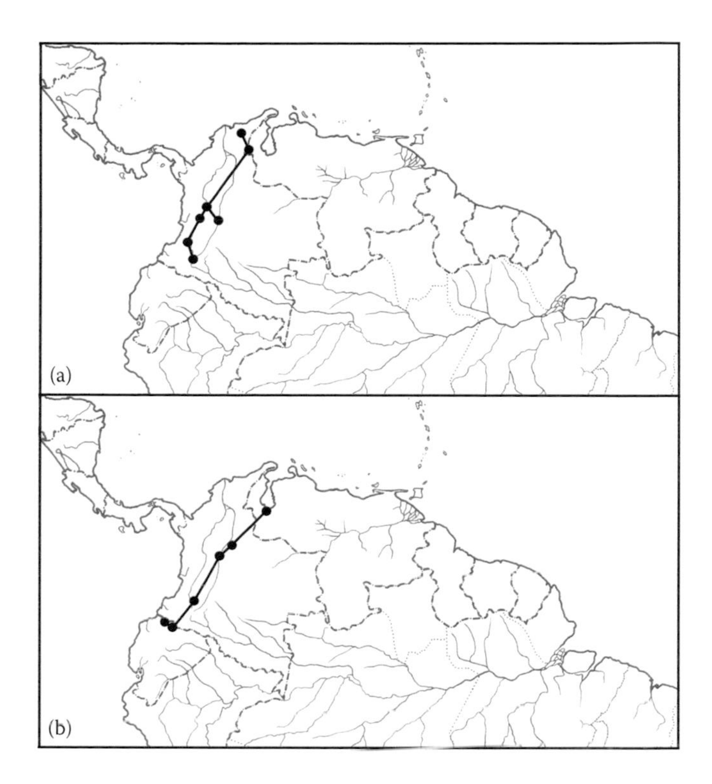

Figure 8.4 Maps with individual tracks in the Páramo province. (a) Strengerianini (Decapoda: Pseudothelphusidae) and (b) *Gigantodax cervicornis* (Diptera: Simuliidae).

Figure 8.5 Paramo vegetation, Belmira, Colombia. (Courtesy of Mario Alberto Quijano.)

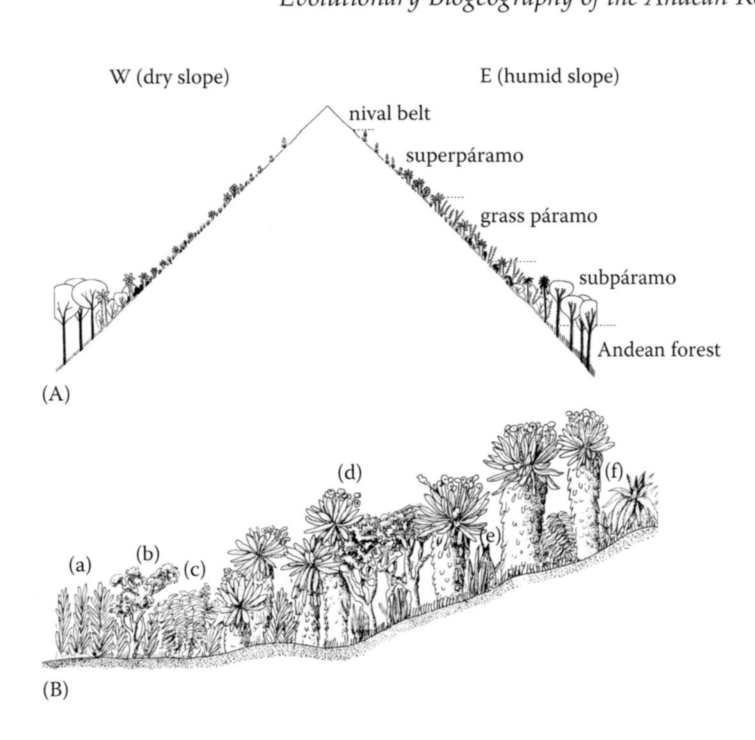

Figure 8.6 Páramo vegetation. (A) Schematic representation of the vegetational zonation (Modified from van der Hammen, T. and Cleef, A.M., Development of the high Andean páramo flora and vegetation. In: Vuilleumier, F. and Monasterio, M. (Eds.), *High Altitude Tropical Biogeography*, Oxford University Press and American Museum of Natural History, New York and Oxford, pp. 153–201, 1986. With permission.) and (B) detail of the grass páramo. (Modified from Rangel, J.O. et al., Tipos de vegetación en Colombia: Una aproximación al conocimiento de la terminología fitosociológica, fitoecológica y de uso común, in Rangel, J.O., Lowy, P.D. and Aguilar, M. (Eds.), *Colombia Diversidad Biótica II: Tipos de vegetación en Colombia*, Instituto de Ciencias Naturales, Universidad Nacional de Colombia, Santafé de Bogotá, pp. 89–381, 1997. With permission.) (a) *Chusquea tessellata*, (b) *Diplostephium schultzii*, (c) *Blechnum loxense*, (d) *Espeletia hartwegiana*, (e) *Neurolepis cf. aperta* and (f) *Puya* sp.

Festuca, Gentiana, Gunnera, Hypericum, Lupinus, Miconia, Paepalanthus and *Rubus* (Cabrera and Willink, 1973; Cleef, 1978, 1981; Monasterio, 1986; Hernández Camacho and Sánchez Páez, 1992; Hernández Camacho et al., 1992a). Studies divided the Páramo province into three altitudinal zones (Figure 8.6a; van der Hammen and Cleef, 1986; Luteyn, 1992; Huber and Riina, 1997; Rangel, 2000a):

1. *Subpáramo, low páramo or ceja andina*: Shrubby transition zone from 3200–3500 m altitude, made up of forest elements from the lower slopes and grass páramo above. Shrubs belong to the genera *Befaria, Brachyotum, Diplostephium, Gaultheria, Gynoxys, Hypericum, Maclenia, Miconia, Pentacalia, Hesperomeles* and *Vaccinium.*

2. *Grass páramo, páramo proper or páramo in the strict sense* (Figure 8.6b): From 3500–4100 m, it has a markedly xeromorphic vegetation. It is composed of buchgrasslands dominated by species of *Calamagostris* and *Festuca,* and dwarf bamboos *(Chusquea)* in wet slopes. It is also rich in shrubs (*Hypericum, Diplostephium, Pentacalia, Pernettya* and *Valeriana*), acaulescent rosette-plants *(Acaena)* and cushion plants *(Werneria).* The most characteristic plants are the species of *Espeletia* and *Espeletiopsis* (Asteraceae), with their columnar, wooly, rosette-plan growth form.
3. *Superpáramo*: Patches situated above 4100 m altitude, with sandy soils, situated between the grass páramo and the snowline subject to regular snowfalls. There are scattered plants belonging to the genera *Draba, Ephedra, Lupinus* and *Senecio.*

Van der Hammen and Cleef (1986) recognized in each of these altitudinal belts a lower and an upper zone, and also described them for each side (west or east) of the cordillera. Restrepo and Duque (1992) analyzed the vegetational types in Llano de Paletara, Colombian Central Cordillera and distinguish eight different types: *frailejones* with *Espeletia hartwegiana, Blechnum loxense* and *Sphagnum sancto-josephense; chuscal* with *Chusquea tessellata* and *Sphagnum sancto-josephense*; swamp with *Carex jamesonii* and *Arthoxanthum odoratum; pajonal* with *Calamagrostis intermedia;* scrubland with *Hypericum luncioides* and *Blechnum loxense;* scrubland with *Ageratina tinifolia;* scrubland with *Diplostephium cf. cinerascens;* and forest with *Escallonia myrtilloides.*

Biotic relationships

The importance given to some tropical elements made Cabrera and Willink (1973) assign the Páramo province to the Neotropical region, whereas their close relationships with other Andean provinces (Rivas-Martínez and Tovar, 1983; Fjeldså, 1982) led other studies (Morrone, 1994a, b, 1996b; Posadas et al., 1997) to place it in the Andean region. Additionally, this province contains numerous north temperate plant genera as *Alnus, Draba, Lupinus, Quercus, Salix, Sanbucus, Valeriana* and *Viburnum* that arrived after the uplift of the northern Andes (Hughes and Eastwood, 2006). The placement of the Páramo province in the South American transition zone (Morrone, 2014a) reflects its conflicting relationships.

Vuilleumier (1986) suggested a close relationship of the Páramo province with the Puna province based on bird taxa. A parsimony analysis of endemicity based on bird taxa (Porzecanski and Cracraft, 2005) postulated a close relationship of this province with the Guatuso-Talamanca, Pantepui and Puna provinces. Track, parsimony and cladistic biogeographical analyses based on insect and plant taxa (Morrone, 1994a,b;

Posadas et al., 1997) suggested that the Páramo province was closely related to the Puna province.

Regionalization

Hernández Camacho et al. (1992b) identified 43 districts in the Colombian portion of this province. I provide in this chapter some synonymies recognizing only 24 Colombian districts based on van der Hammen (1997) and unpublished analyses (Carlos Jiménez-Rivillas, Juan Daza and Mario Alberto Quijano, pers. comm.). Delimitation of Páramo districts in Colombia is still under study, and future studies should identify additional districts for the páramos of Ecuador and Venezuela.

Alto Cauca Highland district

Alto Cauca Highland district—Hernández Camacho et al., 1992b: 111 (regionalization); Morrone, 2014a: 84 (regionalization), 2017: 221 (regionalization).

Central Cordillera Southeastern Subandean district—Hernández Camacho et al., 1992b: 110 (regionalization); Morrone, 2014a: 85 (regionalization), 2017: 222 (regionalization), **syn. nov.**

Definition: Central cordillera, Colombia (Hernández Camacho et al., 1992b). Alto Cauca Highland district corresponds to the Colombian Massif nucleus (van der Hammen, 1997) and belongs to the hydrographic basin of the Magdalena-Cauca rivers.

Alto Patía district

Alto Patía district—Hernández Camacho et al., 1992b: 147 (regionalization); Morrone, 2014a: 84 (regionalization), 2017: 221 (regionalization).

Alto Patía Subandean district—Hernández Camacho et al., 1992b: 110 (regionalization); Morrone, 2014a: 84 (regionalization), 2017: 221 (regionalization), **syn. nov.**

Eastern Nariño Andean Forests district—Hernández Camacho et al., 1992b: 110 (regionalization); Morrone, 2014a: 85 (regionalization), 2017: 222 (regionalization), **syn. nov.**

Nariño-Putumayo Páramos district—Hernández Camacho et al., 1992a: 110 (regionalization); Morrone, 2014a: 85 (regionalization), 2017: 222 (regionalization), **syn. nov.**

Western Nariño Andean Forests district—Hernández Camacho et al., 1992b: 110 (regionalization); Morrone, 2014a: 86 (regionalization), 2017: 222 (regionalization), **syn. nov.**

Definition: Central cordillera, Colombia (Hernández Camacho et al., 1992b). It corresponds to the Troncal Sur of the Central Cordillera (van der Hammen, 1997).

Andalucía district

Andalucía district—Hernández Camacho et al., 1992b: 144 (regionalization); Morrone, 2014a: 84 (regionalization), 2017: 221 (regionalization).

Definition: Eastern cordillera, Colombia (Hernández Camacho et al., 1992b).

Awa district

Awa district—Hernández Camacho et al., 1992b: 110 (regionalization); Morrone, 2014a: 84 (regionalization), 2017: 221 (regionalization).

Definition: Western cordillera, southern Colombia (Hernández Camacho et al., 1992b).

Cañón Chicamocha district

Cañón Chicamocha district—Hernández Camacho et al., 1992b: 143 (regionalization); Morrone, 2014a: 84 (regionalization), 2017: 221 (regionalization).

Definition: Eastern cordillera, northern Colombia (Hernández Camacho et al., 1992b).

Cañón del Cauca district

Cañón del Cauca district—Hernández Camacho et al., 1992b: 111 (regionalization); Morrone, 2014a: 84 (regionalization), 2017: 221 (regionalization).

Definition: Northern part of the Western cordillera, western Colombia (Hernández Camacho et al., 1992b).

Catatumbo Mountains Forest district

Catatumbo Mountains district—Hernández Camacho et al., 1992b: 110 (regionalization); Morrone, 2014a: 85 (regionalization), 2017: 222 (regionalization).

Definition: Northern part of the Eastern cordillera, Colombia (Hernández Camacho et al., 1992b). Catatumbo Mountains Forest district corresponds to the Tamá-Santander and northern Santander areas (van der Hammen, 1997).

Cauca and Valle Western Cordillera Andean Forest district

Cauca and Valle Western Cordillera Andean Forest district—Hernández Camacho et al., 1992b: 149 (regionalization); Morrone, 2014a: 84 (regionalization), 2017: 221 (regionalization).

Cauca-Valle Cordillera Subandean Forests district—Hernández Camacho et al., 1992b: 111 (regionalization); Morrone, 2014a: 85 (regionalization), 2017: 222 (regionalization), **syn. nov.**

Definition: Western Cordillera, northwestern Colombia (Hernández Camacho et al., 1992b).

Cauca Pacific Slope Subandean Forest district

Cauca Pacific Slope Subandean Forest district—Hernández Camacho et al., 1992b: 149 (regionalization); Morrone, 2014a: 84 (regionalization), 2017: 222 (regionalization).

Definition: Western Cordillera, northwestern Colombia (Hernández Camacho et al., 1992b).

Eastern Andean district

Eastern Andean district—Hernández Camacho et al., 1992b: 110 (regionalization); Morrone, 2014a: 85 (regionalization), 2017: 222 (regionalization).

Definition: Eastern Cordillera, northern Colombia (Hernández Camacho et al., 1992b).

Eastern Cordillera Páramos district

Eastern Cordillera Páramos district—Hernández Camacho et al., 1992b: 110 (regionalization); Morrone, 2014a: 85 (regionalization), 2017: 222 (regionalization).

Eastern Cordillera Cloud Forests district—Hernández Camacho et al., 1992b: 144 (regionalization); Morrone, 2014a: 85 (regionalization), 2017: 222 (regionalization), **syn. nov.**

Páramos de la Cordillera Oriental sector (in part) van der Hammen, 1997: 20, **syn. nov.**

Definition: Central portion of the Eastern Cordillera, central Colombia (Hernández Camacho et al., 1992b).

Farallones de Cali district

Farallones de Cali district—Hernández Camacho et al., 1992b: 110 (regionalization); van der Hammen, 1997: 20 (regionalization); Morrone, 2014a: 85 (regionalization), 2017: 222 (regionalization).

Cerro Calima district (in part)—van der Hammen, 1997: 20, **syn. nov.**

Farallones de Cali district (in part)—van der Hammen, 1997: 20, **syn. nov.**

Definition: Eastern Cordillera, northern Colombia (Hernández Camacho et al., 1992b).

Frontino district

Frontino district—Hernández Camacho et al., 1992b: 110 (regionalization); Morrone, 2014a: 85 (regionalization), 2017: 222 (regionalization).

Citará district—Hernández Camacho et al., 1992b: 110 (regionalization); Morrone, 2014a: 85 (regionalization), 2017: 222 (regionalization), **syn. nov.**

Páramos Citará-Tatamá district (in part)—van der Hammen, 1997: 20 (regionalization), **syn. nov.**

Páramos Paramillo-Frontino district (in part) van der Hammen, 1997: 20 (regionalization), **syn. nov.**

Definition: Western Cordillera, western Colombia (Hernández Camacho et al., 1992b).

Paramillo del Sinú district

Paramillo del Sinú district—Hernández Camacho et al., 1992b: 149 (regionalization); Morrone, 2014a: 85 (regionalization), 2017: 222 (regionalization).

Dabeiba district—Hernández Camacho et al., 1992b: 110 (regionalization); Morrone, 2014a: 85 (regionalization), 2017: 222 (regionalization), **syn. nov.**

Páramos Paramillo-Frontino district (in part)—van der Hammen, 1997: 20, **syn. nov.**

Definition: Western Cordillera, western Colombia (Hernández Camacho et al., 1992b).

Páramos Huila-Tolima district

Cauca-Huila Eastern Subandean Forests district—Hernández Camacho et al., 1992b: 110 (regionalization); Morrone, 2014a: 85 (regionalization), 2017: 222 (regionalization), **syn. nov.**

Cauca-Huila-Valle-Tolima Andean Forests district—Hernández Camacho et al., 1992b: 110 (regionalization); Morrone, 2014a: 85 (regionalization), 2017: 222 (regionalization), **syn. nov.**

Cauca-Huila-Valle-Tolima Páramos district—Hernández Camacho et al., 1992b: 110 (regionalization); Morrone, 2014a: 85 (regionalization), 2017: 222 (regionalization), **syn. nov.**

Páramos Huila-Tolima district—van der Hammen, 1997: 20 (regionalization).

Definition: Central Cordillera, Colombia (Hernández Camacho et al., 1992b).

Perijá district

Perijá district—Hernández Camacho et al., 1992b: 139 (regionalization); Morrone, 2014a: 85 (regionalization), 2017: 222 (regionalization).

Perijá Páramos district—Hernández Camacho et al., 1992b: 110 (regionalization); van der Hammen, 1997: 20 (regionalization); Morrone, 2014a: 86 (regionalization), 2017: 222 (regionalization), **syn. nov.**

Southern Perijá district—Hernández Camacho et al., 1992b: 110 (regionalization); Morrone, 2014a: 86 (regionalization), 2017: 222 (regionalization), **syn. nov.**

Definition: Eastern Cordillera, northern Colombia and northwestern Venezuela (Hernández Camacho et al., 1992b).

Quindío Páramo district

Quindío Páramo district—Hernández Camacho et al., 1992b: 148 (regionalization); van der Hammen, 1997: 20 (regionalization); Morrone, 2014a: 86 (regionalization), 2017: 222 (regionalization).

Quindío Andean Forests district—Hernández Camacho et al., 1992b: 110 (regionalization); Morrone, 2014a: 86 (regionalization), 2017: 222 (regionalization), **syn. nov.**

Quindío-Antioquia Central Cordillera Subandean Forests district—Hernández Camacho et al., 1992b: 148 (regionalization); Morrone, 2014a: 86 (regionalization), 2017: 222 (regionalization), **syn. nov.**

Definition: Central Cordillera, Colombia (Hernández Camacho et al., 1992b).

San Agustín district

San Agustín district—Hernández Camacho et al., 1992b: 144 (regionalization); Morrone, 2014a: 86 (regionalization), 2017: 222 (regionalization).

Definition: Eastern Cordillera, Colombia (Hernández Camacho et al., 1992b).

San Juan Cloud Forest district

San Juan Cloud Forest district—Hernández Camacho et al., 1992b: 149 (regionalization); Morrone, 2014a: 86 (regionalization), 2017: 222 (regionalization).

Definition: Western Cordillera, western Colombia (Hernández Camacho et al., 1992b).

San Lucas Mountains district

San Lucas Mountains district—Hernández Camacho et al., 1992b: 147 (regionalization); Morrone, 2014a: 86 (regionalization), 2017: 222 (regionalization).

Definition: Western Cordillera, western Colombia (Hernández Camacho et al., 1992b).

Sierra Nevada district

Sierra Nevada subprovince—Müller, 1973: 20 (regionalization).

Santa Marta Páramo ecoregion—Dinerstein et al., 1995: 101 (ecoregionalization), **syn. nov.**

Páramos de Santa Marta district—van der Hammen, 1997: 20 (ecoregionalization), **syn. nov.**

Sierra Nevada de Santa Marta ecoregion—Carbonó, 2017: 1 (ecoregionalization), **syn. nov.**

Definition: Sierra Nevada de Santa Marta, in the western cordillera, northern Colombia (Hernández Camacho et al., 1992b). Sierra Nevada district is an isolated mountain that rises to 5775 m (Carbonó, 2017).

Endemic taxa: *Raouliopsis* (Asteraceae) and *Obtegomeria* (Lamiaceae; Carbonó, 2017).

Tachira district

Tachira district—Hernández Camacho et al., 1992b: 142 (regionalization); Morrone, 2014a: 86 (regionalization), 2017: 222 (regionalization).

Definition: Eastern Cordillera, northern Colombia and northwestern Venezuela (Hernández Camacho et al., 1992b).

Tolima district

Tolima district—Hernández Camacho et al., 1992b: 145 (regionalization); Morrone, 2014a: 86 (regionalization), 2017: 222 (regionalization).

Tolima Central Cordillera Subandean Forests district—Hernández Camacho et al., 1992b: 110 (regionalization); Morrone, 2014a: 86 (regionalization), 2017: 222 (regionalization), **syn. nov.**

Definition: Western Cordillera, central Colombia (Hernández Camacho et al., 1992b).

Western Cordillera Northern Andean Forests district

Western Cordillera Northern Andean Forests district—Hernández Camacho et al., 1992b: 149 (regionalization); Morrone, 2014a: 86 (regionalization), 2017: 222 (regionalization).

Western Cordillera Eastern Subandean Forests district—Hernández Camacho et al., 1992b: 110 (regionalization); Morrone, 2014a: 86 (regionalization), 2017: 222 (regionalization), **syn. nov.**

Western Cordillera Northern Subandean Forests district—Hernández Camacho et al., 1992b: 149 (regionalization); Morrone, 2014a: 86 (regionalization), 2017: 222 (regionalization), **syn. nov.**

Definition: Western Cordillera, western Colombia (Hernández Camacho et al., 1992b).

Cenocrons

Studies postulated that the biota of the Páramo province derived from different biotic sources (Cleef, 1978, 1981; Vuilleumier, 1986; Sklenář et al., 2011). The most probable scenario suggested the existence of an initial prepáramo flora developing on savanna-like hilltops from Neotropical elements already adapted to a climate with marked pluvial seasonality and/or special edaphic conditions. During or shortly after the final upheaval of the northern Andes, Subantarctic elements that migrated northward along the Andes and by Nearctic elements which crossed the Panama Isthmus southwards since ca. 3 m.y.a., finding temperate-like conditions in a tropical setting (Londoño et al., 2014), gradually invaded the early páramo vegetation or proto-páramo (typical páramo taxa, but floristically poorer than present-day paramo). Additionally, there are numerous Holarctic plant genera, such as *Alnus, Draba, Lupinus, Quercus, Salix, Sambucus, Valeriana* and *Viburnum*, that may have arrived before the existence of the Panama Isthmus and after the uplift of the northern Andes (Antonelli et al., 2009). During the Pleistocene, conditions exposed páramo vegetation to at least 20 events of glacial and interglacial periods from the recurrent upward and downward displacement of the Andean vegetation zones (van der Hammen, 1974).

Sklenář et al. (2011) analyzed distributional and phylogenetic evidence available on several plant genera, finding that half of the Páramo species derived through dispersal from temperate areas, being those from the southern Andes the initial source of immigrants. Taxa of northern temperate origin show currently a higher number of species in the Páramo province, including several genera that underwent adaptive radiations in the area (*Gentianella, Draba, Valeriana, Cerastium* and *Lupinus*).

Case study: Evolutionary biogeography of the Páramo flora

Title: "Angiosperm flora and biogeography of the páramo region of Colombia, northern Andes." (Londoño, C., A. Cleef and S. Madriñán, *Flora*, 209: 81–87, 2014.)

Goal: To propose a biogeographical regionalization of the páramos of Colombia based on plant species.

Location: Colombia.

Methods: Parsimony analysis of endemicity based on plant species distributed in 34 páramo units (Figure 8.7). The study downloaded georeferences specimen data from data bases and selected local floras. The study constructed the data matrix with Mesquite 2.75 (Maddison and Maddison, 2013). The study performed the parsimony analysis with PAUP (Swofford, 2003).

Results: The parsimony analysis yielded three cladograms, which differ in the position of two of the main clades and one páramo complex. The study identified five main clades: (1) Páramos de la Cordillera Oriental province, (2) Páramos del Macizo and Cordillera Central province, (3) Páramos de Antioquia province, (4) Páramos del Norte province and (5) Páramos de la Cordillera Occidental province (Figure 8.7).

Main conclusions: The study treated the five groups of páramos found as biogeographical provinces, which are not strictly equivalent to those

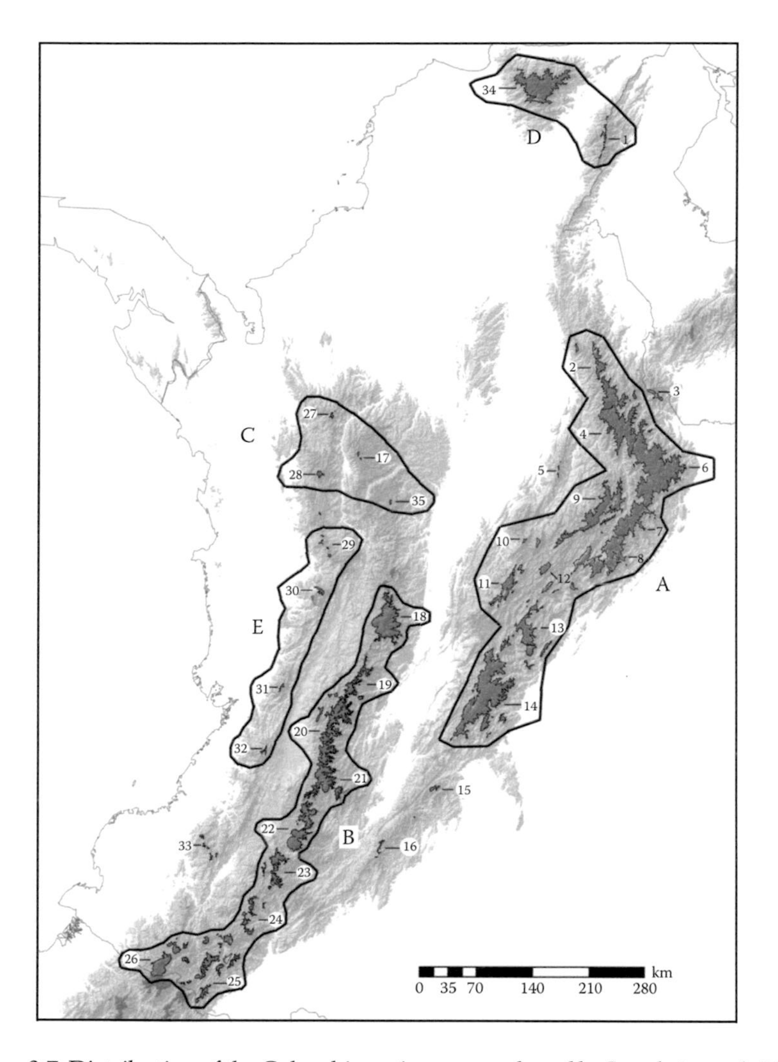

Figure 8.7 Distribution of the Colombian páramos analyzed by Londoño et al. (2014). (Modified from Londoño, C. et al., *Flora*, 209, 81–87, 2014. With permission.) (1) Perijá, (2) Jurisdicciones/Santurbán, (3) Tamá, (4) Almorzadero, (5) Yariguíes, (6) Cocuy, (7) Pisba, (8) Tota/Bijagual/Mamapacha, (9) Guantiva/Rusia, (10) Iguaque/Merchán, (11) Guerrero, (12) Rabanal and Bogotá river, (13) Chingaza, (14) Cruz Verde/Sumapaz, (15) Los Picachos, (16) Miraflores, (17) Belmira, (18) Nevados, (19) Chilí/Barragán, (20) Las Hermosas, (21) Nevado del Huila/Moras, (22) Guanacas/Puracé/Coconucos, (23) Sotará, (24) Doña Juana/Chimayoy, (25) La Cocha/Patascoy, (26) Chiles/Cumbal, (27) Paramillo, (28) Frontino/Urrao, (29) Citará, (30) Tatamá, (31) Duende, (32) Farallones de Cali, (33) Cerro Plateado, (34) Santa Marta, (A) Páramos de la Cordillera Oriental province, (B) Páramos del Macizo and Cordillera Central province, (C) Páramos de Antioquia province, (D) Páramos del Norte province and (E) Páramos de la Cordillera Occidental province.

delimited by previous studies (e.g., van der Hammen, 1997). van der Hammen (1997) originally recognized the Páramos de la Cordillera Oriental, Páramos de la Cordillera Central and Macizo Colombiano and Páramos de la Cordillera Occidental. The results of this analysis proposed the Páramos de Antioquia and Páramos del Norte provinces as new biogeographical units.

Desert province

Desert province—Cabrera and Willink, 1973: 89 (regionalization); Willink, 1988: 206 (regionalization); Huber and Riina, 1997: 272 (glossary); Morrone, 1999: 12 (regionalization); Donato, 2006: 422 (dispersal-vicariance analysis); Ezcurra et al., 2014: 28 (biotic evolution); Morrone, 2014a: 86 (regionalization); del Río et al., 2015: 1294 (track analysis and cladistic biogeography); Morrone, 2015b: 210 (regionalization); Daniel and Vaz-de-Mello, 2016: 1169 (track analysis and regionalization); Morrone, 2017: 222 (regionalization); Brignone et al., 2016: 327 (systematic revision).

Andean Pacific center—Müller, 1973: 100 (regionalization).

Peruvian subcenter—Müller, 1973: 101 (regionalization).

Salares zone—Artigas, 1975: map (regionalization).

Pacific Desert province—Udvardy, 1975: 41 (regionalization); Huber and Riina, 1997: 238 (glossary).

Pacific Coastal Deserts dominion—Ab'Sáber, 1977: map (climate).

Pacific Desert region—Rivas-Martínez and Tovar, 1983: 516 (regionalization).

Peruvian Arid Coastal center—Cracraft, 1985: 68 (areas of endemism).

Peruvian Pacific Desert region—Rivas-Martínez and Navarro, 1994: map (regionalization).

Desert area (in part)—Coscarón and Coscarón-Arias, 1995: 726 (areas of endemism).

Sechura Desert ecoregion—Dinerstein et al., 1995: 105 (ecoregionalization).

Pacific Desert ecoregion—Huber and Riina, 1997: 154 (glossary).

Coastal Peruvian Desert province—Morrone, 2001a: 110 (regionalization), 2001d: 4 (regionalization), 2004b: 46 (regionalization), 2006: 483 (regionalization); Vidal et al., 2009: 161 (endemicity analysis); Urtubey et al., 2010: 506 (track analysis and cladistic biogeography); Hechem et al., 2011: 46 (track analysis).

Desertic Peruvian-Ecuadorean province—Rivas-Martínez et al., 2011: 27 (regionalization).

Hyperdesertic North Peruvian province—Rivas-Martínez et al., 2011: 27 (regionalization).

Hyperdesertic Tropical Chilean-Arequipan province (in part)—Rivas-Martínez et al., 2011: 27 (regionalization).
Hyperdesertic Tropical Pacific region—Rivas-Martínez et al., 2011: 27 (regionalization).
Western Coast Desert province—Baranzelli et al., 2014: 752 (phylogeographic analysis).
Nazca province—Stonis et al., 2016: 561 (systematic revision).

Definition

The Desert province comprises a narrow strip along the Pacific Ocean coast, from northern Peru to northern Chile (Cabrera and Willink, 1973; Morrone, 2001d, 2006, 2014a; Moreira-Muñoz, 2011). Several factors maintain its extreme aridity, including the South Pacific anticyclone, the Humboldt current and a rain shadow created by the Andes (Dillon et al., 2009).

Endemic and characteristic taxa

Morrone (2014a) provided a list of endemic and characteristic taxa. Some examples included *Nolana* spp. (Solanaceae); *Notiobia moffetti* (Coleoptera: Carabidae); *Echemoides aguilari* and *E. penai* (Araneae: Gnaphosidae); *Notiobia moffetti* (Coleoptera: Carabidae); *Galapaganus lacertosus* and *G. squamosus* (Coleoptera: Curculionidae); *Tropidurus peruvianus* (Squamata: Iguanidae); *Amorphochilus schnablii* (Chiroptera: Furipteridae); *Conepatus rex inca* (Figure 8.8; Carnivora: Mephitidae); *Atlapetes nationi* (Passeriformes: Passeridae); and *Rhodopis* (Apodiformes: Trochilidae).

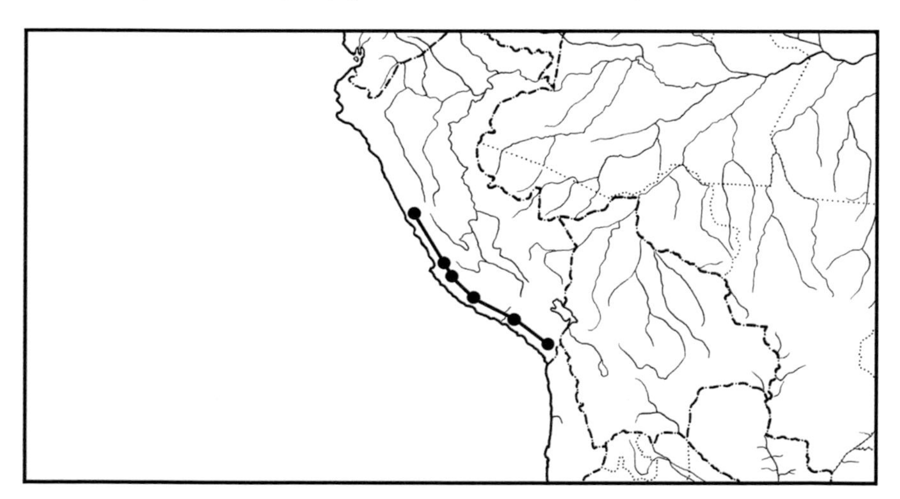

Figure 8.8 Map with the individual track of *Conepatus rex inca* (Carnivora: Mephitidae) in the desert province.

Vegetation

Scarce, permanent vegetation is abundant only near the rivers and the sea; between 1500 and 3000 m tree-like cacti are abundant, among them shrubs and herbs grow when it rains (Cabrera and Willink, 1973). Dominant plant species include *Caesalpinia tinctoria, Diplostephium tacorense, Franseria fruticosa, Inga feuillei, Kageneckia lanceolata, Lemaireocereus cartwrightianus, Paspalum vaginatum, Prosopis chilensis, P. limensis, Salicornia ambigua, Schinus areira, Tillandsia latifolia, T. purpurea, T. straminea* and *Trichocereus peruvianus* (Cabrera and Willink, 1973; Rivas-Martínez and Tovar, 1983; Rivas-Martínez et al., 2011).

Biotic relationships

Rivas-Martínez and Tovar (1983) considered that the Desert province was related to the Chaco province (Chacoan subregion, Neotropical region). Studies assigned the Desert province, assigned in the past to the Neotropical region, to the South American transition zone based on its close biotic links with the Puna and Páramo provinces (Fjeldså, 1992; Morrone and Urtubey, 1997; Posadas et al., 1997).

Regionalization

Cabrera and Willink (1973) have delimited two districts. In addition, I treat five units identified by Lamas (1982) in this chapter as districts.

Arequipa district, ***stat. nov.***

Arequipa unit—Lamas, 1982: 353 (regionalization).

Definition: Area along the western Andes of Arequipa, Moquegua and Tacna, extending south into northern Chile (Lamas, 1982).

Endemic taxon: *Teriocolias zelia kuscheli* (Lepidoptera: Pieridae; Lamas, 1982).

Callao district, ***stat. nov.***

Callao unit—Lamas, 1982: 352 (regionalization).

Definition: This district consists of a narrow strip at the foot of the western Andes, presumably reaching from southern La Libertad in the north to the Pisco valley in the south (Lamas, 1982).

Endemic taxa: *Junonia evarete lima* (Lepidoptera: Nymphalidae); and *Eurema nise stygma* and *Phoebis argante chincha* (Lepidoptera: Pieridae; Lamas, 1982).

Cardonales district

Cardonales district—Cabrera and Willink, 1973: 91 (regionalization); Morrone, 2014a: 87 (regionalization), 2017: 224 (regionalization).

Desertic Peruvian-Ecuadorean province—Rivas-Martínez et al., 2011: 27 (regionalization).

Definition: This district corresponds to the western slopes of the Andes, between 1500 and 3000 m (Cabrera and Willink, 1973; Rivas-Martínez et al., 2011).

Coastal Desert district

Coastal Desert district—Cabrera and Willink, 1973: 91 (regionalization); Morrone, 2014a: 87 (regionalization), 2017: 224 (regionalization).

Definition: This district corresponds to the driest portion of the province, situated in the coast (Cabrera and Willink, 1973).

Mollendo district, ***stat. nov.***

Mollendo unit—Lamas, 1982: 353 (regionalization).

Definition: This district covers from about the Río Grande (Ica) in the north to the northern limit of the Atacama desert in Chile (Lamas, 1982).

Endemic taxon: *Danaus plexippus erippus* (Lepidoptera: Nymphalidae; Lamas, 1982).

Porculla district, ***stat. nov.***

Porculla unit—Lamas, 1982: 353 (regionalization).

Definition: This district includes the western slopes of the northern Andes, from southwestern Ecuador to the Santa valley in Ancash (Lamas, 1982).

Endemic taxa: *Danaus gilippus nivosus* (Lepidoptera: Nymphalidae); *Battus polydanas streckerianus* (Lepidoptera: Papilionidae); and *Eurema nigrocincta* and *Teriocolias zelia mathani* (Lepidoptera: Pieridae; Lamas, 1982).

Surco district, ***stat. nov.***

Surco unit—Lamas, 1982: 353 (regionalization).

Definition: This district covers a narrow strip along the western slopes of Cordillera Occidental, with its northern limit somewhere in western Ancash and possibly extending as far south as southwestern Ayacucho (Lamas, 1982).

Endemic taxa: *Parapedaliodes parepa milvia* (Lepidoptera: Nymphalidae) and *Teriocolias zelia andina* (Lepidoptera: Pieridae; Lamas, 1982).

Cenocrons

Dillon et al. (2009) analyzed the diversification of *Nolana* (Solanaceae) in the Desert province and northern Chile. Based on a molecular phylogenetic analysis, they found that the genus diverged from its sister genus 8.5 m.y.a. and the crown group, dated at 4 m.y.a., corresponded to the time when western South America was suffering increasing aridity. The study considered that the species of the genus from the Desert province derived from northern Chilean ancestors, which dispersed into the province in several episodes and then radiated there.

Case study: Diversification of the plant genus Nolana

Title: "Biogeographic diversification in *Nolana* (Solanaceae), a ubiquitous member of the Atacama and Peruvian Deserts along the western coast of South America." (Dillon, M. O., T. Tu, L. Xie, V. Quipuscoa Silvestre and J. Wen, *Journal of Systematics and Evolution*, 47: 457–476, 2009.)

Goal: To reconstruct the biogeographical diversification for *Nolana* (Solanaceae), a genus of 89 endemic species largely restricted to fog-dependent desert lomas formations of coastal Peru and Chile.

Location: Coastal desert, Peru and Chile.

Methods: The study included 70 species of *Nolana* plus two outgroups in the phylogenetic analysis. The study mapped and generated the distribution and range of each species using 1° intervals of latitude and longitude from a database of over 1200 georeferenced records from herbarium accessions. The study used three chloroplast genes (ndhF, rbcL and atpB) to estimate the divergence time of *Nolana* in a broad phylogenetic framework, including species of Solanaceae and Convolvulaceae. For each region, the study editee and aligned sequences from all primers using Sequencher vers. 4.8 (GeneCodes, Ann Arbor, Michigan). The study initially aligned all sequences for each genomic accession with ClustalX version 1.83 (Thompson et al., 1997) followed by manual adjustments using the program Se-Al v2.0a11 (Rambaut, 2007). The study performed parsimony and Bayesian analyses with the LEAFY second intron sequences and the combined data of four chloroplast markers. The study performed the parsimony analysis with PAUP 4.10b (Swofford, 2003). The study derived the phylogenetic tree for the estimation of the divergence time of *Nolana* from the combined data of three chloroplast genes. The study applied a dispersal-vicariance (DIVA) analysis (Ronquist, 1997) to infer the biogeographical diversification of *Nolana* based on a simplified most parsimonious tree generated by PAUP using a subset of the LEAFY second intron sequences, which includes one outgroup species and 14 species of *Nolana*. The study implemented DIVA analysis using DIVA version 1.1 (Ronquist, 1996).

Results: Both parsimony and Bayesian analyses produced the same cladogram. Using two fossil calibration points from Convolvulaceae and Solanaceae and one estimated age point for the crown group of Convolvulaceae and Solanaceae, the study estimated 8.48 m.y.a. as the divergence time of *Nolana* from its sister *Sclerophylax* and 4.02 m.y.a. as the crown age of *Nolana*. The study dated the divergence of the *N. sessiliflora-N. acuminata-N. baccata* node at 2.88 m.y.a. The study estimated 0.35 m.y.a. as the divergence time between *N. galapagensis* (Islas Galápagos, Ecuador) and *N. adansonii* from mainland Peru. The node for the divergence of *N. tomentella* was 2.0 m.y.a. and the divergence of *N. coelestis* and *N. werdermannii* was 1.25 m.y.a. The DIVA analysis suggested that *Nolana* had its origin in Chile (Figure 8.9) and more than one dispersal event from Chile into Peru, with subsequent radiation (clades D, F), with 18 and 14 species,

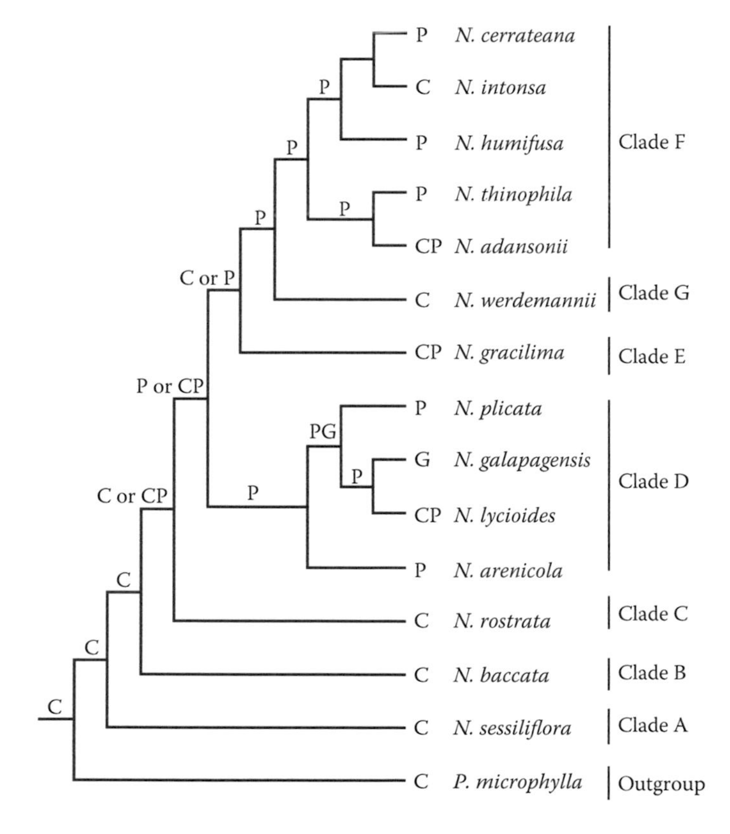

Figure 8.9 Dispersal-vicariance analysis of *Nolana* by Dillon et al. (2009). (Modified from Dillon, M.O. et al., *J. Syst. Evol.*, 47, 457–476, 2009. With permission.)

respectively. The pattern in Chile also suggested a south-to-north dispersal and radiation, beginning with clade B, clade C and ultimately clade G, the small-flowered northern Chilean group. These groups displayed a south-to-north dispersal pattern with highest diversity found around 25°25′S. The observed patterns in the DIVA analysis suggested a south-to-north pattern for *N. galapagensis,* with Peru as the geographical origin or source of Peruvian *N. lycioides,* a species perhaps preadapted to sandy habitats prior to its dispersal to the island chain.

The Bayesian divergence estimates yielded a divergence time of 0.35 m.y.a. Not all distributions followed the south to north pattern as in the case of *N. adansonii, N. gracillima* and *N. lycioides.* Similarly, both *N. intonsa* and *N. tarapacana* were Chilean endemics with obvious phylogenetic connections with the largely Peruvian clade F that included *N. lycioides, N. volcanica* and *N. cerrateana.* Studies have not resolved whether these are the product of vicariance or long-distance dispersal; however, it is evident that short-term climatic fluctuations, such as El Niño events, and longer-term climatic changes associated with glacial cycles have been influential in expansions and contractions in the floras of the Andean Cordillera. Pleistocene glacial cycles caused significant sea level fluctuations, with estimates of lowering between 120 and 230 m, which would have significantly changed the position of the shoreline in relation to that of the present day. This event would have exposed the northern Peruvian continental shelf and would have displaced plant communities, especially from 5° to 13°S. These events provided opportunities for biotic interchange of the adjacent Andean Cordillera and the coastal desert.

Main conclusions: The lomas formations of coastal Peru and Chile are dynamic areas. The continuous deserts appear partitioned with phyletic differences reflected in different histories for different regions of coastal Peru and Chile. *Nolana* stands out as the most widespread genus, with no fewer than 71 coastal endemic species found from northern Peru to southern Chile and 17 species in interior Andean sites. There appears to be a significant barrier to dispersal along the coast between 18°S and 20°S and that barrier reflects in the present-day distribution and phylogeny of *Nolana* species. Given past climate changes in northern Peru, the pattern suggests the possibility of vicariant distributions; however, one cannot rule out long-distance dispersal from the adjacent Andean Cordillera. The lomas formations in southern Peru (13°S–18°S) are home to extensive radiation, with 30 species of *Nolana* recorded from the Department of Arequipa constituting two clades, namely D and F. *Nolana galapagensis* appears to derive from a Peruvian ancestor (clade D) and may have reached the Galápagos Islands at ca. 0.35 m.y.a. Northern

Chilean endemics (clade G), Peruvian endemics (clade F) and clade E with Chilean and Peruvian species have more distant relationships with southern taxa. Although studies still debate the age of the aridity in coastal Peru and Chile, the rapid diversification in such environments appears in separate clades between 2.69 and 2.88 m.y.a. These results suggest that the crown group in *Nolana* is approximately 4.02 m.y.a. and the observed species radiations in *Nolana* are the most specious in the Peruvian and Atacama deserts. The DIVA analysis suggests a Chilean origin for the genus. The diversity in form and function in *Nolana* represents a radiation of several clades into the same geographical areas and environments, but at different times. Partitioning environments occurred with the development of different habits, leaf forms, pubescence types, ecophysiology and phenology reflected in secondary sympatry.

Puna province

Puna formation—Holmberg, 1898: 433 (regionalization).
Puna subregion—Goetsch, 1931: 2 (regionalization).
Puna district—Osgood, 1943: 27 (regionalization).
Puna province—Cabrera, 1951: 52 (regionalization), 1953: 107 (regionalization), 1957: 336 (regionalization), 1958: 200 (regionalization), 1971: 32 (regionalization); Cabrera and Willink, 1973: 87 (regionalization); Udvardy, 1975: 42 (regionalization); Cabrera, 1976: 59 (regionalization); Willink, 1988: 206 (regionalization), 1991: 138 (regionalization); Morrone, 1994a: 190 (parsimony analysis of endemicity); Huber and Riina, 1997: 271 (glossary); Posadas et al., 1997: 2 (parsimony analysis of endemicity); Morrone, 1996a: 108 (regionalization); Katinas et al., 1999: 112 (track analysis); Morrone, 2001a: 111 (regionalization), 2001d: 5 (regionalization); Ojeda et al., 2002: 24 (biotic evolution); Roig-Juñent et al., 2003: 275 (track analysis and cladistic biogeography); Morrone, 2004b: 46 (regionalization); Donato, 2006: 422 (dispersal-vicariance analysis); Morrone, 2006: 483 (regionalization); Alzate et al., 2008: 1252 (track analysis); Aagesen et al., 2009: 309 (endemicity analysis); Urtubey et al., 2010: 506 (track analysis and cladistic biogeography); Hechem et al., 2011: 46 (track analysis); Ferretti et al., 2012: 2 (parsimony analysis of endemicity and track analysis); Kutschker and Morrone, 2012: 543 (track analysis); Campos-Soldini et al., 2013: 16 (track analysis); Ferro, 2013: 324 (biotic evolution); Ezcurra et al., 2014: 28 (biotic evolution); Morrone, 2014a: 87 (regionalization); Amarilla et al., 2015: 8 (biotic evolution); Apodaca et al., 2015a: 96 (regionalization); Biganzoli and Zuloaga, 2015: 339 (floristics); del Río et al., 2015: 1294 (track analysis and cladistic biogeography); Morrone, 2015b: 210 (regionalization); Figueroa and Ratcliffe, 2016: 68 (faunistics); Monckton, 2016:

125 (systematic revision and map); Stonis et al., 2016: 561 (systematic revision); Arana et al., 2017: 421 (shapefiles); Escalante, 2017: 351 (parsimony analysis of endemicity); Martínez et al., 2017: 479 (track analysis and regionalization); Morrone, 2017: 225 (regionalization).

Northern Andean Cordillera region—Peña, 1966a: 5 (regionalization), 1966b: 213 (regionalization).

High Andean province (in part)—Cabrera, 1971: 30 (regionalization); Huber and Riina, 1997: 270 (glossary); Ojeda et al., 2002: 24 (biotic evolution); Aagesen et al., 2009: 309 (endemicity analysis).

Quechua High Andean district—Cabrera, 1971: 30 (regionalization), 1976: 52 (regionalization).

Puna center—Müller, 1973: 92 (regionalization).

Altiplanic zone—Artigas, 1975: map (regionalization).

Puna zone—Artigas, 1975: map (regionalization).

Cuyan Subandean province—Ringuelet, 1975: 107 (regionalization).

Titicaca province—Ringuelet, 1975: 107 (regionalization).

Lake Titicaca province—Udvardy, 1975: 42 (regionalization).

Punas dominion—Ab'Sáber, 1977: map (climate).

Puna region-Rivas—Rivas-Martínez and Tovar, 1983: 516 (regionalization); Huber and Riina, 1997: 284 (glossary).

Austral Andean center—Cracraft, 1985: 65 (areas of endemism).

Punan subregion—Rivas-Martínez and Navarro, 1994: map (regionalization).

Bolivian province—Rivas-Martínez and Navarro, 1994: map (regionalization).

Peruvian province—Rivas-Martínez and Navarro, 1994: map (regionalization).

Puna area—Coscarón and Coscarón-Arias, 1995: 726 (areas of endemism); Apodaca et al., 2015b: 5 (dispersal-vicariance analysis).

Central Andean Dry Puna ecoregion—Dinerstein et al., 1995: 102 (ecoregionalization); Locklin, 2017: 1 (ecoregionalization).

Central Andean Puna ecoregion—Dinerstein et al., 1995: 102 (ecoregionalization); Riveros Salcedo, 2017: 1 (ecoregionalization).

Central Andean Wet Puna ecoregion—Dinerstein et al., 1995: 102 (ecoregionalization); Riveros Salcedo and Locklin, 2017: 1 (ecoregionalization).

Puna ecoregion—Huber and Riina, 1997: 154 (glossary); Burkart et al., 1999: 11 (ecoregionalization).

Arid Puna province—Morrone, 1999: 12 (regionalization).

Central Puna province—Morrone, 1999: 12 (regionalization).

Humid Puna province—Morrone, 1999: 12 (regionalization).

Central Andes area—Porzecanski and Cracraft, 2005: 266 (parsimony analysis of endemicity).

Punean province—Donato, 2006: 422 (dispersal-vicariance analysis).
Andean Cuyan province (in part)—López et al., 2008: 1572 (parsimony analysis of endemicity and regionalization).
Aymara province López et al., 2008: 1574 (parsimony analysis of endemicity and regionalization).
Bolivian-Tucumanan province—Rivas-Martínez et al., 2011: 27 (regionalization).
Mesophytic Punenian province—Rivas-Martínez et al., 2011: 27 (regionalization).
Xerophytic Punenian province—Rivas-Martínez et al., 2011: 27 (regionalization).
Northern Andean area (in part)—Apodaca et al., 2015b: 5 (dispersal-vicariance analysis).
Puneña province—Brignone et al., 2016: 327 (systematic revision).

Definition

The Puna province corresponds to the plateau at approximately 3800–4500 m of altitude, situated from southern Peru to northwestern Argentina between 15°S and 27°S (Peña, 1966a; Cabrera and Willink, 1973; Müller, 1973; Willink, 1988; Rivas-Martínez and Navarro, 1994; Posadas et al., 1997; Burkart et al., 1999; Morrone, 2001d, 2006, 2014a; Palma et al., 2005). The climate is cold and dry. Mean annual precipitation is 150–230 mm (Palma et al., 2005), with rain falling predominately during summer (Aagesen et al., 2009). Rivas-Martínez and Tovar (1983) lumped the High Andean and Mesoandean floors together.

Endemic and characteristic taxa

Morrone (2014a) provided a list of endemic and characteristic taxa. Some examples included *Chuquiraga atacamensis* and *C. kuscheli* (Asteraceae); *Epilobium pedicellare* (Figure 8.10); *Fuchsia austromontana* and *F. tincta* species group (Onagraceae); *Chuquiraga atacamensis* (Asteraceae); *Piper carrapanum, P. edurumglaberimicaule* and *P. tardum* (Piperaceae); *Trachelopachys bidentatus* and *T. tarma* (Araneae: Clubionidae); *Aegla humahuaca, A. intercalata* and *A. jujuyana* (Decapoda: Aeglidae); *Curicta peruviana* (Heteroptera: Nepidae); *Atacamacris* and *Punacris* (Orthoptera: Tristiridae); *Ancognatha erythrodera* and *A. lutea* (Coleoptera: Scarabaeidae; Figueroa and Ratcliffe, 2016); *Pilobalia* (Coleoptera: Tenebrionidae); *Caliadurgus subandinus* (Hymenoptera: Pompilidae); *Akodon lutescens* and *Punomys* spp. (Rodentia: Cricetidae); *Grallaria andicola* (Passeriformes: Formicaridae); and *Rhea pennata garleppi* (Rheiformes: Rheidae).

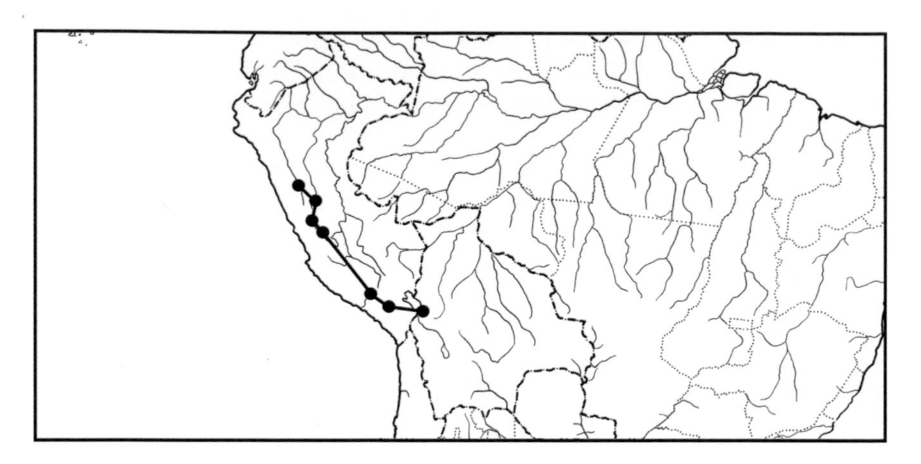

Figure 8.10 Map with the individual track of *Epilobium pedicelare* (Onagraceae) in the Puna province.

Vegetation

The Puna province consists of bush steppes, montane grasslands, low scrubland, trees and herbaceous plants (Figure 8.11; Cabrera, 1957; Cabrera and Willink, 1973; Dinerstein et al., 1995; Martínez Carretero, 1995; Roig and Martínez Carretero, 1998; Roig-Juñent et al., 2003; Palma et al., 2005). Species of Solanaceae, Fabaceae and Asteraceae mainly formed the open bush steppes, accompanied by grass species such as *Jarava leptostachya, Pennisetum chilense, Deyeuxia rigida* and *Aristida antoniana* (Aagesen et al., 2009). Steppe grasses include species of *Festuca* and tribe Stipeae (Palma et al., 2005).

Biotic relationships

Vuilleumier (1986) suggested a close relationship with the Páramo province based on bird taxa. Track, parsimony and cladistic biogeographical analyses based on insect and plant taxa (Morrone, 1994a,b; Posadas et al., 1997) supported this suggestion. A parsimony analysis of endemicity based on bird taxa (Porzecanski and Cracraft, 2005) postulated a close relationship of this province with the Guatuso-Talamanca, Pantepui and Páramo provinces. Two track analyses (Katinas et al., 1999; Mihoč et al., 2006) hypothesized that the Puna province was a node.

Regionalization

Martínez Carretero (1995) identified four districts within the Argentinean portion of this province. One may also treat the five units identified by Lamas (1982) within the Peruvian portion of this province as districts.

Figure 8.11 Vegetation typical of the Puna province, western Peru. (a) Shrubs and herbs and (b) shrubs. (Courtesy of Erick Yábar-Landa.)

Bolivian district

Bolivian district—Martínez Carretero, 1995: 30 (regionalization); Morrone, 2014a: 88 (regionalization), 2017: 227 (regionalization).

Definition: The Bolivian district covers southern Peru, southwestern Bolivia and part of Tucumán, Argentina (Martínez Carretero, 1995).

Cajamarca district, ***stat. nov.***

Cajamarca unit—Lamas, 1982: 355 (regionalization).

Cajamarca district—Morrone, 2017: 227 (regionalization).

Definition: The Cajamarca district covers southern Cajamarca, between the Huancabamba Depression to the north, the upper Marañon to the east and the Cordillera Pelagatos (La Libertad) to the south (Lamas, 1982).

Endemic taxa: *Mutisia viciaefolia* (Asteraceae), *Astragalus uniflorus* (Fabaceae) and *Puya raimondi* (Bromeliaceae; Martínez Carretero, 1995).

Central district

Central district—Martínez Carretero, 1995: 30 (regionalization); Morrone, 2014a: 88 (regionalization), 2017: 227 (regionalization).

Definition: The Central district covers La Rioja and Catamarca, Argentina (Martínez Carretero, 1995).

Endemic taxa: *Ephedra multiflora* (Ephedraceae), *Chuquiraga ruscifolia* (Asteraceae), *Nicotiana longibracteata* (Solanaceae) and *Fabiana bryoides* (Solanaceae; Martínez Carretero, 1995).

Cuyan district

Cuyan district—Martínez Carretero, 1995: 30 (regionalization); Morrone, 2014a: 88 (regionalization), 2017: 227 (regionalization).

Definition: The Cuyan district covers the southern part of the province, in northwestern Mendoza and southwestern San Juan, Argentina (Martínez Carretero, 1995).

Endemic taxa: *Stipa humilis* var. *ruiziana* (Poaceae), *Chuquiraga echegarayi* and *Senecio uspallatensis* (Asteraceae) and *Junellia erinacea* (Verbenaceae; Martínez Carretero, 1995).

*Huancaspata district, **stat. nov.***

Huancaspata unit—Lamas, 1982: 355 (regionalization).

Huancaspata district—Morrone, 2017: 227 (regionalization).

Definition: The Huancaspata district covers the high mountains between the upper Marañon and the upper Huallaga, with its northern limit lying in the border between Amazonas and La Libertad, and its southern limit could be in northwestern Huánuco (Lamas, 1982).

Jujuyan district

Jujuyan district—Martínez Carretero, 1995: 30 (regionalization); Morrone, 2014a: 88 (regionalization), 2017: 227 (regionalization).

Definition: The Jujuyan district covers the western parts of Jujuy and Salta (Argentina) and northern Chile (Martínez Carretero, 1995).

Endemic taxa: *Mutisia saltensis* (Asteraceae), *Stipa arcuata* (Poaceae) and *Opuntia* cfr. *tilcarensis* (Cactaceae; Martínez Carretero, 1995).

*Pasco district, **stat. nov.***

Pasco unit—Lamas, 1982: 355 (regionalization).

Pasco district—Morrone, 2017: 227 (regionalization).

Definition: The Pasco district covers the Puna formation of La Libertad, Ancash, Lima, Pasco, Junín, Huancavelica and northern Ayacucho;

in the north it may reach the Cordillera Pelagatos and its southern limits appears to be at the border between Huancavelica and Ayacucho (Lamas, 1982).

Endemic taxa: *Phulia garleppi* and *P. nannophyes* (Lepidoptera: Pieridae); and *Punargentus lamna* and *Yramea cora* (Lepidoptera: Nymphalidae; Lamas, 1982).

Shimbe district, ***stat. nov.***

Shimbe unit—Lamas, 1982: 355 (regionalization).

Shimbe district—Morrone, 2017: 227 (regionalization).

Definition: The Shimbe district covers the border between Piura and northern Cajamarca, over 3500 m (Lamas, 1982).

Atacama province

Northern Plateaus and Cordilleras region—Reiche, 1905: map (regionalization).

Atacama region—Goetsch, 1931: 1 (regionalization); Urbina-Casanova et al., 2015: 273 (floristics).

Atacaman district—Osgood, 1943: 27 (regionalization).

Desertic zone—Mann, 1960: 16 (regionalization).

High Plateau region—Peña, 1966a: 4 (regionalization), 1966b: 212 (regionalization).

Atacaman region—O'Brien, 1971: 199 (regionalization).

Coquimban district (in part)—Cabrera and Willink, 1973: 91 (regionalization); Huber and Riina, 1997: 146 (glossary).

Chilean subcenter—Müller, 1973: 101 (regionalization).

Atacaman area (in part)—Artigas, 1975: map (regionalization).

Argentinean-Atacaman province (in part)—Rivas-Martínez and Navarro, 1994: map (regionalization).

Atacama Hyperdesert province—Rivas-Martínez and Navarro, 1994: map (regionalization).

Desert area (in part)—Coscarón and Coscarón-Arias, 1995: 726 (areas of endemism).

Atacama Desert ecoregion—Dinerstein et al., 1995: 105 (ecoregionalization); Huber and Riina, 1997: 141 (glossary); Armstrong, 2017: 1 (ecoregionalization).

Atacama province—Morrone, 1999: 12 (regionalization), 2001a: 112 (regionalization), 2001d: 6 (regionalization), 2004b: 46 (regionalization), 2006: 483 (regionalization); Vidal et al., 2009: 161 (endemicity analysis); Urtubey et al., 2010: 506 (track analysis and cladistic biogeography); Hechem et al., 2011: 46 (track analysis); Alfaro et al., 2013: 236 (faunistics); Morrone, 2014a: 88 (regionalization); del Río et al., 2015: 1294 (track analysis and cladistic biogeography); Leivas et al., 2015:

116 (faunistics); Morrone, 2015b: 210 (regionalization); Stonis et al., 2016: 561 (systematic revision); Morrone, 2017: 227 (regionalization).

Atacama Desert region—Roig-Juñent and Domínguez, 2001: 557 (faunistics).

Subtropical Pacific area (in part)—Porzecanski and Cracraft, 2005: 266 (parsimony analysis of endemicity).

Atacama ecoregion—Abell et al., 2008: 408 (ecoregionalization).

Hyperdesertic Tropical Chilean-Arequipan province (in part)—Rivas-Martínez et al., 2011: 27 (regionalization).

Atacama Desert province—Baranzelli et al., 2014: 752 (phylogeographic analysis).

Definition

The Atacama province is situated in northern Chile, between 18° and 26°S (Peña, 1966a; O'Brien, 1971; Cabrera and Willink, 1973; Müller, 1973; Artigas, 1975; Rivas-Martínez and Navarro, 1994; Dinerstein et al., 1995; Morrone, 2001d, 2006, 2014a; Luebert, 2011). Within it, studies identified three physiographic units: the Cordillera de la Costa, which constitutes a narrow coastal plain; the Central Depression, which is a relatively flat area of *ca.* 200 km wide; and the Andean Cordillera, which can reach 4000 m elevation (Cavieres et al., 2002). Studies characterized this province as arid (<50 mm/year) to hyperarid (<5 mm/year) climates (Guerrero et al., 2013; Armstrong, 2017). Some of the most diverse communities of the Atacama province occur in *lomas* supported by winter fogs, which form over cool Pacific Ocean currents (Dinerstein et al., 1995).

Endemic and characteristic taxa

Morrone (2014a) provided a list of endemic and characteristic taxa. Some examples included *Cleome chilensis* (Cleomaceae); *Heliotropium* sect. *Cochranea* (Heliotropiaceae); *Malesherbia arequipana* and *M. tocopillana* (Passifloraceae); *Zephyra* (Tecophilaeaceae); *Chileotracha* (Solifugae: Ammotrechidae); *Chilehexops platnicki* (Araneae: Dipluridae); Elasmoderini (Orthoptera: Tristiridae); *Polymerus atacamensis* and *Rhinacloa azapa* (Heteroptera: Miridae); *Chilicola lickana* (Hymenoptera: Colletidae; Monckton, 2016); *Omalodes atacamanus* (Coleoptera: Histeridae; Leivas et al., 2015); *Listroderes robustior* (Figure 8.12; Coleoptera: Curculionidae); *Gigantodax cortesi*, *Simulium hectorvargasi* and *S. tenuipes* (Diptera: Simuliidae); and *Liolaemus atacamensis* and *L. poconchilensis* (Squamata: Iguanidae). Roig-Juñent and Domínguez (2001) considered that this province harbored a high endemicity (33.7%) of species of Carabidae (Coleoptera).

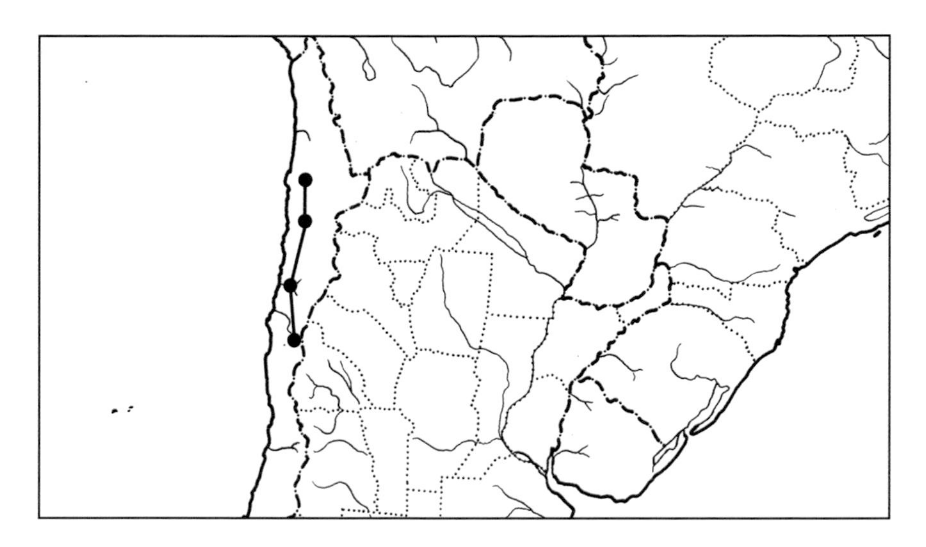

Figure 8.12 Map with the individual track of *Listroderes robustior* (Coleoptera: Curculionidae) in the Atacama province.

Vegetation

The Atacama province consists of desert areas with scarce vegetation, coastal and desert scrubland (Figures 8.13 and 8.14). There are rich communities in the *lomas*, due to the fog that the currents of the Pacific Ocean form during winter (Dinerstein et al., 1995; Luebert and Pliscoff, 2006). Some species rich taxa include *Copiapoa* (Cactaceae), *Heliotropium* sect. *Cochranea* (Heliotropiaceae), *Nolana* and *Solanum* sect. *Regmandra* (Solanaceae) and *Oxalis* sect. *Carnosae* (Oxalidaceae; Luebert, 2011). Some of the most abundant species in the coastal desert are *Adesmia argentea, Atriplex clivicola, Berberis littoralis, Bulnesia chilensis, Croton chilensis, Ephedra breana, Eulychnia iquiquensis, Euphorbia lactiflua, Frankenia chilensis, Gypothamnium pinifolium, Heliotropium eremogenum, H. floridum, H. pychnophyllum, H. stenophyllum, Nolana adansonii, N. lycioides, N. tocopillensis* and *Oxalis gigantea*. In the desert scrub the most abundant species are *Atriplex atacamensis, A. deserticola, Flourensia thurifera, Gymnophyton foliosum, Heliotropium stenophyllum, Huidobria chilensis, H. fruticosa, Monttea chilensis, Nolana leptophylla, Oxyphyllum ulicinum* and *Tessaria absinthioides* (Luebert and Pliscoff, 2006; Armstrong, 2017). In the interior part of the province there are stands of natural and human-induced forests of *Prosopis tamarugo* (Moreira-Muñoz, 2011).

Figure 8.13 Vegetation of the Atacama province, 17 km east of Caldera, northern Chile. (Courtesy of Mario Elgueta.)

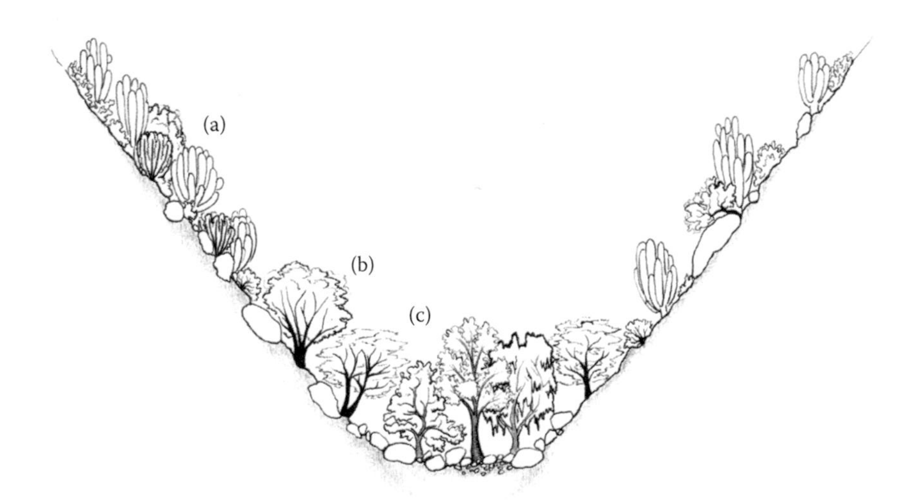

Figure 8.14 Vegetation in the scrub desert of the Atacama province. (Modified from Luebert, F. and P. Pliscoff, *Sinopsis bioclimática y vegetacional de Chile.* Editorial Universitaria, Santiago de Chile, 2006. With permission.) (a) Scrub, (b) xeric forest with *Acacia caven* and (c) ravine forest with *Salix humboldtiana.*

Biotic relationships

O'Brien (1971) considered that the entomofauna of the Atacama province was related to the entomofauna of the Coquimban province (Central Chilean subregion).

Regionalization

Peña (1966a,b), Di Castri (1968) and Artigas (1975) identified five nested units, which Morrone (2014a, 2017) treated as districts. They based their preliminary delimitation on the map of Artigas (1975). I assign the Desventuradas Archipelago in this chapter formally to this province and treat it as a sixth district following Teillier and Vilina (2017).

*Desventuradas Archipelago district, **stat. nov.***

Fernandezian region (in part)—Takhtajan, 1986: 252 (regionalization); Cox, 2001: 519 (regionalization).

Desventuradas Archipelago ecoregion—Teillier and Vilina, 2017: 1 (ecoregionalization).

Definition: This district covers the islands of San Félix (26° 17′ S, 80° 5′ W) and San Ambrosio (26° 21′ S, 79° 53′ W), situated 850 km of continental Chile and 780 km north of the Juan Fernández islands (Kuschel, 1961, 1962; Moreira-Muñoz, 2011; Teillier and Vilina, 2017). San Ambrosio island is larger, measuring roughly 4 km long by 1 km wide (Kuschel, 1961). Both islands are of relatively recent volcanic origin.

Endemic taxa: *Lycapsus* and *Thamnoseris* (Asteraceae); *Sanctambrosia* (Cariophyllaceae); and *Nesocaryum* (Boraginaceae; Moreira-Muñoz, 2011; Teillier and Vilina, 2017). Marticorena (1990) reported a 60.6% of endemic plant species. Urbina-Casanova et al. (2015) considered that 86.9% of the plant taxa are endemic to the islands.

Vegetation: The district consists of sclerophyllous, low scrubland. Most frequent plant species include *Eragrostis peruviana* (Poaceae); *Atriplex chapini* (Amaranthaceae); *Thamnoseris lacerata, Cotula coronopifolia* and *Sonchus asper* (Asteraceae); *Sanctambrosia manicata* (Cariophyllaceae); *Cristaria insularis* (Malvaceae); and *Chenopodium murale* (Chenopodiaceae; Kuschel, 1962; Teillier and Vilina, 2017).

Interior Desert district

Northern Desert region (in part)—Peña, 1966a: 7 (regionalization), 1966b: 214 (regionalization).

Interior Desert region—Di Castri, 1968: 18 (regionalization).

Mediterranean Desertic zone—Artigas, 1975: 20 (regionalization).

Interior Desert district—Morrone, 2014a: 89 (regionalization); Monckton, 2016: 125 (systematic revision and map); Morrone, 2017: 230 (regionalization).

Definition: The Interior Desert district occupies most of Chile between 18°S and 26°S, west of the Andean Cordillera, never reaching 2000 m of altitude (Peña, 1966a).

Vegetation: Luebert and Pliscoff (2006) characterized the interior tropical desert as having very scarce vegetation, except in the places with subterranean aquifers with salty water where scrubs of *Tessaria absinthioides* (Asteraceae) grew. Other vegetation type identified by Luebert and Pliscoff (2006) is the interior tropical desert scrubland dominated by *Nolana leptophylla* (Solanaceae) and *Cistanthe salsoloides* (Montiaceae); other species are *Adesmia atacamensis, Argylia glutinosa, Atriplex deserticola, Cryptantha werdermanniana, Heliotropium chenopodiaceum, Nolana flaccida, Reyesia parviflora* and *Skytanthus acutus.*

Northern Andean district

Northern Andean zone—Artigas, 1975: 20 (regionalization).

Northern Andean district—Morrone, 2014a: 89 (regionalization); Monckton, 2016: 125 (systematic revision and map); Morrone, 2017: 230 (regionalization).

Definition: The North Andean district extends between 18° and 25° 30′ S, from 2000 to 4000 m of altitude, parallel to the previous district (Artigas, 1975).

Vegetation: Within this district, Luebert and Pliscoff (2006) identified three main vegetational types. In the rocky soils of the Andean Cordillera of Tarapacá, *Parastrephia lucida* (Asteraceae) and *Azorella compacta* (Apiaceae) dominate the scrubland; other species are *Adesmia melanthes, Baccharis incarum, Caiophora rahmeri, Festuca orthophylla, Opuntia ignescens, Parastrephia quadrangularis, Senecio nutans* and *Werneria aretioides.* Between 4000 and 4400 m of Tarapacá, *Parastrephia lucida* (Asteraceae) and *Festuca orthophylla* (Poaceae) dominate the scrubland alternating with grasses; other species are *Adesmia patancana, Baccharis incarum, Deyeuxia breviaristata, Nototriche argentea, N. turritella, Perezia ciliosa, Senecio humillimus* and *Werneria glaberrima.* In cordilleran areas of Antofagasta, Tarapacá and northern Atacama, between 4200 and 4900 m, *Adesmia melanthes* (Fabaceae), *Artemisia copa* (Asteraceae) and *Stipa frigida* (Poaceae) dominate the low scrubland; other species are *Baccharis incarum, Cristaria andicola, Descurainia stricta, Fabiana denudata, Haplopappus rigidus, Hoffmanseggia eremophila, Sisymbrium philippianum* and *Stipa frigida.*

Northern Coast district

Northern Littoral region—Reiche, 1905: map (regionalization).

Northern Coast region—Peña, 1966a: 8 (regionalization), 1966b: 214 (regionalization).

Littoral Desert region—Di Castri, 1968: 18 (regionalization).

Littoral Desertic zone—Artigas, 1975: 20 (regionalization).

Northern Coast district—Morrone, 2014a: 89 (regionalization); Monckton, 2016: 125 (systematic revision and map); Morrone, 2017: 230 (regionalization).

Definition: The Northern Coast district covers the narrow littoral area between 20° 30′ to near 25°S, with vegetation supported by the moisture from the fog (Peña, 1966a).

Vegetation: Within this district, Luebert and Pliscoff (2006) identified three main types of tropical desert scrublands. In the low coastal areas (between sea level and 400 m) of Antofagasta and southern Tarapacá, *Nolana adansoni* and *N. lycioides* (Solanaceae) are dominant; other species are *Alona stenophylla, Malesherbia tocopillana, Nolana peruviana* and *Solanum chilense*. Between 400 and 1200 m in the coast of Tarapacá and northern Antofagasta, dominant plants are *Ephedra breana* (Ephedraceae), *Solanum chilense* (Solanaceae) and *Eulychnia iquiquensis* (Cactaceae); other species are *Frankenia chilensis, Nolana sedifolia, Camassia biflora, Oxalis bulbocastanum* and *Leucoryne appendiculata*. Between sea level and 300 m in the coast of central Antofagasta, dominant plants are *Copiapoa boliviana* (Cactaceae), *Heliotropium pycnophyllum* (Boraginaceae) and *Nolana peruviana* (Solanaceae); other species are *Cristaria oxyptera, Dinemandra ericoides, Nolana leptophylla, Polyachyrus fuscus* and *Tetragonia maritima*.

Northern Precordilleran district

Northern Precordilleran region—Peña, 1966a: 7 (regionalization), 1966b: 214 (regionalization).

Highland zone—Artigas, 1975: 20 (regionalization).

Northern Precordilleran zone—Artigas, 1975: 20 (regionalization).

Northern Precordilleran district—Morrone, 2014a: 89 (regionalization); Monckton, 2016: 125 (systematic revision and map); Morrone, 2017: 230 (regionalization).

Definition: The Northern Precordilleran district extends between 18° and 24° 30′ S (Artigas, 1975).

Vegetation: Luebert and Pliscoff (2006) characterized the scrubland as dominated by *Fabiana amulosa* (Solanaceae), *Baccharis boliviensis, Diplostephium meyenii* and *Lophopappus tarapacanus* (Asteraceae); other species are *Adesmia spinossisima, Balbisia microphylla, Cheilanthes pruinata, Junellia seriphioides, Oreocereus leucotrichus, Senna birostris, Stipa pubiflora* and *Tagetes multiflora*.

Tamarugal district

Northern Desert region (in part)—Peña, 1966a: 7 (regionalization), 1966b: 214 (regionalization).

Tamarugal zone—Artigas, 1975: 20 (regionalization).

Tamarugal district—Morrone, 2014a: 89 (regionalization); Monckton, 2016: 125 (systematic revision and map); Morrone, 2017: 230 (regionalization).

Definition: The Tamarugal district occupies a small area between 20°S and 22°S, in the central part of the province (Artigas, 1975).

Vegetation: The district consists of thorn forest dominated by *Prosopis tamarugo* (Fabaceae); other species are *Atriplex atacamensis, Caesalpinia aphylla, Cressa truxillensis, Distichlis spicata, Euphorbia tarapacana, Prosopis alba, P. burkartii, P. strombulifera, Schinus molle, Tagete minuta* and *Tessaria absinthioides* (Luebert and Pliscoff, 2006).

Cenocrons

Luebert (2011) distinguished four floristic elements assimilated to cenocrons: Neotropical (basically from the Desert province), Central Chilean, Transandean (basically from the Chacoan province) and Antitropical (basically from the Nearctic region). Guerrero et al. (2013) analyzed the time of colonization of three plant genera (*Chaetanthera, Nolana* and *Malesherbia*) and one lizard genus (*Liolaemus*) in the Atacama province and other arid areas in western South America and reconstructed the timing of the shifts in climate and the invasion of these taxa to desert areas. They concluded that *Chaetanthera, Malesherbia* and *Liolaemus* might have invaded the region 10 m.y.a, about 20 m.y.a. after the initial onset of regional aridity; whereas *Nolana*, the most diverse of these taxa, colonized it 2 m.y.a.

Cuyan High Andean province

Subandean province (in part)—Fittkau, 1969: 642 (regionalization).

Cuyan High Andean district—Cabrera, 1971: 31 (regionalization), 1976: 55 (regionalization); Huber and Riina, 1997: 145 (glossary).

Southern Andean province (in part)—Udvardy, 1975: 42 (regionalization); Morrone, 1996a: 108 (regionalization); Huber and Riina, 1997: 328 (glossary).

Argentinean-Atacaman province (in part)—Rivas-Martínez and Navarro, 1994: map (regionalization).

Southern Andean Steppe ecoregion—Dinerstein et al., 1995: 102 (ecoregionalization); Olson et al., 2001: 935 (ecoregionalization); Ezcurra, 2017: 1 (ecoregionalization).

Prepuna province—Morrone, 1999: 12 (*non* Cabrera, 1951; regionalization), 2001a: 112 (regionalization), 2006: 483 (regionalization);

Kutschker and Morrone, 2012: 543 (track analysis); Morrone, 2014a: 89 (regionalization), 2015b: 210 (regionalization).

Andean Cuyan province (in part)—López et al., 2008: 1572 (parsimony analysis of endemicity and regionalization).

Mediterranean Andean province—Rivas-Martínez et al., 2011: 27 (regionalization).

Southern Andean area—Apodaca et al., 2015b: 5 (dispersal-vicariance analysis).

Cuyan High Andean province—Morrone and Ezcurra, 2016: 287 (regionalization); Arana et al., 2017: 421 (shapefiles); Morrone, 2017: 230 (regionalization).

Definition

The Cuyan High Andean province comprises the eastern slopes of the Andes of Catamarca, La Rioja, San Juan, Mendoza and northern Neuquén provinces in Argentina and limiting areas in central Chile, between 27°S and 38°S (Cabrera, 1971; Morrone and Ezcurra, 2016). South of this area, fragments extend as isolated islands on high elevations of the Valdivian Forest and Magellanic Forest provinces, and north of this area on the high mountains of the Puna province (Cabrera, 1976). Relief is abrupt, its northern part harbors the highest peaks of the Andes: Mt. Ojos del Salado (6863 m), Mt. Pissis (6858 m), Mt. Aconcagua (6960 m) and Mt. Tupungato (6800 m); whereas in the south elevation decreases to less than 3000 m (Ezcurra, 2017). Climate is dry and very cold at high elevations. The lowest latitudinal limits of vegetation range from 3500 m in the north to 1800 m in the south, and the highest altitudinal limits are at about 5000 m in the north and 3000 m in the south (Ezcurra, 2017).

Endemic and characteristic taxa

Morrone and Ezcurra (2016) and Ezcurra (2017) provided a list of endemic and characteristic taxa. Some examples include *Adesmia pinifolia* (Fabaceae); *Chuquiraga echegarayi, Huarpea andina, Nassauvia cumingii* and *Senecio uspallatensis* (Asteraceae); *Azorella cryptantha* (Figure 8.15a) and *Laretia acaulis* (Apiaceae); *Lithodraba mendocinensis* (Brassicaceae); *Jaborosa laciniata* (Solanaceae); *Oxychloe bisexualis* (Juncaceae); *Mauryius* (Scorpiones: Bothriuridae; Ojanguren-Affilastro and Mattoni, 2017); *Alsodes nodosus* (Alsodidae); *Homonota andicola, Liolaemus andinus, L. buergeri, L. fitzgeraldi* and *L. ruibalis* (Squamata: Iguanidae); *Euneomys chinchilloides noei* (Rodentia: Cricetidae); *Geositta isabellina* (Passeriformes: Furnariidae); *Sicalis auriventris* (Passeriformes: Emberizidae); and *Cyanoliseus patagonus andinus* (Figure 8.15b; Psittaciformes: Psittacidae).

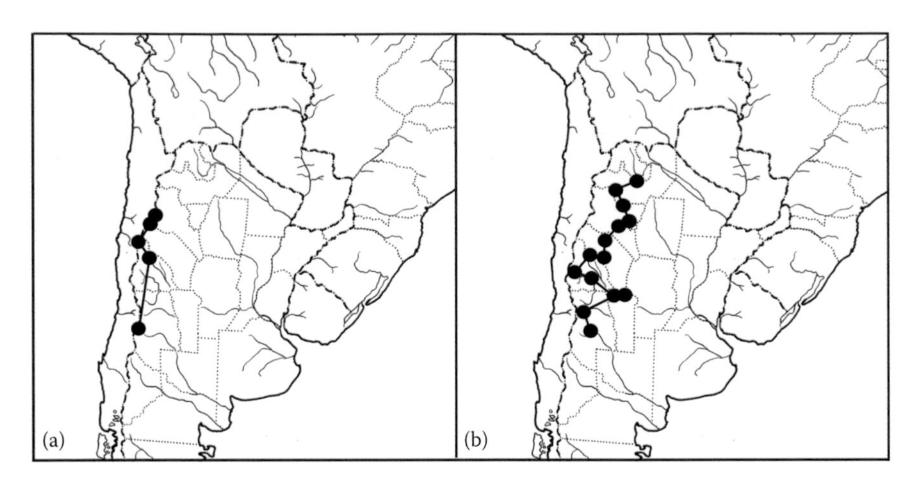

Figure 8.15 Maps with individual tracks in the Cuyan High Andean province. (a) *Azorella cryptantha* (Apiaceae) and (b) *Cyanoliseus patagonus andinus* (Psittacidae).

Vegetation

The Cuyan High Andean province consists of grass steppes and low vegetation of bushes and cushion bushes adapted to snow (Figure 8.16; Cabrera, 1971, 1976; Cabrera and Willink, 1973; Ezcurra, 2017). Dominant plant species include *Adesmia obovata, A. pinifolia, A. subterranea, Azorella cryptantha, Berberis empetrifolia, Chuquiraga echegaragyi, C. ruscifolia, Menonvillea cuneata, Nassauvia lagascae, Papostipa tenuisima, P. chrysophylla, Plantago barbata, Poa holciformis, Senecio uspallatensis* and *Werneria pygmaea* (Cabrera, 1971, 1976; Cabrera and Willink, 1973; Ezcurra, 2017).

Biotic relationships

Studies assigned the Cuyan High Andean province, formerly included in the Neotropical region, to the South American transition zone based on its close biotic links with other transitional provinces (Fjeldså, 1992; Morrone and Urtubey, 1997; Posadas et al., 1997). One example of this pattern is *Caenocrypticoides triplehorni* (Coleoptera: Tenebrionidae), closely related to species of the Atacaman and Desert provinces (Flores and Pizarro-Araya, 2004).

Cenocrons

Based on the analysis of palynological and geological evidence, Simpson (1983) concluded that the flora of this province —as well as the flora of other High Andean habitats—were relatively young, derived from older elements of lower altitudes that radiated in the area when the Andes uplifted

Figure 8.16 Representative plant species of the Cuyan High Andean province, Mendoza, Argentina. (a) *Argylia uspallatensis* (Bignoniaceae) and (b) *Nastanthus ventosus* (Calyceraceae). (Courtesy of Roberto Kiesling.)

in the Neogene-Pleistocene. In contrast, Ezcurra (2017) indicated that the monotypic and phylogenetically isolated genera *Huarpea* (Asteraceae) and *Lithodraba* (Brassicaceae) were relicts of an ancient Paleogene flora, but other genera with important diversification in the area (e.g., *Senecio*, Asteraceae) had a relatively younger origin, possibly related with Quaternary climatic fluctuations. Finally, genera with Subantarctic (e.g., *Ourisia*, Scrophulariaceae) or Arctic (e.g., *Ribes*, Grossulariaceae) affinities may have dispersed to the area when these high elevation environments appeared in the Neogene (Ezcurra, 2017).

Monte province

Monte formation—Holmberg, 1898: 419 (regionalization).

Monte area—Hauman, 1920: 54 (regionalization); Hueck, 1957: 40 (regionalization), 1966: 3 (regionalization); Roig-Juñent, 1994b: 183 (cladistic biogeography); Coscarón and Coscarón-Arias, 1995: 726 (areas of endemism).

Monte province—Hauman, 1931: 60 (regionalization); Soriano, 1950: 33 (regionalization); Cabrera, 1951: 41 (regionalization), 1953: 107 (regionalization); Morello, 1955: 386 (regionalization); Cabrera, 1958: 200 (regionalization); Morello, 1958: 131 (regionalization); Cabrera, 1971: 22 (regionalization); Cabrera and Willink, 1973: 77 (regionalization); Cabrera, 1976: 36 (regionalization); Udvardy, 1975: 41 (regionalization); Stange et al., 1976: 78 (biotic evolution); Rivas-Martínez and Navarro, 1994: map (regionalization); Huber and Riina, 1997: 228 (glossary); León et al., 1998: 135 (regionalization); Roig, 1998: 136 (regionalization); Morrone, 1999: 9 (regionalization); Zuloaga et al., 1999: 37 (floristics); Marino et al., 2001: 115 (parsimony analysis of endemicity); Morrone, 2000a: 61 (regionalization), 2001a: 94 (regionalization); Roig-Juñent et al., 2001: 78 (areas of endemism); Ojeda et al., 2002: 24 (biotic evolution); Roig-Juñent et al., 2003: 275 (track analysis and cladistic biogeography); Morrone, 2004b: 46 (regionalization), 2006: 483 (regionalization); Spinelli et al., 2006: 302 (parsimony analysis of endemicity); Abraham et al., 2009: 145 (geography); Roig et al., 2009: 164 (regionalization); Urtubey et al., 2010: 506 (track analysis and cladistic biogeography); Hechem et al., 2011: 46 (track analysis); Sérsic et al., 2011: 477 (phylogeographic analysis); Ferretti et al., 2012: 2 (parsimony analysis of endemicity and track analysis); Campos-Soldini et al., 2013: 16 (track analysis); Baranzelli et al., 2014: 752 (phylogeographic analysis); Carpintero, 2014: 339 (faunistics); Ezcurra et al., 2014: 28 (biotic evolution); Ferretti et al., 2014a: 1089 (track and endemicity analysis); Flores and Cheli, 2014: 283 (faunistics); Grismado and Ramírez, 2014: 144 (faunistics); Morrone, 2014a: 90 (regionalization); Amarilla et al., 2015: 8 (biotic evolution); Apodaca et al., 2015a: 88 (regionalization); Biganzoli and Zuloaga, 2015: 342 (floristics); del Río et al., 2015: 1294 (track analysis and cladistic biogeography); Ferretti, 2015: 3 (dispersal-vicariance analysis); Morrone, 2015b: 210 (regionalization); del Valle Elías and Aagesen, 2016: 161 (endemicity analysis); Ruiz et al., 2016: 385 (track analysis); Stonis et al., 2016: 561 (systematic revision); Agrain et al., 2017: 73 (faunistics); Arana et al., 2017: 421 (shapefiles); Martínez et al., 2017: 479 (track analysis and regionalization); Morrone, 2017: 232 (regionalization).

Xerophyllous Forests area—Parodi, 1934: 171 (regionalization).

Central Xerophyllous Forest area—Castellanos and Pérez-Moreau, 1944: 382 (regionalization).

Western Monte area—Parodi, 1945: 130 (regionalization); Bölcke, 1957: 2 (regionalization).

Central province—Soriano, 1949: 198 (regionalization).

Central or Subandean dominion—Ringuelet, 1961: 160 (biotic evol. and regionalization).

Monte center—Müller, 1973: 146 (regionalization).

Monte with Cactaceae dominion—Ab'Sáber, 1977: map (climate).
Argentine Monte ecoregion—Dinerstein et al., 1995: 99 (ecoregionalization); Dellafiore, 2017c: 1 (ecoregionalization).
Monte de Sierras y Bolsones ecoregion—Burkart et al., 1999: 13 (ecoregionalization).
Argentina Monte province—Rivas-Martínez et al., 2011: 27 (regionalization).
Mar Chiquita-Salinas Grandes ecoregion—Abell et al., 2008: 408 (ecoregionalization).

Definition

The Monte province comprises southern Bolivia and central Argentina, between 24°S and 44°S, from Jujuy to northeastern Chubut (Morello, 1955, 1958; Cabrera, 1958, 1971; Cabrera and Willink, 1973; Müller, 1973; Stange et al., 1976; Willink, 1988, 1991; Roig-Juñent, 1994b; Coscarón and Coscarón-Arias, 1995; Morrone, 2000a, 2006, 2014a; Roig-Juñent et al., 2001; Abraham et al., 2009; Roig et al., 2009; del Valle Elías and Aagesen, 2016). Altitudinally, the province ranges from sea-level to 3500 m, depending on latitude (del Valle Elías and Aagesen, 2016). To the west it shows a transition with the Cuyan High Andean province, as the elevation increases, and to the east it shows a transition with the Pampa province, of the Chacoan dominion (Dellafiore, 2017c). Based on geomorphological features, studies divided it into the High Monte (mountainous areas in the north) and the Low Monte (piedmonts, hills and desert valleys in the central and southern areas; Olson et al., 2001). Studies more recently treated the High Monte, previously recognized as the Prepuna province (e.g., Cabrera, 1951, 1971; Morello, 1958), as a district of the Monte province (Morrone and Ezcurra, 2016).

Endemic and characteristic taxa

Morrone (2014a), del Valle Elías and Aagesen (2016) and Carbonó (2017) provided lists of endemic taxa. Some examples include *Ephedra boelckei* (Ephedraceae); *Hickenia* (Asclepiadaceae); *Denmoza* and *Tephrocactus* (Cactaceae); *Chuquiraga rosulata* (Asteraceae); *Halophytum* (Halophytaceae); *Monttea aphylla* (Scrophullariaceae); *Larrea divaricata, L. cuneifolia* and *L. nitida* (Zygophyllaceae); *Urophonius brachycentrus* and *Timogenes* (Scorpiones: Bothriuridae); *Bradynobaenus subandinus* (Hymenoptera: Bradynobaenidae); *Amblycerus caryoboriformis* (Coleoptera: Chrysomelidae); *Cyrtomon hirsutus, Enoplopactus lizeri* (Figure 8.17a), *Listroderes bruchi, Enoplopactus catamarcensis, Mendozella* and *Pantomorus luteipes* (Coleoptera: Curculionidae); *Epitragella, Megelenophorus* and *Phrynocarenum* (Coleoptera: Tenebrionidae); *Catadacus* (Hymenoptera: Ichneumonidae); *Scaptodactyla* (Hymenoptera: Mutillidae);

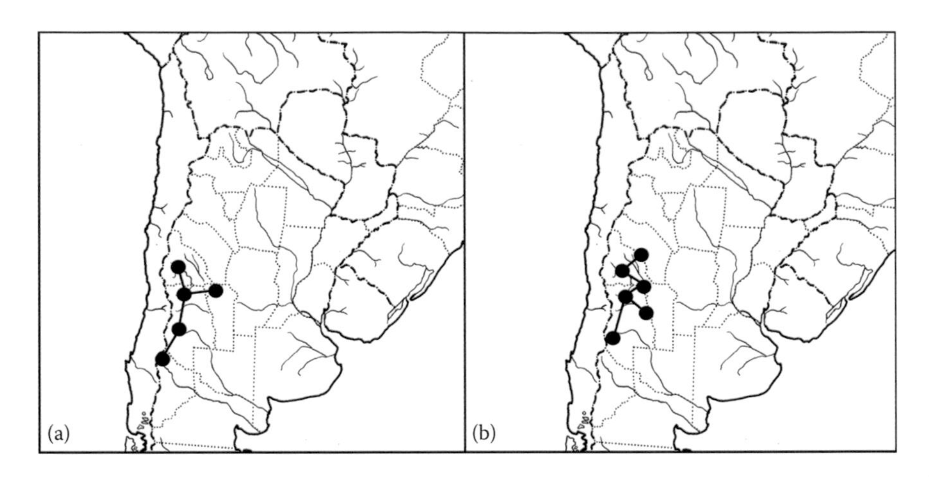

Figure 8.17 Maps with individual tracks in the Monte province. (a) *Enoplopactus lizeri* (Coleoptera: Curculionidae) and (b) *Bothrops ammodytoides* (Squamata: Viperidae).

Papipappus (Orthoptera: Ommexechidae); *Burmeisteriellus* and *Vulcanocanthon seminulum* (Coleoptera: Scarabaeidae); *Bothrops ammodytoides* (Figure 8.17b; Squamata: Viperidae); *Pseudasthenes patagonica* (Passeriformes: Furnariidae); *Phrygilus carbonarius, Poospiza ornata* and *Sicalis mendozae* (Passeriformes: Thraupidae); *Chlamyphorus truncatus* (Xenarthra: Dasypodidae); and *Tympanoctomys barrerae* (Rodentia: Octodontidae).

Vegetation

The Monte province consists of open scrubland, with species of the genera *Aloysia, Parkinsonia, Capparis* and the Zygophyllaceae genera *Larrea, Bulnesia* and *Plectrocarpa* (Figure 8.18; Cabrera and Willink, 1973; Roig-Juñent et al., 2003; Roig et al., 2009; del Valle Elías and Aagesen, 2016). The most characteristic community is the *jarillal*, with *Larrea cuneifolia, L. divaricata* and *L. nitida*, that develops in pockets of sandy or rocky-sandy soils (Dellafiore, 2017c). Additionally, there are edaphic communities such as woods of *Prosopis* scrubs and *Baccharis salicifolia* and *Tessaria dodonaefolia* in humid places (Carbonó, 2017). Dominant plant species include *Allenrolfea vaginata, Atriplex lampa, Baccharis salicifolia, Bougainvillea spinosa, Bulnesia retama, B. schikendantzii, Cassia aphylla, C. rigida, Cercidium praecox, Fabiana patagonica, Larrea cuneifolia, L. divaricata, L. nitida, Mimosa ephedroides, Monttea aphylla, Plectrocarpa rougesii, Portulaca* spp., *Prosopis alpataco, P. chilensis, P. flexuosa, P. torquata, Proustia cuneifolia, Tessaria dodoneaefolia* and *Zuccagnia punctata* (Morello, 1955; Cabrera, 1971, 1976; Cabrera and Willink, 1973; Roig, 1998; Roig-Juñent et al., 2001). Plants similar to those

Figure 8.18 Plant species characteristic of the Monte province, Argentina. (a) *Cereus forbesii* (Cactaceae), (b) *Aspidosperma* sp. (Apocynaceae) and (c) *Polylepis tarapacana* (Rosaceae). (Courtesy of Sergio Roig-Juñent [a] and Roberto Kiesling [b and c].)

in the North American dry areas belong to the genera *Acacia*, *Caesalpinia*, *Cassia*, *Cercidium*, *Larrea*, *Mimosa* and *Prosopis* (Graham, 2004).

Biotic relationships

Studies suggested that the Monte represents a biotically impoverished Chaco (Willink, 1991), especially when considering its plant species (e.g., Cabrera and Willink, 1973), but this province harbor several endemic insect taxa (Willink, 1991). According to a cladistic biogeographical analysis based on beetle and plant taxa (Morrone, 1993a), the Monte province was closely related to the Chaco province.

In addition to its historical relationship with the Chacoan provinces, the Monte province shows biotic similarities with the Sonora province of the Nearctic region (Clark, 1979; Willink, 1988; Morello, 1984; Roig et al., 2009; Wen and Ickert-Bond, 2009). Solbrig et al. (1977) and Wen and Ickert-Bond (2009) reviewed the hypotheses that postulated to explain these similarities: usually involving long-distance dispersal as well as vicariance and parallel evolution due to adaptations to arid environments from closely related ancestors. Examples of the latter included Bruchinae (Coleoptera: Chrysomelidae) associated with Fabaceae shrubs in both areas (Stange et al., 1976) and weevils (Coleoptera: Curculionidae) of the *Sibinia sulcifer* species group (Clark, 1979).

Regionalization

Roig et al. (2009), Morrone (2015b) and Morrone and Ezcurra (2016) identified four districts (Figure 8.19).

Eremean district

Eremean district—Roig et al., 2009: 166 (regionalization); Morrone, 2014a: 90 (regionalization), 2017: 235 (regionalization).
Uspallata-Callingasta Valley area—Roig-Juñent et al., 2001: 87 (regionalization).

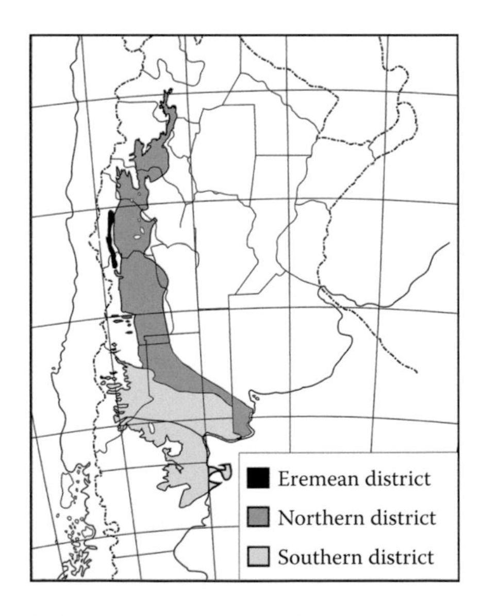

Figure 8.19 Map with three districts of the Monte province (Prepuna district not represented). (Modified from Roig, F.A. et al., *J. Arid Environ.*, 73, 164–172, 2009. With permission.)

Definition: The Eremean district is a narrow strip along the length of the high pre-Andean valleys of Mendoza and San Juan (Roig et al., 2009). Annual precipitation is between 50 and 100 mm, sometimes lower (Roig et al., 2009).

Endemic taxa: *Maihueniopsis clavarioides* (Cactaceae); *Cistanthe densiflora* (Montiaceae); *Larrea divaricata* subsp. *monticellii* (Zygophyllaceae); *Nyctelia plicatipennis, Physogaster chechoi* and *P. longipilis* (Coleoptera: Tenebrionidae); and *Homonota andicola* (Squamata: Phyllodactylidae; Roig et al., 2009).

Northern district

Central area—Roig-Juñent et al., 2001: 87 (regionalization).

Northern area—Roig-Juñent et al., 2001: 86 (regionalization).

Northern district—Roig et al., 2009: 168 (regionalization); Morrone, 2014a: 90 (regionalization), 2017: 234 (regionalization).

Central subdistrict—Roig et al., 2009: 168 (regionalization).

Pampa subdistrict—Roig et al., 2009: 168 (regionalization).

Tucumán-Salta subdistrict—Roig et al., 2009: 168 (regionalization).

Definition: The Northern district encompasses the largest extension of the province, from Salta and Tucumán to southern Mendoza (Roig et al., 2009). Studies characterize it by subtropical rainfall, with an annual precipitation of 200–400 mm (Roig et al., 2009).

Endemic taxa: *Ephedra boelckei* (Ephedraceae); *Heliotropium curassavicum* var. *fruticulosum* (Boraginaceae); *Prosopis alpataco* (Fabaceae); *Mimodromius proseni* (Coleoptera: Carabidae); *Enoplopactus catamarcensis* and *E. sanjuaninus* (Coleoptera: Curculionidae); *Glyphoderus monticola* (Coleoptera: Scarabaeidae); *Entomoderes infernalis, E. pustulosus, E. subauratus* and *Thylacoderes sphaericus* (Coleoptera: Tenebrionidae); *Liolaemus anomalus* and *L. scapularis* (Squamata: Iguanidae); and *Leiosaurus catamarcencis* (Squamata: Leiosauridae; Roig-Juñent et al., 2001; Roig et al., 2009).

Prepuna district

Prepuna province—Cabrera, 1951: 40 (regionalization), 1953: 107 (regionalization), 1958: 200 (regionalization); Morello, 1958: 117 (regionalization); Cabrera, 1971: 21 (regionalization); Cabrera and Willink, 1973: 76 (regionalization); Cabrera, 1976: 34 (regionalization); Aagesen et al., 2009: 295 (endemicity analysis).

Patagónica province—Brignone et al., 2016: 327 (systematic revision), Ferretti, 2012: 219 (endemicity analysis).

Cardonales de Laderas subregion—Burkart, 1999: 14 (regionalization).

Monte de Sierras y Bolsones ecoregion (in part)—Burkart, 1999: 13 (regionalization).

High Monte area (in part)—Abraham et al., 2009: 145 (regionalization).

Andean province—Roig et al., 2009: 164 (regionalization).
Prepuneña province—Brignone et al., 2016: 334 (systematic revision).
Prepuna district—Morrone and Ezcurra, 2016: 2 (regionalization); Morrone, 2017: 234 (regionalization).
Definition: The Prepuna district extends from southern Bolivia to northwestern Argentina (Jujuy to La Rioja provinces), from about 1000 to 3500 m (Morrone and Ezcurra, 2016). The district corresponds to the High Monte (Olson et al., 2011).
Endemic taxon: *Entomoderes zupai* (Coleoptera: Tenebrionidae; Flores and Roig-Juñent, 1997).

Southern district

Península de Valdés area—Roig-Juñent et al., 2001: 87 (regionalization).
Southern area—Roig-Juñent et al., 2001: 87 (regionalization).
Southern district—Roig et al., 2009: 168 (regionalization); Morrone, 2014a: 91 (regionalization), 2017: 235 (regionalization).
Northern Patagonia subdistrict—Roig et al., 2009: 168 (regionalization).
Península de Valdés and Punta Ninfas subdistrict—Roig et al., 2009: 165 (regionalization).
San Jorge Gulf and Punta Ninfas or Atlantic Shrub subdistrict—Roig et al., 2009: 168 (regionalization).
Southern Patagonian subdistrict—Roig et al., 2009: 168 (regionalization).
Definition: The Southern district lies south of the Colorado river (Roig et al., 2009). Annual rainfall ranges between 100 and 500 mm (Roig et al., 2009).
Endemic taxa: *Aylacophora deserticola, Chuquiraga avellanedae* and *C. rosulata* (Asteraceae); *Maihuenia patagonica* (Cactaceae); *Tetraglochin caespitosum* (Rosaceae); *Barypus dentipennis, B. schajovskoyi, Mimodromius fleissi* and *M. straneoi* (Coleoptera: Carabidae); *Leptynoderes fuscula, Nyctelia dorsata, Patagonogenius collaris and Psectrascelis convexipennis* (Coleoptera: Tenebrionidae); *Amphisbaena angustifrons* (Squamata: Amphisbaenidae); and *Liolaemus donosobarrosi* and *L. melanops* (Squamata: Iguanidae; Roig et al., 2009; Roig-Juñent et al., 2001).
Vegetation: León et al. (1998) characterized the shrub steppe of this district as dominated by *Larrea divaricata, L. cuneifolia* and *L. nitida* (Zygophyllaceae); other species are *Bouganvillaea spinosa, Bromus tectorum, Chuquiraga* spp., *Condalia microphyla, Lycium* spp., *Monthea aphylla, Perezia recurvata, Plantago patagonica, Prosopidastrum globossum, Prosopis strombuligera, Schinus barbatus, S. polygamus* and *Verbena alatocarpa*.

Cenocrons

The entomofauna of the Monte province has basically Chacoan affinities, showing also Patagonian and Subantarctic elements (Willink, 1988, 1991; Ellenrieder, 2001; Roig-Juñent et al., 2001, 2003; Marino et al., 2001; Roig et al., 2009). The Patagonian and Subantarctic affinities are particularly clear in its southernmost districts (Willink, 1991).

Roig et al. (2009) identified four cenocrons in the Monte province based on the phylogenetic relationships and the distribution of the taxa analyzed:

Paleorelict taxa with related groups in other continents: Taxa originated before the break-up of Gondwana. They included Zygophyllaceae, *Prosopis,* Daesiidae (Solifuga) and Cicindinae (Coleoptera: Carabidae).

Taxa with American related groups: Taxa that evolved in South America after the break-up of Gondwana. They included Cactaceae and *Larrea* (Zygophyllaceae).

Endemic South American desert taxa: Taxa with their sister taxa in mesic or humid tropical areas of South America. Examples included the beetle tribes Cnemalobini (Coleoptera: Carabidae) and Eucranini (Scarabaeidae).

Neoendemic desert taxa: Taxa that underwent rapid speciation in desert environments. One example was the lizard genus *Pristidactylus,* which originated in Subantarctic Chile.

Comechingones province

Comechingones province—Martínez et al., 2017: 486 (track analysis and regionalization); Arana et al., 2017: 421 (shapefiles).

Definition

The Comechingones province covers the mountain areas in central Argentina (Córdoba and San Luis provinces) situated between 29°S and 33°S, at an altitude above 1000 (Martínez et al., 2017).

Endemic and characteristic taxa

Martínez et al. (2017) provided a list of endemic and characteristic taxa. Some examples included *Isoetes hieronymii* (Isoetaceae); *Alternanthera pumila* (Amaranthaceae); *Grindelia globulariifolia, Hypochaerisncaespitosa* and *Senecio retanensis* (Asteraceae); *Berberis hieronymi* (Berberidaceae); *Aa achalensis* (Orchidaceae); *Escallonia cordobensis* (Escalloniaceae); *Nassella nidulans* and *Poa stuckertii* (Poaceae); *Apurimacia dolichocarpa* (Fabaceae); *Rhinella achalensis* and *Melanophryniscus estebani* (Anura: Bufonidae); *Pristydactilus achalensis* (Squamata: Leiosauridae); *Asthenes modesta cordobae, Cinclodes atacamensis schocolatinus, C. comechingonus* and *C. oustaleti olrogi* (Passeriformes: Furnariidae); *Catamenia inornata cordobensis and Phrygilus plebejus naroskyi* (Passeriformes: Thraupidae); and *Agriornis montana fumosus* (Passeriformes: Tyrannidae).

Vegetation

The Comechingones province consists of high altitude grasslands and steppes (Figure 8.20) with predominance of species of *Festuca* and *Nassella*

Figure 8.20 Vegetation of the Comechingones province, Córdoba, Argentina. (a) High altitude grassland and (b) high altitude steppe. (Courtesy of Marcelo Arana.)

(Poaceae). There are also woodlands dominated by *Polylepis australis* (Rosacae) and *Maytenus boaria* (Celastraceae; Martínez et al., 2017).

Biotic relationships

A track analysis based on plant and vertebrate taxa (Martínez et al., 2017) identified complex biotic relationships with provinces from the Andean region (e.g., Patagonian, Magellanic Forest, Valdivian Forest, Maule, Magellanic Forest, Santiagan and Coquimban), the Neotropical region (e.g., Chaco, Yungas and Parana Forest) and the South American transition zone (e.g., Puna, Monte, and Páramo). Based on these results, Martínez et al. (2017) assigned this province formally to the South American transition zone.

References

Aagesen, L., C. A. Szumik, F. O. Zuloaga and O. Morrone. 2009. Quantitative biogeography in the South America highlands–recognizing the Altoandina, Puna and Prepuna through the study of Poaceae. *Cladistics*, 25: 295–310.

Abell, R., M. L. Thieme, C. Revenga, M. Bryer, M. Kottelat, N. Bogutskaya, B. Coad et al. 2008. Freshwater ecoregions of the world: A new map of biogeographic units for freshwater biodiversity conservation. *BioScience*, 58: 493–514.

Abraham, E., H. F. del Valle, F. Roig. L. Torres, J. O. Ares, F. Coronato and R. Godagnone. 2009. Overview of the geography of the Monte Desert biome (Argentina). *Journal of Arid Environments*, 73: 144–153.

Abrahamovich, A., N. B. Díaz and J. J. Morrone. 2004. Distributional patterns of the Neotropical and Andean species of the genus *Bombus* (Hymenoptera: Apidae). *Acta Zoológica Mexicana (nueva serie)*, 20: 99–117.

Ab'Sáber, A. N. 1977. Os domínios morfoclimáticos na América do Sul. Primeira aproximaçao. *Geomorfologia*, 52: 1–21.

Absolon, B. A., V. Gallo and L. S. Avilla. 2016. Distributional patterns of living ungulates (Mammalia: Cetartiodactyla and Perissodactyla) of the Neotropical region, the South American transition zone and Andean region. *Journal of South American Earth Sciences*, 71: 63–70.

Acosta, L. E. and E. A. Maury. 1998a. Scorpiones. In: Morrone, J. J. and Coscarón, S. (Eds.), *Biodiversidad de Artrópodos Argentinos: Un enfoque Biotaxonómico*, Ediciones Sur, La Plata, Argentina, pp. 545–559.

Acosta, L. E. and E. A. Maury. 1998b. Opiliones. In: Morrone, J. J. and Coscarón, S. (Eds.), *Biodiversidad de Artrópodos Argentinos: Un enfoque biotaxonómico*, Ediciones Sur, La Plata, Argentina, pp. 569–580.

Agrain, F. A., M. L. Chamorron, N. Cabrera, D. Sassi and S. Roig-Juñent. 2017. A comprehensive guide to the Argentinian case-bearer beetle fauna (Coleoptera, Chrysomelidae, Camptostomata). *ZooKeys*, 677: 11–88.

Alfaro, F. M., J. Pizarro-Araya, L. Letelier and J. Cepeda. 2013. Distribución geográfica de los ortópteros (Insecta: Orthoptera) presentes en las provincias biogeográficas de Atacama y Coquimbo (Chile). *Revista de Geografía Norte Grande*, 56: 235–250.

Alfaro, F. M., J. Pizarro-Araya and J. Mondaca. 2014. A new insular record of *Germarostes (Germarostes) posticus* (Germar) (Coleoptera: Hybosoridae: Ceratocanthinae) for the Chilean transitional costal desert. *The Coleopterists Bulletin*, 68: 387–390.

Allen, J. A. 1892. The geographical distribution of North American mammals. *Bulletin of the American Museum of Natural History*, 4: 199–243.
Almirón, A., M. Azpelicueta, J. Casciotta and A. López Cazorla. 1997. Ichthyogeographic boundary between the Brazilian and Austral subregions in South America, Argentina. *Biogeographica*, 73: 23–30.
Alzate, F., M. A. Quijano-Abril and J. J. Morrone. 2008. Panbiogeographical analysis of the genus *Bomarea* (Alstroemeriaceae). *Journal of Biogeography*, 35: 1250–1257.
Amarilla, L. D., A. M. Anton, J. O. Chiapella, M. M. Manifesto, D. F. Angulo and V. Sosa. 2015. *Munroa argentina*, a grass of the South American transition zone, survived the Andean uplift, aridification and glaciations of the Quaternary. *PLoS One*, 10: 1–21.
Amorim, D. S. and M. R. S. Pires. 1996. Neotropical biogeography and a method for maximum biodiversity estimation. In: Bicudo, C. E. M. and N. A. Menezes (Eds.), *Biodiversity in Brazil: A first approach*. CNPq, São Paulo, Brazil, pp. 183–219.
Amorim, D. S., C. M. D. Santos and S. S. de Oliveira. 2009. Allochronic taxa as an alternative model to explain circumantarctic distributions. *Systematic Entomology*, 34: 2–9.
Amorim, D. S. and H. S. Tozoni. 1994. Phylogenetic and biogeographic analysis of the Anisopodoidea (Diptera, Bibionomorpha), with an area cladogram for intercontinental relationships. *Revista Brasileira de Entomologia*, 38: 517–543.
Antonelli, A., J. A. Nylander, C. Person and I. Sanmartín. 2009. Tracing the impact of the Andean uplift on Neotropical plant evolution. *Proceedings of the National Academy of Sciences*, 106: 9749–9754.
Apodaca, M. J., J. V. Crisci and L. Katinas. 2015a. Las provincias fitogeográficas de la República Argentina: Definición y sus principales áreas protegidas. In: Casas, R. R. and G. F. Albarracín (Eds.), *El deterioro del suelo y del ambiente en la Argentina*, Fundación Ciencia, Educación y Cultura, Buenos Aires, Argentia.
Apodaca, M. J., J. V. Crisci and L. Katinas. 2015b. Andean origin and diversification of the genus *Perezia*, an ancient lineage of Asteraceae. *Smithsonian Contributions to Botany*, 102: 1–28.
Arana, M. D., G. A. Martínez, A. J. Oggero, E. S. Natale and J. J. Morrone. 2017. Map and shapefile of the biogeographic provinces of Argentina. *Zootaxa*, 4341: 420–422.
Armstrong, S. 2017. Western South America: Northwestern Chile. In: *Terrestrial Ecoregions of Latin America and the Caribbean*, World Wildlife Fund. WWF NT1303. https://www.worldwildlife.org/ecoregions/nt1303. Accessed November 20, 2017.
Artigas, J. N. 1975. Introducción al estudio por computación de las áreas zoogeográficas de Chile continental basado en la distribución de 903 especies de animales terrestres. *Gayana, Miscelánea*, 4: 1–25.
Ashworth, A. C. and G. Kuschel. 2003. Fossil weevils (Coleoptera: Curculionidae) from latitude 85°S Antarctica. *Palaeogeography, Palaeoclimatology, Palaeoecology*, 191: 191–202.
Axelrod, D. I., M. T. K. Arroyo and P. H. Raven. 1991. Historical development of the temperate vegetation in the Americas. *Revista Chilena de Historia Natural*, 64: 413–446.

Avise, J. C. 2000. *Phylogeography: The History and Formation of Species*. Harvard University Press, Cambridge, MA.

Baeza, C. M., T. F. Stuessy and C. Marticorena. 2002. Notes on the Poaceae of Robinson Crusoe (Juan Fernández) Islands, Chile. *Brittonia*, 54: 154–163.

Bailey, R. G. 1998. *Ecoregions: The Ecosystem Geography of the Oceans and Continents*. Springer-Verlag, New York.

Baranzelli, M. C., L. A. Johnson, A. Cosacov and A. N. Sérsic. 2014. Historial and ecological divergence among populations of *Monttea chilensis* (Plantaginaceae), an endemic endangered shrub bordering the Atacama Desert, Chile. *Evolutionary Ecology*, 28: 751–774.

Barnard, J. L. and C. M. Barnard. 1983. *Freshwater Amphipoda of the World*. Hayfield Associates, Mount Vernon, VA.

Barriga, J. E. and D. E. Cepeda. 2017. Nueva especie de *Stenomela* Erichson (Coleoptera: Chrysomelidae) para Chile. *Revista Chilena de Entomología*, 42: 17–22.

Bartholomew, J. G., W. E. Clark and P. H. Grimshaw. 1911. *Atlas of Zoogeography: A Series of Maps Illustrating the Distribution of over Seven Hundred Families, Genera, and Species of Existing Animals*. Edinburgh Geographical Institute, Edinburgh, Scotland.

Bernardello, G., G. J. Anderson, T. F. Stuessy and D. J. Crawford. 2006. The angiosperm flora of the Archipelago Juan Fernandez (Chile): Origin and dispersal. *Canadian Journal of Botany*, 84: 1266–1281.

Bernardello, G. and T. F. Stuessy. 2017. Island group off the coast on central Chile in the Pacific Ocean. In: *Terrestrial Ecoregions of Latin America and the Caribbean*, World Wildlife Fund. WWF NT0401. https://www.worldwildlife.org/ecoregions/nt0401. Accessed November 20, 2017.

Biganzoli, F. and F. Zuloaga. 2015. Análisis de la diversidad de la familia Poaceae de la región austral de América del Sur. *Rodriguésia*, 66: 337–351.

Birnie, J. F. and D. E. Roberts. 1986. Evidence of tertiary forest in the Falkland Islands (Islas Malvinas). *Palaeogeography, Palaeoclimatology, Palaeoecology*, 55: 45–53.

Blöcher, R. and J. P. Frahm. 2002. A comparison of the moss floras of Chile and New Zealand: Studies in austral temperate rain forest bryophytes 17. *Tropical Bryology*, 21: 81–92.

Bölcke, O. 1957. La situación forrajera argentina. *IDIA*, 113: 1–36.

Bouckaert, R., J. Heled, D. Kühnert, T. Vaughan, C. H. Wu, D. Xie, M. A. Suchard, A. Rambaut and A. J. Drummond. 2014. BEAST2: A software platform for Bayesian evolutionary analysis. *PloS Computational Biology*, 10: e1003537.

Brignone, N. F., S. S. Denham and R. Pozner. 2016. Synopsis of the genus *Atriplex* (Amaranthaceae, Chenopodioideae) for South America. *Australian Systematic Botany*, 29: 324–357.

Brinkhurst, R. O. and B. J. M. Jamieson. 1971. *Aquatic Oligochaeta of the World*. Oliver and Boyd, Edinburgh, Scotland.

Brion, C. and C. Ezcurra. 2017. South America: Chile and Argentina. In: *Terrestrial Ecoregions of Latin America and the Caribbean*, World Wildlife Fund. WWF NT0402. https://www.worldwildlife.org/ecoregions/nt0402. Accessed November 20, 2017.

Bryant, D. and V. Moulton. 2004. Neighbor-Net: An agglomerative method for the construction of phylogenetic networks. *Molecular Biology and Evolution*, 21: 255–265.

Buckland, P. C. and P. M. Hammond. 1997. The origins of the biota of the Falkland Islands and South Georgia. *Quaternary Proceedings*, 5: 59–69.

Blyth, E. 1871. A suggested new division of the Earth into zoological regions. *Nature*, 3: 427–429.

Boyer, S. L., R. M. Clouse, L. R. Benavides, P. Sharma, P. J. Schwendiger, I. Karunathna and G. Giribet. 2007. Biogeography of the world: A case study from cyphophtalmid Opiliones, a globally distributed group of arachnids. *Journal of Biogeography*, 34: 2070–2085.

Brooks, D. R. 1985. Historical ecology: A new approach to studying the evolution of ecological associations. *Annals of the Missouri Botanical Garden*, 72: 660–680.

Brooks, D. R. 1990. Parsimony analysis in historical biogeography and coevolution: Methodological and theoretical update. *Systematic Zoology*, 39: 14–30.

Broughton, D. A. and J. H. McAdam. 2005. A checklist of the native vascular flora of the Falkland Islands (Islas Malvinas): New information on the species present, their ecology, status and distribution. *Journal of the Torrey Botanical Society*, 132: 115–148.

Burkart, R., N. O. Bárbaro, R. O. Sánchez and D. A. Gómez. 1999. *Eco-regiones de la Argentina*. Administración de Parques Nacionales, Buenos Aires, Argentina.

Cabrera, A. and J. Yepes. 1940. *Mamíferos Sud-americanos (vida, costumbres y descripción)*. Historia Natural Ediar, Buenos Aires, Argentina.

Cabrera, A. L. 1951. Territorios fitogeográficos de la República Argentina. *Boletín de la Sociedad Argentina de Botánica*, 4: 21–65.

Cabrera, A. L. 1953. Esquema fitogeográfico de la República Argentina. *Revista del Museo de la Ciudad Eva Perón, Botánica*, 8: 87–168.

Cabrera, A. L. 1957. La vegetación de la Puna argentina. *Revista de Investigaciones Agrícolas*, 11: 317–412.

Cabrera, A. L. 1958. Fitogeografía en la Argentina. *Suma de Geografía*, 3: 101–207.

Cabrera, A. L. 1971. Fitogeografía de la República Argentina. *Boletín de la Sociedad Argentina de Botánica*, 14: 1–42.

Cabrera, A. L. 1976. Regiones fitogeográficas argentinas. In: Kugler, W. F. (Ed.), *Enciclopedia Argentina de Agricultura y Jardinería*. II, ACME, Buenos Aires, Argentina, pp. 1–85.

Cabrera, A. L. and A. Willink. 1973. *Biogeografía de América Latina*. Monografía 13, Serie de Biología, OEA, Washington, DC.

Campos-Soldini, M. P., M. G. Del Río and S. A. Roig-Juñent. 2013. Análisis panbiogeográfico de las especies de *Epicauta* Dejean, 1834 (Coleoptera: Meloidae) en América del Sur austral. *Revista de la Sociedad Entomológica Argentina*, 72: 15–25.

Carbonó, E. 2017. Northern South America: Northern Colombia. In: *Terrestrial Ecoregions of Latin America and the Caribbean*, World Wildlife Fund. WWF NT1007. https://www.worldwildlife.org/ecoregions/nt1007. Accessed November 20, 2017.

Carpintero, D. L. 1998. Miridae. In: Morrone, J. J. and S. Coscarón (Eds.), *Biodiversidad de artrópodos argentinos: Un enfoque biotaxonómico*, Ediciones Sur, La Plata, Argentina, pp. 144–150.

Carpintero, D. L. 2014. Enicocephalomorpha y Dipsocoromorpha. In: Roig-Juñent, S., L. E. Claps and J. J. Morrone (Eds.), *Biodiversidad de Artrópodos Argentinos: Volumen 3*, Facultad de Ciencias Naturales e Instituto Miguel Lillo, San Miguel de Tucumán, Argentina, pp. 335–340.

Carpintero, D. L. and S. I. Montemayor. 2008. Revision of the Cantacaderinae (Heteroptera, Tingidae) from Argentina and Chile, with the description of a new species of Cantacaderini. *Deutsche Entomologische Zeitung*, 55: 109–116.

Carrara, R. and G. E. Flores. 2013. Endemic tenebrionids (Coleoptera: Tenebrionidae) from the Patagonian steppe: A preliminary identification of areas of microendemism and richness hotspots. *Entomological Science*, 16: 100–111.

Carrara, R. and G. E. Flores. 2015. Endemic epigean tenebrionids (Coleoptera: Tenebrionidae) from the Andean region: Exploring the Patagonian-diversification hypothesis. *Zootaxa*, 4007: 47–62.

Carvalho, C. J. B. de and M. S. Couri. 2010. Biogeografia de Muscidae (Insecta, Diptera) da América do Sul. In: Carvalho, C. J. B. and E. A. B. Almeida (Eds.), *Biogeografia da América do Sul: Padroes e processos*. Editora Roca Limitada, São Paulo, Brazil, pp. 277–298.

Casagranda, M. D., S. Roig-Juñent and C. Szumik. 2009. Endemismo a diferentes escalas espaciales: Un ejemplo con Carabidae (Coleoptera: Insecta) de América del Sur austral. *Revista Chilena de Historia Natural*, 82: 17–42.

Castellanos, A. and R. A. Pérez-Moreau. 1944. Los tipos de vegetación de la República Argentina. *Monografías del Instituto de Estudios Geográficos, Universidad Nacional de Tucumán*, 4: 1–154.

Caviedes, C. N. and A. Iriarte. 1989. Migration and distribution of reodents in central Chile since the Pleistocene: The palaeogeographic evidence. *Journal of Biogeography*, 16: 181–187.

Cavieres, L. A., M. T. K. Arroyo, P. Posadas, C. Marticorena, O. Mattei, R. Rodríguez, F. A. Squeo and G. Arancio. 2002. Identification of priority areas for conservation in an arid zone: Application of parsimony analysis of endemicity in the vascular flora of the Antofagasta region, northern Chile. *Biodiversity and Conservation*, 11: 1301–1311.

Cecca, F., J. J. Morrone and M. C. Ebach. 2011. Biogeographical convergence and time-slicing: Concepts and methods in cladistic biogeography. In: Upchurch, P., A. McGowan and C. Slater (Eds.), *Palaeogeography and Palaeobiogeography: Biodiversity in Space and Time*, Systematics Association Special Volume, CRC Press, Boca Raton, FL, pp. 1–12.

Cekalovic, T. 1974. Divisiones biogeográficas de la XII Región Chilena (Magallanes). *Boletín de la Sociedad de Biología de Concepción*, 48: 297–314.

Chani-Posse, M. 2010. Revision of the southern South American species of *Philonthus* Stephens (Coleoptera: Staphylinidae). *Zootaxa*, 2595: 1–70.

Chani-Posse, M. 2014. An illustrated key to the New World genera of Philonthina Kirby (Coleoptera: Staphylinidae), with morphological, taxonomical and distributional notes. *Zootaxa*, 3755: 62–86.

Charrier, A. 2012. Los anfibios de Chile en medio de la tormenta global: Perspectivas y proyecciones para su conservación. *Revista de Conservación Ambiental*, 2: 3–5.

Cione, A. L., G. M. Gasparini, E. Soibelzon, L. H. Soibelzon and E. P. Tonni. 2015. *The Great American Biotic Interchange: A South American Perspective*. Springer Briefs in Earth System Sciences, Dordrecht, the Netherlands.

Clark, W. E. 1979. Taxonomy and biogeography of weevils of the genus *Sibinia* Germar (Coleoptera: Curculionidae) associated with *Prosopis* (Leguminosae: Mimosoideae) in Argentina. *Proceedings of the Entomological Society of Washington*, 81: 153–170.

Clarke, C. B. 1892. On biologic regions and tabulation areas. *Philosophical Transactions of the Royal Society of London*, 183: 371–387.

Cleef, A. M. 1978. Characteristics of Neotropical páramo vegetation and its Subantarctic relations. In: Troll, C. and W. Lauer (Eds.), *Geoecological Relations between the Southern Temperate Zone and the Tropical Mountains. Erdwissenschaftliche Forschung*, 11: 365–390.

Cleef, A. M. 1981. The vegetation of the páramos of the Colombian Cordillera Oriental. *Dissertationes Botanicae*, 61: 1–320.

Clement, M., D. Posada and K. A. Crandall. 2000. TCS: A computer program to estimate gene genealogies. *Molecular Ecology*, 9: 1657–1659.

Coelho, L. A., C. Molineri, D. A. Dos Santos and P. S. F. Ferreira. 2016. Biogeography and areas of endemism of *Prepops* Reuter (Heteroptera: Miridae). *Revista de Biología Tropical*, 64: 17–31.

Collins, W. D., M. Blackmon, C. Bitz, G. Bonan and C. Bretherton. 2004. The community climate system model: CCSM3. *Journal of Climate*, 19: 2122–2143.

Cook, L. G. and M. D. Crisp. 2005. Not so ancient: The extant crown group of *Nothofagus* represents a post. Gondwanan radiation. *Proceedings of the Royal Society B: Biological Sciences*, 272: 2535–2544.

Cosacov, A., L. A. Johnson, V. Paiaro, A. A. Cocucci, F. E. Córdoba and A. N. Sérsic. 2013. Precipitation rather than temperature influenced the phylogeography of the endemic shrub *Anarthrophyllum desideratum* in the Patagonian steppe. *Journal of Biogeography*, 40: 168–182.

Coscarón, S. and C. L. Coscarón-Arias. 1995. Distribution of Neotropical Simuliidae (Insecta, Diptera) and its areas of endemism. *Revista de la Academia Colombiana de Ciencias*, 19: 717–732.

Coulleri, J. P. and M. S. Ferrucci. 2012. Biogeografía histórica de *Cardiospermum* y *Urvillea* (Sapindaceae) en América: Paralelismos geográficos e históricos con los bosques secos estacionales neotropicales. *Boletín de la Sociedad Argentina de Botánica*, 47: 103–117.

Cox, C. B. C. 2001. The biogeographic regions reconsidered. *Journal of Biogeography*, 28: 511–523.

Cox, C. B. C. and P. D. Moore. 2005. *Biogeography: An Ecological and Evolutionary Approach*, 7th ed. Blackwell Science, Oxford, UK.

Cracraft, J. 1985. Historical biogeography and patterns of differentiation within the South American avifauna: Areas of endemism. *Ornithological Monographs*, 36: 49–84.

Craw, R. C., J. R. Grehan and M. J. Heads. 1999. *Panbiogeography: Tracking the History of Life*. Oxford Biogeography Series 11, Oxford University Press, New York.

Crawford, D. J., T. F. Stuessy, M. B. Cosner, D. H. Haines, M. Silva and M. Baeza. 1992. Evolution of the genus *Dendroseris* (Asteraceae: Lactuceae) on the Juan Fernandez islands: Evidence from chloroplast and ribosomal DNA. *Systematic Botany*, 17: 676–682.

Crisci, J. V., M. M. Cigliano, J. J. Morrone and S. Roig-Juñent. 1991. Historical biogeography of southern South America. *Systematic Zoology*, 40: 152–171.

Crisci, J. V., M. S. de la Fuente, A. A. Lanteri, J. J. Morrone, E. Ortiz Jaureguizar, R. Pascual and J. L. Prado. 1993. Patagonia, Gondwana Occidental (GW) y Oriental (GE), un modelo de biogeografía histórica. *Ameghiniana*, 30: 104.

Cuatrecasas, J. 1986. Speciation and radiation of the Espeletiinae in the Andes. In: Vuilleumier, F. and M. Monasterio. (Eds.), *High Altitude Tropical Biogeography*, Oxford University Press and American Museum of Natural History, New York, pp. 267–303.

Curler, G. R., J. K. Moulton and R. I. Madriz. 2015. Redescription of *Aposycorax chilensis* (Tonnoir) (Diptera, Psychodidae, Sycorinae) with the first identification of a blood meal for the species. *Zootaxa*, 4048: 114–126.

Daniel, G. M. and F. Z. Vaz-de-Mello. 2016. Biotic components of dung beetles (Insecta: Coleoptera: Scarabaeidae: Scarabaeinae) from Pantanal–Cerrado border and its implications for Chaco regionalization. *Journal of Natural History*, 50: 1159–1173.

Daniels, L. D. and T. V. Veblen. 2000. ENSO effects on temperature and precipitation of the Patagonian-Andean region: Implications for biogeography. *Physical Geography*, 21: 223–243.

Darlington, P. J. Jr. 1957. *Zoogeography: The Geographical Distribution of Animals*. John Wiley & Sons, New York.

Davis, S. D., V. H. Heywood and A. C. Hamilton. 1997. *Centres of Plant Diversity: A Guide and Strategy for Their Conservation. Volume 3. The Americas*. IUCN Publications Unit, Cambridge, UK.

De Candolle, A. P. 1820. Géographie botanique. In: Levrault, F. G. (Ed.), *Dictionnaire des Sciences Naturelles*, Imprimeur du Roi, Paris, France, pp. 359–422.

Dellafiore, C. 2017a. Southern South America: Southern Argentina and southeastern Chile. In: *Terrestrial Ecoregions of Latin America and the Caribbean*, World Wildlife Fund. WWF NT0805. https://www.worldwildlife.org/ecoregions/nt0805. Accessed November 20, 2017.

Dellafiore, C. 2017b. Southern South America: Southern Argentina and Chile. In: *Terrestrial Ecoregions of Latin America and the Caribbean*, World Wildlife Fund. WWF NT0804. https://www.worldwildlife.org/ecoregions/nt0804. Accessed November 20, 2017.

Dellafiore, C. 2017c. Southern South America: Southern Argentina, streching northward. In: *Terrestrial Ecoregions of Latin America and the Caribbean*, World Wildlife Fund. WWF NT0802. https://www.worldwildlife.org/ecoregions/nt0802. Accessed November 20, 2017.

Del Río, M. G., J. J. Morrone and A. A. Lanteri. 2015. Evolutionary biogeography of South American weevils of the tribe Naupactini (Coleoptera: Curculionidae). *Journal of Biogeography*, 42: 1293–1304.

Delachaux, E. A. S. 1920. Las regiones físicas de la República Argentina. *Revista del Museo de La Plata*, 15: 102–131.

del Valle Elías, G. and L. Aagesen. 2016. Areas of vascular plants endemism in the Monte desert (Argentina). *Phytotaxa*, 266: 161–251.

Díaz Gómez, J. M. 2009. Historical biogeography of *Phymaturus* (Iguania: Liolaemidae) from Andean and Patagonian South America. *Zoologica Scripta*, 38: 1–7.

Di Castri, F. 1968. Esquisse écologique du Chili. In: Delamare Debouteville, C. and Rappoport, E. (Eds.), *Biologie de L'Amérique Australe. Vol. 4*. Editions du Centre National de la Recherche Scientifique, Paris, France, pp. 6–52.

Diels, L. 1908. *Pflanzengeographie*. Göschensche Verlagshandlung, Leipzig, Germany.

Dillon, M. O., T. Tu, L. Xie, V. Quipuscoa Silvestre and J. Wen. 2009. Biogeographic diversification in *Nolana* (Solanaceae), a ubiquitous member of the Atacama and Peruvian Deserts along the western coast of South America. *Journal of Systematics and Evolution*, 47: 457–476.

Dinerstein, E., D. M. Olson, D. J. Graham, A. L. Webster, S. A. Primm, M. P. Bookbinder and G. Ledec. 1995. *A Conservation Assessment of the Terrestrial Ecoregions of Latin America and the Caribbean*. The World Bank, Washington, DC.

Domínguez, M. C., S. Roig-Juñent, J. J. Tassin, F. C. Ocampo and G. E. Flores. 2006. Areas of endemism of the Patagonian steppe: An approach based on insect distributional patterns using endemicity analysis. *Journal of Biogeography*, 33: 1527–1537.

Domínguez Díaz, E., D. Vega-Valdés, O. Dollenz, R. Villa-Martínez, J. C. Aravena, J. M. Henríquez and C. Muñoz-Escobar. 2015. Flora y vegetación de turberas de la Región de Magallanes. In: Domínguez, E. and D. Vega-Valdés (Eds.), Funciones y servicios ecosistémicos de las turberas en Magallanes, Colección de libros INIA 33, Instituto de Investigaciones Agropecuarias, Centro Regional Kampenaike, Punta Arenas, Chile, pp. 149–195.

Donato, M. 2006. Historical biogeography of the family Tristiridae (Orthoptera: Acridomorpha) applying dispersal-vicariance analysis. *Journal of Arid Environments*, 66: 421–434.

Donato, M., P. Posadas, D. R. Miranda-Esquivel, E. Ortiz Jaureguizar and G. Cladera. 2003. Historical biogeography of the Andean region: Evidence from Listroderina (Coleoptera: Curculionidae: Rhytirrhinini) in the context of the South American geobiotic scenario. *Biological Journal of the Linnean Society*, 80: 339–352.

Donoso, C. 1996. Ecology of *Nothofagus* forests in Central Chile. In: Veblen, T. T., R. S. Hill and J. Read (Eds.), *The Ecology and Biogeography of Nothofagus forests*, Yale University Press, London, UK, pp. 271–292.

Doyle, J. J. and J. A. Doyle. 1987. A rapid DNA isolation procedure for small quantities of fresh leaf tissue. *Phytochemistry Bulletin*, 19: 11–15.

Drude, O. 1884. *Die Florenreiche der Erde: Darstellung der gegenwärtigen Verbreitungsverhältnisse der Pflanzen: Ein Beitrag zur vergleichenden Erdkunde*. Petermanns Mitteilungen, Justus Perthes, Gotha, Germany.

Drude, O. 1890. *Handbuch der Pflanzengeographie*. Bibliothek Geographischer Handbücher, Verlag J. Engelhorn, Stuttgart, Germany.

Durante, S. P. and A. Abrahamovich. 2002. New leafcutting bee species of the subgenus *Megachile (Dasymegachile)* (Hymenoptera, Megachilidae) from Magellanic Forest province, in Patagonia Argentina. *Transactions of the American Entomological Society*, 128: 361–366.

Ebach, M. C., A. C. Gill, A. Kwan, S. T. Ahyong, D. J. Murphy and G. Cassis. 2013. Towards an Australian Bioregionalisation Atlas: A provisional area taxonomy of Australia's biogeographical regions. *Zootaxa*, 3619: 315–342.

Ebach, M. C., J. J. Morrone, L. R. Parenti and A. L. Viloria. 2008. International code of area nomenclature. *Journal of Biogeography*, 35: 1153–1157.

Echeverría-Londoño, S. and D. R. Miranda-Esquivel. 2011. MartiTracks: A geometrical approach for identifying geographical patterns of distribution. *PLoS One*, 6: 1–7.

Echeverry, A. and J. J. Morrone. 2010. Parsimony analysis of endemicity as a panbiogeographical tool: An analysis of Caribbean plant taxa. *Biological Journal of the Linnean Society*, 101: 961–976.

Elgueta, M. 2013. *Geosphaeropterus*, nuevo género de Tropiphorini (Coleoptera: Curculionidae) de Chile, con descripción de tres nuevas especies. *Boletín del Museo Nacional de Historia Natural, Chile*, 62: 203–217.

Ellenrieder, N. von. 2001. Species composition and distribution of the Argentinean Aeshnidae (Odonata: Anisoptera). *Revista de la Sociedad Entomológica Argentina*, 60: 39–60.

Engler, A. 1879. *Versuch Einer Entwicklungsgeschichte der Pflanzenwelt, Insbesondere der Florengebiete Seit der Tertiärperiode. Vol 1. Die Extratropischen Gebiete der Nördlichen Hemisphäre*. Verlag von W. Engelmann, Leipzig, Germany.

Engler, A. 1882. *Versuch Einer Entwicklungsgeschichte der Pflanzenwelt, Insbesondere der Florengebiete Seit der Tertiärperiode. Vol. 2. Die Extratropischen Gebiete der Südlichen Hemisphäre und die Tropischen Gebiete*. Verlag von W. Engelmann, Leipzig, Germany.

Engler, A. 1899. *Die Entwickelung der Pflanzengeographie in den Letzten Hundert Jahren und Weitere Aufgaben Derselben. Wissenschaftliche Beiträge zum Gedächtnis der Hundertjährigen Wiederkehr des Antritts von Alexander von Humboldt's Reise nach Amerika*. Gesellschaft für Erdkunde, Berlin, Germany.

Erséus, C. and R. Grimm. 2002. A new species of *Ainudrilus* (Tubificidae) from South Georgia, and other Subantarctic freshwater oligochaetes. *Hydrobiologia*, 468: 77–81.

Escalante, T. 2009. Un ensayo sobre regionalización biogeográfica. *Revista Mexicana de Biodiversidad*, 80: 551–560.

Escalante, T. 2017. A natural regionalization of the world based on primary biogeographical homology of terrestrial mammals. *Biological Journal of the Linnean Society*, 120: 349–362.

Escalante, T., M. Linaje, P. Illoldi-Rangel, M. Rivas, P. Estrada, F. Neira and J. J. Morrone. 2009. Ecological niche models and patterns of richness and endemism of the southern Andean genus *Eurymetopum* (Coleoptera, Cleridae). *Revista Brasileira de Entomologia*, 53: 379–385.

Excoffier, L., G. Laval and S. Schneider. 2005. Arlequin ver 3.01. An integrated software package for population genetics data analysis. *Evolutionary Bioinformatics Online*, 1: 47–60.

Ezcurra, C. 2017. Southern South America: Western Argentina into Chile. In: *Terrestrial Ecoregions of Latin America and the Caribbean*, World Wildlife Fund. WWF NT1008. https://www.worldwildlife.org/ecoregions/nt1008. Accessed November 20, 2017.

Ezcurra, C., A. C. Premoli, C. P. Souto, M. A. Aizen, M. Arbetman, P. Mathiasen, M. C. Acosta and P. Quiroga. 2014. La vegetación de la región Andino-Patagónica tiene su historia. In: Raffaele, E., M. De Torres Curth, C. L. Morales and T. Kitzberger (Eds.), *Ecología e Historia Natural de la Patagonia andina: Un cuarto de Siglo de Investigación em Biogeografía, Ecología y Conservación*, Fundación de Historia Natural Felix de Azara, Buenos Aires, Argentina, pp. 19–36.

Farris, J. S. 1988. *Hennig86 reference. Version 1.5*. Published by the author, Port Jefferson, New York.

Felsenstein, J. 1981. Evolutionary trees from DNA sequences: A maximum likelihood approach. *Journal of Molecular Evolution*, 17: 368–376.

Fergnani, P. N., Sackmann, P. and Ruggiero, A. 2013. The spatial variation in ant species composition and functional groups across the Subantarctic-Patagonian transition zone. *Journal of Insect Conservation*, 17: 295–305.

Ferretti, N. 2015. Cladistic reanalysis and historical biogeography of the genus *Lycinus* Thorell, 1894 (Araneae: Mygalomorphae: Nemesiidae) with description of two new species from westerm Argentina. *Zoological Studies*, 54: 1–15.

Ferretti, N., A. González and F. Pérez-Miles. 2012. Historical biogeography of mygalomorph spiders from the peripampasic orogenic arc based on track analysis and PAE as a panbiogeographical tool. *Systematics and Biodiversity*, 10: 179–193.

Ferretti, N., A. González and F. Pérez-Miles. 2014a. Identification of priority areas for conservation in Argentina: Quantitative biogeography insights from mygalomorph spiders (Araneae: Mygalomorphae). *Journal of Insect Conservation*, 18: 1087–1096.

Ferretti, N., F. Pérez-Miles and A. González. 2014b. Historical relationships among Argentinean biogeographic provinces based on mygalomorph spider distribution data (Araneae: Mygalomorphae). *Studies on Neotropical Fauna and Environment*, 49: 1–10.

Ferro, I. 2013. Rodent endemism, turnover and biogeographical transitions on elevation gradients in the northwestern Argentinian Andes. *Mammalian Biology*, 78: 322–331.

Ferro, I. and J. J. Morrone. 2014. Biogeographic transition zones: A search for conceptual synthesis. *Biological Journal of the Linnean Society*, 113: 1–12.

Figueroa, L. and B. C. Ratcliffe. 2016. A new species of *Ancognatha* Erichson (Coleoptera: Scarabaeidae: Dynastinae: Cyclocephalini) from Peru, with distributions of Peruvian *Ancognatha* species. *The Coleopterists Bulletin*, 70: 65–72.

Fittkau, E. J. 1969. The fauna of South America. In: Fittkau, E., Illies, J. J., Klinge, H., Schwabe, G. H. and Sioli, H. (Eds.), *Biogeography and Ecology in South America, 2*. Junk, The Hague, pp. 624–650.

Fjeldså, J. 1992. Biogeographic patterns and evolution of the avifauna of relict high-altitude woodlands of the Andes. *Steenstrupia*, 18: 9–62.

Fleming, C. A. 1987. Comments on Udvardy's biogeographical realm Antarctica. *Journal of the Royal Society of New Zealand*, 17: 195–200.

Flint, O. S. 1989. Studies of Neotropical caddisflies, XXXIX: The genus *Smicridea* in the Chilean subregion (Trichoptera: Hydropsychidae). *Smithsonian Contributions to Zoology*, 472: 1–45.

Flores, G. E. 2000. Systematic revision and cladistic analysis of the Neotropical genera *Mitragenius* Solier, *Auladera* Solier and *Patagonogenius* gen. n. (Coleoptera: Tenebrionidae). *Entomologica Scandinavica*, 30: 361–396.

Flores, G. E. 2004. Systematic revision and cladistic analysis of the Patagonian genus *Platesthes* (Coleoptera: Tenebrionidae). *European Journal of Entomology*, 101: 591–608.

Flores, G. E. and G. H. Cheli. 2014. Two new species of *Nyctelia* (Coleoptera: Tenebrionidae) from Argentinean Patagonia with zoogeographical and ecological. *Zootaxa*, 3765: 279–287.

Flores, G. E. and J. Pizarro-Araya. 2004. *Caenocrypticoides triplehorni* new species, the first record of Caenocrypticini (Coleoptera: Tenebrionidae) in Argentina, with cladistic analysis of the genus. *Annales Zoologici*, 54: 721–728.

Flores, G. E. and M. Chani-Posse. 2005. *Patagonopraocis*, a new genus of Praocini from Patagonia (Coleoptera: Tenebrionidae). *Annales Zoologici*, 55: 575–581.

Flores, G. E. and J. Pizarro-Araya. 2012. Systematic revision of the South American genus *Praocis* Escholtz, 1829 (Coleoptera: Tenebrionidae). Part 1: Introduction and subgenus *Praocis* s. str. *Zootaxa*, 3336: 1–35.

Flores, G. E. and S. Roig-Juñent. 1997. Systematic revision of the Neotropical genus *Entomoderes* Solier (Coleoptera: Tenebrionidae). *Entomologica Scandinavica*, 28: 141–162.

Flores, G. E. and S. Roig-Juñent. 2001. Cladistic and biogeographic analyses of the Neotropical genus *Epipedonota* Solier (Coleoptera: Tenebrionidae), with conservation considerations. *Journal of the New York Entomological Society*, 109: 309–336.

Flores, G. E. and P. Vidal. 2000. Cladistic analysis of the Chilean genus *Callyntra* Solier (Coleoptera: Tenebrionidae), with description of a new species. *Journal of the New York Entomological Society*, 108: 187–204.

Flores, G. E. and P. Vidal. 2001. Systematic revision and redefinition of the Neotropical genus *Epipedonota* Solier (Coleoptera: Tenebrionidae), with descriptions of eight new species. *Insect Systematics and Evolution*, 32: 1–43.

Giribet, G. and G. D. Edgecombe. 2006. The importance of looking at small-scale patterns when inferring Gondwanan biogeography: A case study of the centipede *Paralamyctes* (Chilopoda, Lithobiomorpha, Henicopidae). *Biological Journal of the Linnean Society*, 89: 65–78.

Glasby, C. J. 2005. Polychaete distribution patterns revisited: An historical explanation. *Marine Ecology*, 26: 235–245.

Godley, E. J. 1960. The botany of southern Chile in relation to New Zealand and the Subantarctic. *Proceedings of the Royal Society of London, Series B-Biological Sciences*, 152: 457–475.

Goetsch, W. 1931. Estudios sobre zoogeografía chilena. *Boletín de la Sociedad de Biología de Concepción*, 5: 1–19.

Goin, F. J., M. O. Woodburne, A. N. Zimicz, G. M. Martin and L. Chornogubsky. 2015. *A brief history of South American Metatherians: Evolutionary contexts and intercontinental dispersals*. Springer Briefs in Earth System Sciences, Dordrecht, the Netherland.

Goloboff, P. 1998. *NONA ver. 2.0*. Available at http://www.cladistics.com/about_nona.htm.

Goloboff, P. 2004. *NDM/ VNDM programs ver. 1.5*. Available at www.zmuc.dk - / public/phylogeny/Endemism/.

Goloboff, P., J. S. Farris and K. C. Nixon. 2008. TNT, a free program for phylogenetic analysis. *Cladistics*, 24: 774–786.

González, G. 2014. Coccinelidae. In: Roig-Juñent, S., Claps L. E. and Morrone J. J. (Eds.), *Biodiversidad de Artrópodos Argentinos: Volumen 3*, Facultad de Ciencias Naturales e Instituto Miguel Lillo, San Miguel de Tucumán, Argentina, pp. 509–530.

Good, R. 1947. *The Geography of the Flowering Plants*. Longman, London, UK.

Gradstein, S. R. 1998. Hepatic diversity in the Neotropical páramos. In: *Proceedings of the VI Congreso Latinoamericano de Botánica (1994), Monographs in Systematic Botany from the Missouri Botanical Garden 68*, St. Louis, Missouri, pp. 69–85.

Graham, A. 2004. *A Natural History of the New World: The Ecology and Evolution of Plants in the Americas*. The University of Chicago Press, Chicago, IL.

Granara de Willink, M. C. 2014. Pseudococcidae. In: Roig-Juñent, S., L. E. Claps and J. J. Morrone (Eds.), *Biodiversidad de Artrópodos Argentinos: Volumen 3*, Facultad de Ciencias Naturales e Instituto Miguel Lillo, San Miguel de Tucumán, Argentina, pp. 249–259.

Green, A. J. A. 1974. Oniscoidea (terrestrial Isopoda). In: Williams, W. D. (Ed.), *Biogeography and Ecology in Tasmania*. Junk, The Hague, the Netherlands, pp. 229–249.

Gressitt, J. L. 1970. Subantarctic entomology and biogeography. *Pacific Insects Monographs*, 23: 295–374.

Grismado, C. J. and N. I. Platnick. 2008. Review of the South American spider genus *Platnickia* (Araneae, Zodaridae). *American Museum Novitates*, 3625: 1–19.

Grismado, C. J. and M. J. Ramírez. 2014. Diguetidae. In: Roig-Juñent, S., L. E. Claps and J. J. Morrone (Eds.), *Biodiversidad de artrópodos argentinos: Volumen 3*, Facultad de Ciencias Naturales e Instituto Miguel Lillo, San Miguel de Tucumán, Argentina, pp. 141–145.

Guerrero, P. C., M. Rosas, M. T. K. Arroyo and J. J. Wiens. 2013. Evolutionary lag times and recent origin of the biota of an ancient desert (Atacama-Sechura). *Proceedings of the National Academy of Sciences*, 110: 11469–11474.

Hartley, A. J. and G. Chong. 2002. Late Pliocene age for the Atacama Desert: Implications for the desertification of western South America. *Geology*, 30: 43–46.

Hasumi, H. and S. Emori. 2004. *K-1 coupled GCM(MIROC) description*. Center for Climate System Research, University of Tokyo, Tokyo, Japan.

Hauman, L. 1920. Ganadería y geobotánica en la Argentina. *Revista del Centro de Estudiantes de Agronomía y Veterinaria de la Universidad de Buenos Aires*, 102: 45–65.

Hauman, L. 1931. Esquisse phytogéographique de l'Argentine subtropicale et de ses relations avec la géobotanique sud-américaine. *Bulletin de la Société Royale de Botanique de Belgique*, 64: 20–79.

Heads, M. 2006. Panbiogeography of *Nothofagus* (Nothofagaceae): Analysis of the main species massings. *Journal of Biogeography*, 33: 1056–1075.

Hechem, V., L. Acheritobehere and J. J. Morrone. 2011. Patrones de distribución de las especies de *Cynanchum, Diplolepis* y *Tweedia* (Apocynaceae: Asclepiadoideae) de America del Sur austral. *Revista de Geografía Norte Grande*, 48: 45–60.

Hechem, V., A. Padró and J. J. Morrone. 2015. Patrones distribucionales de la flora vascular de la estepa patagónica y su relevancia para la regionalización biogeográfica. *Darwiniana*, 3: 5–20.

Heilprin, A. 1887. *The Geographical and Geological Distribution of Animals*. International Scientific Series, London, UK.

Hernández Camacho, J., A. Hurtado Guerra, R. Ortiz Quijano and T. Walschburger. 1992a. Centros de endemismo en Colombia. In: Halffter, G. (Ed.), *La Diversidad Biológica de Iberoamérica*, Acta Zoológica Mexicana. Vol. Esp., 1992, Cyted-D, Programa Iberoamericano de Ciencia y Tecnología para el Desarrollo, Instituto de Ecología, A. C., Xalapa, Mexico, pp. 175–190.

Hernández Camacho, J., A. Hurtado Guerra, R. Ortiz Quijano and T. Walschburger. 1992b. Unidades biogeográficas de Colombia. In: Halffter, G. (Ed.), *La Diversidad Biológica de Iberoamérica*, Acta Zoológica Mexicana. Volumen Especial, 1992, Cyted-D, Programa Iberoamericano de Ciencia y Tecnología para el Desarrollo, Instituto de Ecología, A. C., Xalapa, Mexico, pp. 105–151.

Hernández Camacho, J. and H. Sánchez Páez. 1992. Biomas terrestres de Colombia. In: Halffter, G. (Ed.), *La diversidad biológica de Iberoamérica*, Acta Zoológica Mexicana, Vol. Esp., 1992, Cyted-D, Programa Iberoamericano de Ciencia y Tecnología para el Desarrollo, Instituto de Ecología, A. C., Xalapa, Mexico, pp. 153–173.

Hershkovitz, P. 1969. The evolution of mammals on southern continents. VI: The Recent mammals of the Neotropical region: A zoogeographic and ecological review. *Quarterly Review of Biology*, 44: 1–70.

Hijmans, R. J., S. E. Cameron, J. L. Parra, G. Jones and A. Jarvis. 2005. Very high resolution interpolated climate surfaces for global land areas. *International Journal of Climatology*, 25: 1965–1978.

Himes, C. M. T., M. H. Gallardo and G. J. Kenagy. 2008. Historical biogeography and post-glacial recolonization of South American temperate rain forest by the relictual marsupial *Dromiciops gliroides*. *Journal of Biogeography*, *35:* 1415–1424.

Holdgate, M. W. 1960. Vegetation and soils in the south Chilean islands. *Journal of Ecology*, 49: 559–580.

Holmberg, E. L. 1898. La flora de la República Argentina. *Segundo Censo de la República Argentina*, 1: 385–474.

Hooker, J. D. 1844–1860. *The Botany of the Antarctic Voyage of H. M. Discovery Ships Erebus and Terror, in the Years 1839–1843. I Flora Antarctica, II Flora Nova-Zelandiae, III Flora Tasmaniae*, L. Reeve, London, UK.

Huber, B. A. 2014. Pholcidae. In: Roig-Juñent, S., L. E. Claps and J. J. Morrone (Eds.), *Biodiversidad de artrópodos argentinos: Volumen 3*, Facultad de Ciencias Naturales e Instituto Miguel Lillo, San Miguel de Tucumán, Argentina, pp. 131–140.

Huber, O. and R. Riina. 1997. *Glosario ecológico de las Américas. Vol. I. América del Sur: Países hispanoparlantes*. UNESCO, Paris, France.

Hueck, K. 1957. Las regiones forestales de Sudamérica. *Boletín del Instituto Forestal Latinoamericano de Investigación y Capacitación* (Mérida), 2: 1–40.

Hueck, K. 1966. *Die Wälder Südamerikas*. Fischer, Stuttgart, Germany.

Hughes, C. and R. Eastwood. 2006. Island radiation on a continental scale: Exceptional rates of plant diversification after uplift of the Andes. *Proceedings of the National Academy of Science*, 103: 10334–10339.

Huson, D. H. and D. Bryant. 2006. Application of phylogenetic networks in evolutionary studies. *Molecular Biology and Evolution*, 23: 254–267.

Huxley, T. H. 1868. On the classification and distribution of Alectoromorphae and Heteromorphae. *Proceedings of the Zoological Society of London*, 1868: 294–319.

Jeannel, R. 1938. Les Migadopides (Coleoptera, Adephaga), une lignée subantarctique. *Revue Française d'Entomologie*, 5: 1–55.

Jeannel, R. 1942. *La genèse des faunes terrestres: Élements de biogéographie*. Presses Universitaires de France, Paris, France.

Jerez, V. and C. Muñoz-Escobar. 2015. Coleópteros y otros insectos asociados a turberas del páramo magallánico en la Región de Magallanes, Chile. In: Domínguez, E. and D. Vega-Valdés (Eds.), Funciones y servicios ecosistémicos de las turberas en Magallanes, Colección de libros INIA 33, Instituto de Investigaciones Agropecuarias, Centro Regional Kampenaike, Punta Arenas, Chile, pp. 199–225.

José de Paggi, S. 1990. Ecological and biogeographical remarks on the rotifer fauna of Argentina. *Revue d'Hydrobiologie Tropicale*, 23: 297–311.

Katinas, L., J. J. Morrone and J. V. Crisci. 1999. Track analysis reveals the composite nature of the Andean biota. *Australian Systematic Botany*, 47: 111–130.

Kirby, W. F. 1872. On the geographical distribution of the diurnal Lepidoptera as compared with that of birds. *Journal of the Linnean Society of London*, 11: 431–439.

Klassa, B. and C. M. D. Santos. 2015. Areas of endemism in the Neotropical region based on the geographical distribution of Tabanomorpha (Diptera: Brachycera). *Zootaxa*, 4058: 519–534.

Kotov, A. A., A. Y. Sinev and V. L. Berrios. 2010. The Cladocera (Crustacea: Branchiopoda) of six high altitude water bodies in the North Chilean Andes, with discussion of Andean endemism. *Zootaxa*, 2430: 1–66.

Kreft, H. and Jetz, W. 2010. A framework for delineating biogeographical regions based on species distributions. *Journal of Biogeography*, 37: 2029–2053.

Kuschel, G. 1960. Terrestrial zoology in southern Chile. *Proceedings of the Royal Society of London, Series B*, 152: 540–550.

Kuschel, G. 1961. Composition and relationship of the terrestrial faunas of Easter, Juan Fernandez, Desventuradas, and Galapágos [sic] Islands. In: Anon. (Ed.), Tenth Pacific Science Congress, Pacific Science Association, Honolulu, Hawaii, pp. 79–95.

Kuschel, G. 1962. Zur Naturgeschichte der Insel San Ambrosio (Islas Desventuradas, Chile). 1. Reisebreicht, geographische Verhältnisse und Pflanzenverbreitung. *Arkiv för Botanik*, 4: 413–419.

Kuschel, G. 1964. Problems concerning an Austral region. In: Gressitt, J. L., Lindroth, C. H., Fosberg, F. R., Fleming, C. A. and Turbott, E. G. (Eds.), *Pacific Basin Biogeography: A Symposium, 1963 [1964]*. Bishop Museum Press, Honolulu, Hawaii, pp. 443–449.

Kuschel, G. 1969. Biogeography and ecology of South American Coleoptera. In: Fittkau, E., Illies, J. J., Klinge, H., Schwabe, G. H. and Sioli, H. (Eds.), *Biogeography and Ecology in South America, 2*. Junk, The Hague, the Netherlands, pp. 709–722.

Kuschel, G. and B. M. May. 1996. Discovery of Palophaginae (Coleoptera: Megalopodidae) on *Araucaria araucana* in Chile and Argentina. *New Zealand Entomologist*, 19: 1–13.

Kutschker, A. and J. J. Morrone. 2012. Distributional patterns of the species of *Valeriana* (Valerianaceae) in southern South America. *Plant Systematics and Evolution*, 298: 535–547.

Lachmuth, S., W. Durka and F. Schurr. 2010. The making of a rapid plant invader: Genetic diversity and differentiation in the native and invaded range of *Senecio inaequidens*. *Molecular Ecology*, 19: 3952–3967.

Lamas, C. J. E., S. S. Nihei, A. M. Cunha and M. S. Couri. 2014. Phylogeny and biogeography of *Heterostylum* (Diptera: Bombyliidae): Evidence for an ancient Caribbean diversification model. *Florida Entomologist*, 97: 952–966.

Lamas, G. 1982. A preliminary zoogeographical division of Peru, based on butterfly distributions (Lepidoptera, Papilionoidea). In: Prance, T. P. (Ed.), *Biological Diversification in the Tropics*, Columbia University Press, New York, pp. 336–357.

Lavery, A. H. 2017. Annotated checklist of the spiders, harvestmen, and pseudoscorpions of the Falkland Islands and South Georgia. *Arachnology*, 17: 210–218.

Leivas, F. W. T., N. Degallier and L. M. Almeida. 2015. New species of *Omalodes* and redefinition of the tribe Omalodini (Coleoptera: Histeridae: Histerinae). *Zootaxa*, 3925: 109–119.

León, R. J. C., D. Bran, M. Collantes, J. M. Paruelo and A. Soriano. 1998. Grandes unidades de vegetación de la Patagonia extra andina. *Ecología Austral*, 8: 125–144.

Liebherr, J. K., J. W. M. Marris, R. M. Emberson, P. Syrett and S. Roig-Juñent. 2011. *Orthoglymma wangapeka* gen.n., sp.n. (Coleoptera: Carabidae: Broscini): A newly discovered relict from the Buller Terrane, north-western South Island, New Zealand, corroborates a general pattern of Gondwanan endemism. *Systematic Entomology*, 36: 395–414.

Liria, J. 2008. Sistemas de información geográfica y análisis espaciales: Un método combinado para realizar estudios panbiogeográficos. *Revista Mexicana de Biodiversidad*, 79: 281–284.

Locklin, C. 2017. Southeastern South America: Central Chile. In: *Terrestrial Ecoregions of Latin America and the Caribbean*, World Wildlife Fund. WWF NT1201. https://www.worldwildlife.org/ecoregions/nt12001. Accessed November 20, 2017.

Londoño, C., A. Cleef and S. Madriñán. 2014. Angiosperm flora and biogeography of the páramo region of Colombia, northern Andes. *Flora*, 209: 81–87.

Lopes-Andrade, C. 2010. The first record of *Orthocis* Casey (Coleoptera: Ciidae) from the Andean region, with the description of a distinctive new species. *Zoological Science*, 27: 830–833.

López, H. L., R. C. Menni, M. Donato and A. M. Miquelarena. 2008. Biogeographical revision of Argentina (Andean and Neotropical regions): An analysis using freshwater fishes. *Journal of Biogeography*, 35: 1564–1579.

Lorentz, P. G. 1876. Cuadro de la vegetación de la República Argentina. In: Napp, R. (Ed.), *La República Argentina*, Buenos Aires, Argentina, pp. 77–136.

Löwenberg-Neto, P. 2015. Andean region: A shapefile of Morrone's (2015) biogeographical regionalization. *Zootaxa*, 3985: 600.

Löwenberg-Neto, P. and C. J. B. de Carvalho. 2009. Areas of endemism and spatial diversification of the Muscidae (Insecta: Diptera) in the Andean and Neotropical regions. *Journal of Biogeography*, 36: 1750–1759.

Löwenberg-Neto, P., C. J. B. de Carvalho and J. A. F. Diniz-Filho. 2008. Spatial congruence between biotic history and species richness of Muscidae (Diptera, Insecta) in the Andean and Neotropical regions. *Journal of Zoological Systematics and Evolutionary Research*, 46: 374–380.

Lücking, R., V. Wirth, L. I. Ferraro and M. E. S. Cáceres. 2003. Foliicolous lichens from Valdivian temperate rain forest of Chile and Argentina: Evidence of an austral element, with the description of seven new taxa. *Global Ecology and Biogeography*, 12: 21–36.

Luebert, F. 2011. Hacia una fitogeografía histórica del desierto de Atacama. *Revista de Geografía Norte Grande*, 50: 105–133.

Luebert, F. and P. Pliscoff. 2006. *Sinopsis bioclimática y vegetacional de Chile*. Editorial Universitaria, Santiago de Chile, Chile.

Luteyn, J. L. 1992. Paramos: Why study them? In: Balslev, H. and J. L. Luteyn (Eds.), *Paramo: An Andean Ecosystem under Human Influence*, Academic Press, London, UK, pp. 1–14.

Lydekker, B. A. 1896. *A Geographical History of Mammals*. Cambridge University Press, Cambridge, UK.

Maddison, W. P. and D. R. Maddison. 2013. *Mesquite: A Modular System for Evolutionary Analyses. Version 2.75*. http://mesquiteproject.org.

Magallón, S. A. 2004. Dating lineages: Molecular and paleontological approaches to the temporal framework of clades. *International Journal of Plant Sciences*, 165: S7–S21.

Mann, G. 1960. Regiones biogeográficas de Chile. *Investigaciones Zoológicas Chilenas*, 6: 15–49.

Marino, P. I., G. R. Spinelli and P. Posadas. 2001. Distributional patterns of species of Ceratopogonidae (Diptera) in southern South America. *Biogeographica*, 77: 113–122.

Marshall, J. E. A. 1994. The Falkland Islands: A key element in Gondwana paleogeography. *Tectonics*, 13: 499–514.

Marticorena, C. 1990. Contribución a la estadística de la flora vascular de Chile. *Gayana, Botánica*, 47: 85–113.

Martin, G. M. 2010. Geographic distribution and historical occurrence of *Dromiciops gliroides* Thomas (Metatheria: Microbiotheria). *Journal of Mammalogy*, 91: 1025–1035.

Martin, G. M. 2011. Geographic distribution of *Rhyncholestes raphanurus* Osgood, 1924 (Paucituberculata: Caenolestidae), an endemic marsupial of the Valdivian Temperate Rainforest. *Australian Journal of Zoology*, 59: 118–126.

Martínez, G. A., M. D. Arana, A. J. Oggero and E. S. Natale. 2017. Biogeographical relationships and new regionalisation of high-altitude grasslands and woodlands of the central Pampean Ranges (Argentina), based on vascular plants and vertebrates. *Australian Systematic Botany*, 29: 473–488.

Martínez Carretero, E. 1995. La Puna argentina: Delimitación general y división en distritos florísticos. *Boletín de la Sociedad Argentina de Botánica*, 31: 27–40.

Marvaldi, A. E. and M. S. Ferrer. 2014. Belidae. In: Roig-Juñent, S., L. E. Claps and J. J. Morrone (Eds.), *Biodiversidad de Artrópodos Argentinos: Volumen 3*, Facultad de Ciencias Naturales e Instituto Miguel Lillo, San Miguel de Tucumán, Argentina, pp. 531–539.

Mattick, F. 1964. Übersicht über die Florenreiche und Florengebiete der Erde. In: Melchior, H. (Ed.), *Engler's Syllabus der Pflanzenfamilien*, 12 ed. vol. 2. Angiospermen, Gerbrüder Borntraeger, Berlin, Germany, pp. 626–629.

Maury, E. A. 1979. Apuntes para una zoogeografía de la escorpiofauna argentina. *Acta Zoológica Lilloana*, 35: 703–719.

McDowall, R. M. 2005. Falkland Islands biogeography: Converging trajectories in the South Atlantic. *Journal of Biogeography*, 32: 49–62.

Melo, M. C. and Faúndez, E. I. 2011. Synopsis of the genus *Empicoris* (Hemiptera: Heteroptera: Reduviidae) in Chile. *Acta Entomologica Musei Nationalis Pragae*, 51: 11–20.

Mello-Leitão, C. de. 1937. *Zoo-geografia do Brasil*. Biblioteca Pedagógica Brasileira, Brasiliana, São Paulo.

Mello-Leitão, C. de. 1939. Les arachnides et la zoogéographie de l'Argentine. *Physis* (Buenos Aires), 17: 601–630.

Mello-Leitão, C. de. 1943. Los alacranes y la zoogeografía de Sudamérica. *Revista Argentina de Zoogeografía*, 2: 125–131.

Menu-Marque, S., J. J. Morrone and C. Locascio. 2000. Distributional patterns of the South American species of *Boeckella* (Copepoda: Centropagidae): A track analysis. *Journal of Crustacean Biology*, 20: 262–272.

Mercado-Salas, N. F., C. Pozo, J. J. Morrone and E. Suárez-Morales. 2012. Distributional patterns of the American species of the freshwater genus *Eucyclops* (Copepoda: Cyclopoida). *Journal of Crustacean Biology*, 32: 457–464.

Mihoč, M. A. K., J. J. Morrone, M. A. Negrito and L. A. Cavieres. 2006. Evolución de la serie *Mycrophyllae* (*Adesmia*, Fabaceae) en la Cordillera de los Andes: Una perspectiva biogeográfica. *Revista Chilena de Historia Natural* 79: 389–404.

Monasterio, M. 1986. Adaptive strategies of *Espeletia* in the Andean desert páramo. In: Vuilleumier, F. and M. Monasterio (Eds.), *High Altitude Tropical Biogeography*, Oxford University Press and American Museum of Natural History, New York, pp. 49–80.

Monckton, S. K. 2016. A revision of *Chilicola (Heteroediscelis)*, a subgenus of xeromelissine bees (Hymenoptera, Colletidae) endemic to Chile: Taxonomy, phylogeny, and biogeography, with descriptions of eight new species. *ZooKeys*, 591: 1–144.

Monrós, F. 1958. Consideraciones sobre la fauna del sur de Chile y revisión de la tribus Stenomelini (Coleoptera, Chrysomelidae). *Acta Zoológica Lilloana*, 15: 143–153.

Moore, D. M. 1968. The vascular flora of the Falkland Islands. *British Antarctic Survey Scientific Reports*, 60: 1–202.

Morain, S. A. 1984. *Systematic and Regional Biogeography*. Van Nostrand Reinhold Company, New York.

Moreira-Muñoz, A. 2007. The Austral floristic realm revisited. *Journal of Biogeography*, 34: 1649–1660.

Moreira-Muñoz, A. 2011. *Plant Geography of Chile*. Springer, New York.

Moreira-Muñoz, A. 2014. Central Chile ecoregion. In: Hobohm, C. (Ed.), *Endemism in Vascular Plants*. Springer Science, Dordrecht, the Netherlands, pp. 221–233.

Moreira-Muñoz, A. and S. Elórtegui Francioli. 2014. Juan Fernández archipelago. In: Hobohm, C. (Ed.), *Endemism in Vascular Plants*. Springer Science, Dordrecht, the Netherlands, pp. 165–181.

Morello, J. 1955. Estudios botánicos en las regiones áridas de la Argentina. II. *Revista de Agricultura del Noroeste Argentino*, 1: 385–524.

Morello, J. 1958. La provincia biogeográfica del Monte. *Opera Lilloana*, 2: 1–155.

Morello, J. 1984. *Perfil Ecológico de Sudamérica: Características Estructurales de Sudamérica y su Relación con Espacios Semejantes del Planeta*. Ediciones Cultura Hispánica, Instituto de Cooperación Iberamericana, Barcelona.

Morrone, J. J. 1992. Revisión sistemática, análisis cladístico y biogeografía histórica de los géneros *Falklandius* Enderlein y *Lanteriella* gen. nov. (Coleoptera: Curculionidae). *Acta Entomológica Chilena*, 17: 157–174.

Morrone, J. J. 1993a. Cladistic and biogeographic analyses of the weevil genus *Listroderes* Schoenherr (Coleoptera: Curculionidae). *Cladistics*, 9: 397–411.

Morrone, J. J. 1993b. Revisión sistemática de un nuevo género de Rhytirrhinini (Coleoptera: Curculionidae), con un análisis biogeográfico del dominio Subantártico. *Boletín de la Sociedad de Biología de Concepción*, 64: 121–145.

Morrone, J. J. 1994a. Distributional patterns of species of Rhytirrhinini (Coleoptera: Curculionidae) and the historical relationships of the Andean provinces. *Global Ecology and Biogeography Letters*, 4: 188–194.
Morrone, J. J. 1994b. Systematics, cladistics, and biogeography of the Andean weevil genera *Macrostyphlus*, *Adioristidius*, *Puranius*, and *Amathynetoides*, new genus (Coleoptera: Curculionidae). *American Museum Novitates*, 3104: 1–63.
Morrone, J. J. 1994c. Systematics of the Patagonian genus *Acrostomus* Kuschel (Coleoptera: Curculionidae). *Annals of the Entomological Society of America*, 87: 403–411.
Morrone, J. J. 1996a. The biogeographical Andean subregion: A proposal exemplified by Arthropod taxa (Arachnida, Crustacea, and Hexapoda). *Neotropica*, 42: 103–114.
Morrone, J. J. 1996b. Distributional patterns of the South American Aterpini (Coleoptera: Curculionidae). *Revista de la Sociedad Entomológica Argentina*, 55: 131–141.
Morrone, J. J. 1996c. The South American weevil genus *Rhyephenes* (Coleoptera: Curculionidae; Cryptorhynchinae). *Journal of the New York Entomological Society*, 104: 1–20.
Morrone, J. J. 1997. Cladistics of the New World genera of Listroderina (Coleoptera: Curculionidae: Rhytirrhinini). *Cladistics*, 13: 247–266.
Morrone, J. J. 1999. Presentación preliminar de un nuevo esquema biogeográfico de América del Sur. *Biogeographica*, 75: 1–16.
Morrone, J. J. 2000a. Delimitation of the Central Chilean subregion and its provinces, based mainly on Arthropod taxa. *Biogeographica*, 76: 97–106.
Morrone, J. J. 2000b. Biogeographic delimitation of the Subantarctic subregion and its provinces. *Revista del Museo Argentino de Ciencias Naturales, Nueva Serie*, 2: 1–15.
Morrone, J. J. 2001a. A proposal concerning formal definitions of the Neotropical and Andean regions. *Biogeographica*, 77: 65–82.
Morrone, J. J. 2001b. *Biogeografía de América Latina y el Caribe. Vol. 3*. Manuales & Tesis SEA, Sociedad Entomológica Aragonesa, Zaragoza, Spain, 148 p.
Morrone, J. J. 2001c. Review of the biogeographic provinces of the Patagonian subregion. *Revista de la Sociedad Entomológica Argentina*, 60: 1–8.
Morrone, J. J. 2001d. Toward a formal definition of the Paramo-Punan subregion and its provinces. *Revista del Museo Argentino de Ciencias Naturales, Nueva Serie*, 3: 1–12.
Morrone, J. J. 2002. Biogeographic regions under track and cladistic scrutiny. *Journal of Biogeography*, 29: 149–152.
Morrone, J. J. 2004a. Panbiogeografía, componentes bióticos y zonas de transición. *Revista Brasileira de Entomologia*, 48: 149–162.
Morrone, J. J. 2004b. La zona de transición Sudamericana: Caracterización y relevancia evolutiva. *Acta Entomológica Chilena*, 28: 41–50.
Morrone, J. J. 2006. Biogeographic areas and transition zones of Latin America and the Caribbean Islands based on panbiogeographic and cladistic analyses of the entomofauna. *Annual Review of Entomology*, 51: 467–494.
Morrone, J. J. 2009. *Evolutionary Biogeography: An Integrative Approach with case Studies*. Columbia University Press, New York.

Morrone, J. J. 2010. América do Sul e geografia da vida: Comparação de algumas propostas de regionalização. In: Carvalho, C. J. B. de and E. A. B. Almeida (Eds.), *Biogeografia da América do Sul: Padroes e processos*. Editora Roca Limitada, São Paulo, pp. 14–40.
Morrone, J. J. 2011. Island evolutionary biogeography: Analysis of the weevils (Coleoptera: Curculionidae) of the Falkland Islands (Islas Malvinas). *Journal of Biogeography*, 38: 2078–2090.
Morrone, J. J. 2014a. Biogeographical regionalisation of the Neotropical region. *Zootaxa*, 3782: 1–110.
Morrone, J. J. 2014b. Cladistic biogeography of the Neotropical region: Identifying the main events in the diversification of the terrestrial biota. *Cladistics*, 30: 202–214.
Morrone, J. J. 2015a. Track analysis beyond panbiogeography. *Journal of Biogeography*, 42: 413–425.
Morrone, J. J. 2015b. Biogeographical regionalisation of the Andean region. *Zootaxa*, 3936: 207–236.
Morrone, J. J. 2015c. Biogeographic regionalisation of the world: A reappraisal. *Australian Systematic Botany*, 28: 8190.
Morrone, J. J. 2017. *Neotropical Biogeography: Regionalization and Evolution*. CRC Press, Boca Raton, FL.
Morrone, J. J. and C. Ezcurra. 2016. On the Prepuna biogeographic province: A nomenclatural clarification. *Zootaxa*, 4132: 287–289.
Morrone, J. J., L. Katinas and J. V. Crisci. 1997. A cladistic biogeographic analysis of Central Chile. *Journal of Comparative Biology*, 2: 25–41.
Morrone, J. J. and P. Posadas. 2005. Falklands: Facts and fiction. *Journal of Biogeography*, 32: 2183–2187.
Morrone, J. J., S. Roig-Juñent and J. V. Crisci. 1994. Cladistic biogeography of terrestrial Sub-Antarctic beetles (Insecta: Coleoptera) from South America. *National Geographic Research and Exploration*, 10: 104–115.
Morrone, J. J., S. Roig-Juñent and G. C. Flores. 2002. Delimitation of biogeographic districts in central Patagonia (southern South America), based on beetle distributional patterns (Coleoptera: Carabidae and Tenebrionidae). *Revista del Museo Argentino de Ciencias Naturales, nueva serie*, 4: 1–6.
Morrone, J. J. and E. Urtubey. 1997. Historical biogeography of the northern Andes: A cladistic analysis based on five genera of Rhytirrhinini (Coleoptera: Curculionidae) and *Barnadesia* (Asteraceae). *Biogeographica*, 73: 115–121.
Müller, P. 1973. *The Dispersal Centres of Terrestrial Vertebrates in the Neotropical Realm: A Study in the Evolution of the Neotropical Biota and Its Native Landscapes*. Junk, The Hague.
Müller, P. 1986. *Biogeography*. Harper and Rows, New York.
Murray, A. 1866. *The Geographical Distribution of Mammals*. Day and Son Limited, London.
Myers, N., R. A. Mittermeier, C. G. Mittermeier, G. A. B. da Fonseca and J. Kent. 2000. Biodiversity hotspots for conservation priorities. *Nature*, 403: 853–858.
Nelson, G. and P. Y. Ladiges. 1991a. Three-area statements: Standard assumptions for biogeographic analysis. *Systematic Zoology*, 40: 470–485.
Nelson, G. and P. Y. Ladiges. 1991b. *TAS (MSDos Computer Program)*. Published by the authors, New York.
Nelson, G. and P. Y. Ladiges. 1995. *TASS*. Published by the authors, New York.

Nelson, G. and P. Y. Ladiges. 1996. Paralogy in cladistic biogeography and analysis of paralogy-free subtrees. *American Museum Novitates*, 3167: 1–58.

Nelson, G. and N. I. Platnick. 1981. *Systematics and Biogeography: Cladistics and Vicariance*. Columbia University Press, New York.

Newbigin, M. 1950. *Plant and Animal Geography*. Methuen, London.

Nixon, K. C. 1999. WinClada ver. 1.0000. Ithaca, New York: Published by the author. Available at http://www.cladistics.com/about_winc.htm.

O'Brien, C. W. 1971. The biogeography of Chile through entomofaunal regions. *Entomological News*, 82: 197–207.

Ocampo, F. and J. J. Morrone. 1999. Generic synopsis of the Subantarctic and Central Chilean Molytinae (Coleoptera: Curculionidae). *Neotropica*, 45: 21–29.

Ojanguren-Affilastro, A. A. and C. I. Mattoni. 2017. *Mauryius* n. gen. (Scorpiones: Bothriuridae), a new Neotropical scorpion genus. *Arthropod Systematics and Phylogeny*, 75: 125–139.

Ojeda, R. A., C. E. Borghi and V. G. Roig. 2002. Mamíferos de Argentina. In: G. Ceballos and J. A. Simonetti (Eds.), *Diversidad y conservación de mamíferos neotropicales*, Conabio and UNAM, Mexico City, Mexico, pp. 23–63.

Olson, D. M., E. Dinerstein, E. D. Wikramanayake, N. D. Burgess, G. V. N. Powell, E. C. Underwood, J. A. D'Amico et al. 2001. Terrestrial ecoregions of the world: A new map of life on Earth. *BioScience*, 51: 933–938.

Omad, G. 2014. Two new species of *Didicrum* Enderlein (Diptera, Psychodidae, Psychodinae) from Argentinean Patagonia. *Zootaxa*, 3794: 565–574.

Orfila, R. N. 1941. Apuntaciones ornitológicas para la zoogeografía neotropical. I. Distrito Sabánico. *Revista Argentina de Zoogeografía*, 1: 85–92.

Osgood, W. H. 1943. The mammals of Chile. *Publications of the Field Museum of Natural History, Zoological Series*, 30: 1–268.

Padró, A. 2017. Caracterización de la zona Altoandina Patagónica mediante análisis de trazos. *Graduate Thesis, Facultad de Ciencias Naturales*, Universidad Nacional de la Patagonia San Juan Bosco, Esquel.

Page, R. D. M. 1989. Component *user's manual. Release 1.5*. Published by the author, Auckland.

Page, R. D. M. 1993. Component *user's manual. Release 2.0*. The Natural History Museum, London.

Page, R. D. M. 1994. Maps between trees and cladistic analysis of historical associations among genes, organisms, and areas. *Systematic Biology*, 43: 58–77.

Page, R. D. M. 1995. Parallel phylogenies: Reconstructing the history of host–parasite assemblages. *Cladistics*, 10: 155–173.

Paggi, S. J. de. 1990. Ecological and biogeographical remarks on the rotifer fauna of Argentina. *Revue d'Hydrobiologie Tropicale*, 23: 297–311.

Palma, R. E., P. A. Marquet and D. Boric-Bargetto. 2005. Inter- and intraspecific phylogeography of small mammals in the Atacama desert and adjacent areas of northern Chile. *Journal of Biogeography*, 32: 1931–1941.

Papadopoulou, A., A. G. Jones, P. M. Hammond and A. P. Vogler. 2009. DNA taxonomy and phylogeography of beetles of the Falkland Islands (Islas Malvinas). *Molecular Phylogenetics and Evolution*, 53: 935–947.

Parenti, L. R. and M. C. Ebach. 2009. *Comparative Biogeography: Discovering and Classifying Biogeographical Patterns of a Dynamic Earth*. University of California Press, Berkeley, CA.

Parodi, L. R. 1934. Las plantas indígenas no alimenticias cultivadas en la Argentina. *Revista Argentina de Agronomía*, 1: 165–212.
Parodi, L. R. 1945. Las regiones fitogeográficas argentinas y sus relaciones con la industria forestal. In: Verdoorn, F. (Ed.), *Plants and Plant Science in Latin America*. Waltham, MA, pp. 127–132.
Paulson, D. R. 1979. Odonata. In: Hurlbert, S. H. (Ed.), *Biota Acuática de Sudamérica austral*, San Diego State University, San Diego, CA, pp. 170–171.
Peña, L. E. 1966a. A preliminary attempt to divide Chile into entomofaunal regions, based on the Tenebrionidae (Coleoptera). *Postilla*, 97: 1–17.
Peña, L. E. 1966b. Ensayo preliminar para dividir Chile en regiones entomofaunísticas, basadas especialmente en la familia Tenebrionidae (Col.). *Revista Universitaria, Universidad Católica de Chile*, 28–29: 209–220.
Pérez, M. E. and P. Posadas. 2006. Cladistics and redescription of *Hybreoleptops* Kuschel (Coleoptera: Curculionidae: Entiminae) with the description of two new species from the Central Chilean subregion. *Journal of Natural History*, 40: 1775–1791.
Peterson, P. M., K. Romaschenko and G. Johnson. 2010. A classification of the Chloridoideae (Poaceae) based on multi-gene phylogenetic trees. *Molecular Phylogenetics and Evolution*, 55: 580–598.
Porzecanski, A. L. and J. Cracraft. 2005. Cladistic analysis of distributions and endemism (CADE): Using raw distributions of birds to unravel the biogeography of the South American aridlands. *Journal of Biogeography*, 32: 261–275.
Posada, D. and K. A. Crandall. 1998. MODELTEST: Testing the model of DNA substitution. *Bioinformatics*, 14: 817–818.
Posadas, P. 2008. A preliminar overview of species composition and geographical distribution of Malvinian weevils (Insecta: Coleoptera: Curculionidae). *Zootaxa*, 1704: 1–26.
Posadas, P. 2012. Species composition and geographic distribution of Fuegian Curculionidae (Coleoptera: Curculionoidea). *Zootaxa*, 3303: 1–36.
Posadas, P., J. M. Estévez and J. J. Morrone. 1997. Distributional patterns and endemism areas of vascular plants in the Andean subregion. *Fontqueria*, 48: 1–10.
Posadas, P., D. R. Miranda and J. V. Crisci. 2001. Using phylogenetic diversity measures to set priorities in conservation: An example from southern South America. *Consevation Biology*, 15: 1325–1344.
Posadas, P. and J. J. Morrone. 2001. Biogeografía cladística de la subregión Subantática: Un anáilsis basado en taxones de la familia Curculionidae (Insecta: Coleoptera). In: Llorente, J. and J. J. Morrone (Eds.), Introducción a la biogeografía en Latinoamérica: Conceptos, teorías, métodos y aplicaciones, Las Prensas de Ciencias, UNAM, Mexico City, pp. 267–271.
Posadas, P. and J. J. Morrone. 2003. Biogeografía histórica de la familia Curculionidae (Insecta: Coleoptera) en las subregiones Subantártica y Chilena Central. *Revista de la Sociedad Entomológica Argentina*, 62: 71–80.
Posadas, P. and J. J. Morrone. 2004. A new species of *Antarctobius* Fairmaire from Islas Malvinas (Coleoptera: Curculionidae: Cyclominae). *Insect systematics and Evolution*, 35: 353–359.
Procheş, Ş. and S. Ramdhani. 2012. The world's zoogeographical regions confirmed by cross-taxon analyses. *BioScience*, 62: 260–270.

Quijano-Abril, M. A., R. Callejas-Posada and D. R. Miranda-Esquivel. 2006. Areas of endemism and distribution patterns for Neotropical *Piper* species (Piperaceae). *Journal of Biogeography*, 33: 1266–1278.

Rambaut, A. 2007. *Se-al version2.0a11*. Available from http://tree.bio.ed.ac.uk/software/seal/.

Rangel, J. O. 2000a. La región parammuna y franja aledaña en Colombia. In: Rangel, J. O. (Ed.), *Colombia Diversidad Biótica III: La región de vida paramuna de Colombia*, Universidad Nacional de Colombia, Santafé de Bogotá, pp. 1–23.

Rangel, J. O. 2000b. La diversidad beta: Tipos de vegetación. In: Rangel, J. O. (Ed.), *Colombia Diversidad Biótica III: La región de vida paramuna de Colombia*, Universidad Nacional de Colombia, Santafé de Bogotá, pp. 658–718.

Rangel, J. O., P. D. Lowy, M. Aguilar and A. Garzón. 1997. Tipos de vegetación en Colombia: Una aproximación al conocimiento de la terminología fitosociológica, fitoecológica y de uso común. In: Rangel, J. O., P. D. Lowy and M. Aguilar (Eds.), *Colombia Diversidad Biótica II: Tipos de vegetación en Colombia*. Instituto de Ciencias Naturales, Universidad Nacional de Colombia, Santafé de Bogotá, pp. 89–381.

Rapoport, E. H. 1968. Algunos problemas biogeográficos del Nuevo Mundo con especial referencia a la región Neotropical. In: Delamare Debouteville, C. and E. H. Rapoport (Eds.), *Biologie de l'Amerique Australe. Vol. 4*. CNRS, Paris, pp. 55–110.

Reiche, C. 1905. La distribución geográfica de las compuestas de la flora de Chile. *Anales del Museo Nacional de Chile*, 17: 5–44.

Restrepo, C. and A. Duque. 1992. Tipos de vegetación del llano de Paletara. Cordillera central Colombia. *Caldasia*, 17: 21–34.

Ribeiro, G. C. and A. Eterovic. 2011. Neat and clear: 700 species of crane flies (Diptera: Tipulomorpha) link southern South America and Australasia. *Systematic Entomology*, 36: 754–767.

Ribichich, A. M. 2002. El modelo clásico de la fitogeografía de Argentina: Un análisis crítico. *Interciencia*, 27: 669–675.

Ringuelet, R. A. 1955a. Vinculaciones faunísticas de la zona boscosa del Nahuel Huapi y el dominio zoogeográfico Australcordillerano. *Notas del Museo de La Plata, Zoología*, 18: 21–121.

Ringuelet, R. A. 1955b. Ubicación zoogeográfica de las Islas Malvinas. *Revista del Museo de La Plata (nueva serie), Zoología*, 6: 419–464.

Ringuelet, R. A. 1961. Rasgos fundamentales de la zoogeografía de la Argentina. *Physis* (Buenos Aires), 22: 151–170.

Ringuelet, R. A. 1975. Zoogeografía y ecología de los peces de aguas continentales de la Argentina y consideraciones sobre las áreas ictiológicas de América del Sur. *Ecosur*, 2: 1–122.

Ringuelet, R. A. 1978. Dinamismo histórico de la fauna brasílica en la Argentina. *Ameghiniana*, 15: 255–262.

Riveros Salcedo, J. C. 2017. South America: Argentina, Bolivia, and Peru. In: *Terrestrial Ecoregions of Latin America and the Caribbean*, World Wildlife Fund. WWF NT1002. https://www.worldwildlife.org/ecoregions/nt1002. Accessed November 20, 2017.

Riveros Salcedo, J. C. and C. Locklin. 2017. Western South America: Peru and Bolivia. In: *Terrestrial Ecoregions of Latin America and the Caribbean*, World Wildlife Fund. WWF NT1003. https://www.worldwildlife.org/ecoregions/nt1003. Accessed November 20, 2017.

Rivas-Martínez, S. and G. Navarro. 1994. *Mapa Biogeográfico de Suramérica*. Published by the authors, Madrid.

Rivas-Martínez, S., G. Navarro, Á. Penas and M. Costa. 2011. Biogeographic map of South America. A preliminary survey. *International Journal of Geobotanical Research*, 1: 21–40.

Rivas-Martínez, S. and O. Tovar. 1983. Síntesis biogeográfica de los Andes. *Collectanea Botanica* (Barcelona), 14: 515–521.

Robinson, G. S. 1984. *Insects of the Falkland islands*. British Museum (Natural History), London, UK.

Roig, F. A. 1998. La vegetación de la Patagonia. In: Correa, M. N. (Ed.), *Flora Patagónica, Tomo VIII(1)*, INTA, Colección Científica, Buenos Aires, Argentina, pp. 48–166.

Roig, F. A. and E. Martínez Carretero. 1998. La vegetación puneña en la provincia de Mendoza, Argentina. *Phytocoenologia*, 28: 565–608.

Roig, F. A., S. Roig-Juñent and V. Corbalán. 2009. Biogeography of the Monte desert. *Journal of Arid Environments*, 73: 164–172.

Roig-Juñent, S. 1994a. Las especies chilenas de *Cnemalobus* Guérin-Méneville 1838 (Coleoptera: Carabidae: Cnemalobini). *Revista Chilena de Entomología*, 21: 5–30.

Roig-Juñent, S. 1994b. Historia biogeográfica de América del Sur austral. *Multequina* (Mendoza), 3: 167–203.

Roig-Juñent, S. 1995. Revisión sistemática de los Creobina de América del Sur (Coleoptera: Carabidae: Broscini). *Acta Entomológica Chilena*, 19: 51–74.

Roig-Juñent, S. and M. C. Domínguez. 2001. Diversidad de la familia Carabidae (Coleoptera) en Chile. *Revista Chilena de Historia Natural*, 74: 549–571.

Roig-Juñent, S., M. C. Domínguez, G. E. Flores and C. Mattoni. 2006. Biogeographic history of South American arid lands: A view from its arthropods using TASS analysis. *Journal of Arid Environments*, 66: 404–420.

Roig-Juñent, S. and G. E. Flores. 2001. Historia biogeográfica de las áreas áridas de América del Sur austral. In: Llorente Bousquets, J. and Morrone, J. J. (Eds.), *Introducción a la Biogeografía en Latinoamérica: Teorías, Conceptos, Métodos y Aplicaciones*. Las Prensas de Ciencias, UNAM, Mexico City, Mexico, pp. 257–266.

Roig-Juñent, S., G. Flores, S. Claver, G. Debandi and A. Marvaldi. 2001. Monte desert (Argentina): Insect biodiversity and natural areas. *Journal of Arid Environments*, 47: 77–94.

Roig-Juñent, S., G. E. Flores and C. Mattoni. 2003. Consideraciones biogeográficas de la Precordillera (Argentina), con base en artrópodos epígeos. In: Morrone, J. J. and J. Llorente (Eds.), *Una perspectiva Latinoamericana de la Biogeografía*, Las Prensas de Ciencias, UNAM, Mexico City, Mexico, pp. 275–288.

Roig-Juñent, S., M. F. Tognelli and J. J. Morrone. 2008. Aspectos biogeográficos de los insectos de la Argentina. In: Claps, L. E., G. Debandi y and S. Roig-Juñent (Eds.), *Biodiversidad de Artrópodos Argentinos, Vol. 2*, Sociedad Entomológica Argentina, San Miguel de Tucumán, Argentina, pp. 11–29.

Rojas-Parra, C. A. 2007. Una herramienta automatizada para realizar análisis panbiogeográficos. *Biogeografía*, 1: 31–33.

Romano, G. M. 2017. A high resolution shapefile of the Andean biogeographical region. *Data in Brief*, 13: 230–232.

Romano, G. M., E. V. Ruiz, B. E. Lechner, A. G. Greslebin and J. J. Morrone. 2017. Track analysis of agaricoid fungi of the Patagonian forests. *Australian Systematic Botany*, 29: 440–446.

Ronquist, F. 1996. DIVA, Version 1.1. Computer program and manual available by from Upssala University at ftp.uu.se or ftp.systbot.uu.se.

Ronquist, F. 1997. Dispersal-vicariance analysis: A new approach to the quantification of historical biogeography. *Systematic Biology*, 46: 195–203.

Ronquist, F. 2002. TreeFitter, version 1.2. Software available from http://morphbank.ebc.uu.se/TreeFitter.

Ronquist, F. and J. P. Huelsenbeck. 2003. MrBayes 3: Bayesian phylogenetic inference under mixed models. *Bioinformatics*, 19: 1572–1574.

Rosen, B. R. 1988. From fossils to earth history: Applied historical biogeography. In: Myers, A. A. and P. S. Giller (Eds.), *Analytical Biogeography: An Integrated Approach to the Study of Animal and Plant Distributions*, Chapman and Hall, New York, pp. 437–481.

Ruggiero, A. and C. Ezcurra. 2003. Regiones y transiciones biogeográficas: Complementariedad de los análisis en biogeografía histórica y ecológica. In: Morrone, J. J. and J. Llorente (Eds.), *Una Perspectiva Latinoamericana de la Biogeografía*, Las Prensas de Ciencias, Mexico City, Mexico, pp. 141–154.

Ruiz, E., D. J. Crawford, T. F. Stuessy, F. González, R. Samuel, J. Becerra and M. Silva. 2004. Phylogenetic relationships and genetic divergence among species of *Berberis, Gunnera, Myrceugenia* and *Sophora* of the Juan Fernández Islands (Chile) and their continental progenitors based on isozymes and nrITS sequences. *Taxon*, 53: 321–332.

Ruiz, E. V., G. M. Romano and J. J. Morrone. 2016. Track analysis of Oribatid mites (Acari: Oribatida) of the Subantarctic subregion of South America. *Zootaxa*, 4127: 383–392.

Sánchez Osés, C. and R. Pérez-Hernández. 1998. Revisión histórica de las subdivisiones biogeográficas de la región Neotropical, con especial énfasis en Suramérica. *Montalbán*, 31: 169–210.

Sánchez Osés, C. and R. Pérez-Hernández. 2005. Historia y tabla de equivalencias de las propuestas de subdivisiones biogeográficas de la región Neotropical. In: Llorente Bousquets, J. and J. J. Morrone (Eds.), *Regionalización Biogeográfica en Iberoamérica y tópicos Afines—Primeras Jornadas Biogeográficas de la Red Iberoamericana de Biogeografía y Entomología Sistemática (RIBES XII.I-CYTED)*. Las Prensas de Ciencias, UNAM, Mexico City, Mexico, pp. 495–508.

Sanders, R. W., T. F. Stuessy, C. Marticorena and M. Silva. 1987. Phytogeography and evolution of *Dendroseris* and *Robinsonia*, tree-Compositae of the Juan Fernandez islands. *Opera Botanica*, 92: 195–215.

Sanmartín, I. and F. Ronquist. 2004. Southern Hemisphere biogeography inferred by event-based models: Plant versus animal patterns. *Systematic Biology*, 53: 216–243.

Schaefer, S. 2011. The Andes: Riding the tectonic uplift. In: Albert, J. S. and Reis, R. E. (Eds.), *Historical Biogeography of Neotropical Freshwater Fishes*, University of California Press, Los Angeles, CA, pp. 259–278.

Sclater, P. L. 1858. On the general geographic distribution of the members of the class Aves. *Proceedings of the Linnean Society of London, Zoology*, 2: 130–145.

Sclater, W. L. 1894. The geography of mammals. I. Introductory. *The Geographical Journal*, 3: 95–105.

Sclater, W. L. and P. L. Sclater. 1899. *The Geography of Mammals*. Kegan Paul, Trench, Trübner and Co., London, UK.

Segovia, R. A. and J. J. Armesto. 2015. The Gondwanan legacy in South American biogeography. *Journal of Biogeography*, 42: 209–217.
Sérsic, A. N., A. Cosacov, A. A. Cocucci, L. A. Johnson, R. Pozner, L. J. Ávila, J. W. Sites Jr. and M. Morando. 2011. Emerging phylogeographical patterns of plants and terrestrial vertebrates from Patagonia. *Biological Journal of the Linnean Society*, 103: 475–494.
Shannon, R. C. 1927. Contribución a los estudios de las zonas biológicas de la República Argentina. *Revista de la Sociedad Entomológica Argentina*, 4: 1–14.
Sick, W. D. 1969. Geographical substance. *Monographiae Biologicae*, 19: 449–474.
Simpson, B. B. 1983. An historical phytogeography of the High Andean flora. *Revista Chilena de Historia Natural*, 36: 109–122.
Sklenář, P., E. Dušková and H. Balslev. 2011. Tropical and temperate: Evolutionary history of páramo flora. *Botanical Review*, 77: 71–108.
Skottsberg, C. 1905. Some remarks upon the geographical distribution of vegetation in the colder Southern Hemisphere. *Ymer*, 4: 402–427.
Skottsberg, C. 1920. Derivation of the flora and fauna of Juan Fernandez and Easter Island. In: Skottsberg, C. (Ed.), *The natural history of Juan Fernandez and Easter Island. Vol. 1*, Almqvist and Wiksells, Uppsala, Sweden, pp. 193–438.
Smith, C. 2017. Southern South America: Chile and Argentina. In: *Terrestrial Ecoregions of Latin America and the Caribbean*, World Wildlife Fund. WWF NT0404. https://www.worldwildlife.org/ecoregions/nt0404. Accessed November 20, 2017.
Smith, C. H. 1983. A system of world mammal faunal regions. I. Logical and statistical derivation of the regions. *Journal of Biogeography*, 10: 455–466.
Soares, E. D. G. and C. J. B. de Carvalho. 2005. Biogeography of *Palpibracus* (Diptera: Muscidae): An integrative study using panbiogeography, parsimony analysis of endemicity, and component analysis. In: Llorente, J. and J. J. Morrone (Eds.), *Regionalización Biogeográfica en Iberoamérica y Tópicos Afines*, Las Prensas de Ciencias, UNAM, Mexico City, Mexico, pp. 485–494.
Solbrig, O. T., W. F. Blair, F. A. Enders, A. C. Hulse, J. H. Hunt, M. A. Mares, J. Neff, D. Otte, B. B. Simpson and C. S. Tomoff. 1977. The biota: The dependent variable. In: Orians, G. H. and O. T. Solbrig (Eds.), *Convegent Evolution in Warm Deserts*, US/IBP Synthesis Series 3, Dowden, Hutchinson and Ross, Stroudsburg, Pennsylvania, PA, pp. 50–66.
Solervicens, J. and M. Elgueta. 1994. Insectos de follaje de bosques pantanosos del Norte Chico, centro y sur de Chile. *Revista Chilena de Entomología*, 21: 135–164.
Soriano, A. 1949. El límite entre las provincias botánicas Patagónica y Central en el territorio del Chubut. *Lilloa*, 20: 193–202.
Soriano, A. 1950. La vegetación del Chubut. *Revista Argentina de Agricultura*, 17: 30–66.
Soriano, A. 1956. Los distritos florísticos de la provincia Patagónica. *Revista de Investigación Agrícola*, 10: 323–347.
Soriano, A. 1983. Deserts and semi-deserts of Patagonia. In: West, N. E. (Ed.), *Temperate Deserts and Semi-deserts*, Elsevier Publishing Company, Amsterdam, the Netherlands, pp. 423–460.
Soto, E. M., F. M. Labarque, F. S. Ceccarelli, M. A. Arnedo, J. Pizarro-Araya and M. J. Ramírez. 2017. The life and adventures of an eight-legged castaway: Colonization and diversification of *Philisca* ghost spiders on Robinson Crusoe Island (Araneae, Anyphaenidae). *Molecular Phylogenetics and Evolution*, 107: 132–141.

Spinelli, G. R., P. I. Marino and P. Posadas. 2006. The Patagonian species of the genus *Atrichopogon* Kieffer, with a biogeographic analysis based on Forcipomyiinae (Diptera: Ceratopogonidae). *Insect Systematics and Evolution*, 37: 301–324.

Stange, L. A., A. L. Terán and A. Willink. 1976. Entomofauna de la provincia biogeográfica del Monte. *Acta Zoológica Lilloana*, 32: 73–119.

Steenis, C. G. and G. Van. 1972. *Nothofagus*, key genus to plant geography. *Blumea*, 19: 65–98.

Stonis, J. R., A. Remeikis, A. Diškus and V. Gerulaitis. 2016. The Ando-Patagonian *Stigmella magnispinella* group (Lepidoptera, Lepticulidae) with description of new species from Ecuador, Peru and Argentina. *Zootaxa*, 4200: 561–579.

Stuessy, T. F., D. J. Crawford and C. Marticorena. 1990. Patterns of phylogeny in the endemic vascular flore of the Juan Fernandez islands, Chile. *Systematic Botany*, 15: 338–346.

Stuessy, T. F., R. W. Sanders and M. Silva. 1984. Phytogeography and evolution of the flora of the Juan Fernandez islands: A progress report. In: Radowsky, F. J., P. H. Raven and S. H. Sohmer (Eds.), *Biogeography of the Tropical Pacific: Proceedings of a Symposium*, Association of Systematics Collections and Bernice P. Bishop Museum, Lawrence, Kansas, pp. 55–69.

Sturm, H. 1990. Contribución al conocimiento de las relaciones entre los frailejones (Espeletiinae, Asteraceae) y los animales en la región del páramo andino. *Revista de la Academia Colombiana de Ciencias*, 17: 667–685.

Swofford, D. L. 2003. *PAUP*: Phylogenetic Analysis Using Parsimony (*and other methods). Version 4.* Sinauer Associates, Sunderland, MA. http://paup.csit.fsu.edu/.

Szumik, C. A., F. Cuezzo, P. A. Goloboff and A. E. Chalup. 2002. An optimality criterion to determine areas of endemism. *Systematic Biology*, 51: 806–816.

Szumik, C. A. and P. Goloboff. 2004. Areas of endemism: An improved optimality criterion. *Systematic Biology*, 53: 968–977.

Takayama, K., P. López-Sepúlveda, J. Greimler, D. J. Crawford, P. Peñailillo, M. Baeza, E. Ruiz et al. 2014. Relationships and genetic consequences of contrasting modes of speciation among endemic species of *Robinsonia* (Asteraceae, Senecioneae) of the Juan Fernández Archipelago, Chile, based on AFLPs and SSRs. *New Phytologist*, 205: 415–428.

Takhtajan, A. 1986. *Floristic Regions of the World.* University of California Press, Berkeley, CA.

Teillier, S. and Y. Vilina. 2017. Islands off the coast of central Chile in the Pacific Ocean. In: *Terrestrial Ecoregions of Latin America and the Caribbean*, World Wildlife Fund. WWF NT0403. https://www.worldwildlife.org/ecoregions/nt0403. Accessed November 20, 2017.

Templeton, A. R., K. A. Crandall and C. F. Sing. 1992. A cladistic analysis of phenotypic associations with haplotypes inferred from restriction endonuclease mapping and DNA sequence data. III. Cladogram estimation. *Genetics*, 132: 619–633.

Thompson, J. D., T. J. Gibson, F. Plewniak, F. Jeanmougin and D. G. Higgins. 1997. The Clustal_X windows interface: Flexible strategies for multiple sequence alignment aided by quality analysis tools. *Nucleic Acids Research*, 25: 4876–4882.

Treviranus, G. R. 1803. *Biologie, Oder Philosophie der ebenden Natur. 2 volumes*. Röwer, Göttingen, Germany.

Troncoso, A. and E. J. Romero. 1998. Evolución de las comunidades florísticas en el extremo sur de Sudamérica durante el Cenofítico. *Monographs in Systematic Botany from the Missouri Botanical Garden*, 68: 149–172.

Udvardy, M. D. F. 1975. A classification of the biogeographical provinces of the world. *International Union for Conservation of Nature and Natural Resources Occasional Paper 18*, Morges, Switzerland.

Udvardy, M. D. F. 1987. The biogeographical realm Antarctica: A proposal. *Journal of the Royal Society of New Zealand*, 17: 187–200.

Urbina-Casanova, R., P. Saldivia and R. A. Scherson. 2015. Consideraciones sobre la sistemática de las familias y los géneros de plantas vasculares endémicos de Chile. *Gayana Botánica*, 72: 272–295.

Urra, F. 2017a. *Coritta attemborughi* sp. nov., nueva especie de Oecophoridae (Lepidoptera: Gelechioidea) de Chile central. *Biodiversity and Natural History*, 3: 29–33.

Urra, F. 2017b. Una nueva especie de *Muna* Clarke (Lepidoptera: Depressariidae) de Chile Central. *Revista Chilena de Entomología*, 42: 29–33.

Urtubey, E., T. F. Stuessy, K. Tremetsberger and J. J. Morrone. 2010. The South American biogeographic transition zone: An analysis from Asteraceae. *Taxon*, 59: 505–509.

van der Hammen, T. 1974. The Pleistocene changes of vegetation and climate intropical South America. *Journal of Biogeography*, 1: 3–26.

van der Hammen, T. 1997. Páramos. In: Chaves, M. E. and N. Arango (Eds.), *Informe Nacional sobre el Estado de la Biodiversidad Volumen I*, Instituto de Investigación de Recursos Biológicos Alexander von Humboldt, Ministerio del Medio Ambiente, Santafé de, Bogotá, pp. 10–37.

van der Hammen, T. and A. M. Cleef. 1986. Development of the High Andean páramo flora and vegetation. In: Vuilleumier, F. and M. Monasterio (Eds.), *High Altitude Tropical Biogeography*, Oxford University Press and American Museum of Natural History, New York, pp. 153–201.

Vidal, M. A., E. R. Soto and A. Veloso. 2009. Biogeography of Chilean herpetofauna: Distributional patterns of species richness and endemism. *Amphibia-Reptilia*, 30: 151–171.

Villagrán, C. and L. F. Hinojosa. 1997. Historia de los bosques del sur de Sudamérica. II. Análisis fitogeográfico. *Revista Chilena de Historia Natural*, 70: 241–267.

Vivallo, F. 2013. Revision of the bee subgenus *Centris* (*Wagenknechtia*) Moure, 1950 (Hymenoptera: Apidae: Centridini). *Zootaxa*, 3683: 501–537.

Vuilleumier, F. 1985. Forest birds of Patagonia: Ecological geography, speciation, endemism, and faunal history. *Ornithological Monographs*, 36: 255–304.

Vuilleumier, F. 1986. Origins of the tropical avifaunas of the high Andes. In: Vuilleumier, F. Y. M. Monasterio (Eds.), *High Altitude Tropical Biogeography*, Oxford University Press and American Museum of Natural History, New York, pp. 586–622.

Wallace, A. R. 1876. *The Geographical Distribution of Animals. Vol. I & II*. Harper and Brothers, New York.

Wen, J. and S. M. Ickert-Bond. 2009. Evolution of the Madrean-Tethyan disjunctions and the North and South American amphitropical disjunctions in plants. *Journal of Systematics and Evolution*, 47: 331–348.

Wiley, E. O. 1988. Vicariance biogeography. *Annual Review of Ecology and Systematics*, 19: 513–542.

Willink, A. 1988. Distribution patterns of Neotropical insects with special reference to the Aculeate Hymenoptera of southern South America. In: Heyer, W. R. and P. E. Vanzolini (Eds.), *Proceedings of a Workshop on Neotropical Distribution Patterns*, Academia Brasileira de Ciencias, Rio de Janeiro, Brazil, pp. 205–221.

Willink, A. 1991. Contribución a la zoogeografía de los insectos argentinos. *Boletín de la Academia de Ciencias, Córdoba*, 59: 125–147.

Zuckerland, E. and L. Pauling. 1965. Molecular disease, evolution and genetic heterogeneity. In: Kasha, M. and B. Pullman (Eds.), *Horizons in Biochemistry*, Academic Press, London, pp. 189–225.

Zuloaga, F. O., O. Morrone and D. Rodríguez. 1999. Análisis de la biodiversidad en plantas vasculares de la Argentina. *Kurtziana*, 27: 17–167.

Index

Note: Page numbers followed by f refer to figures.

A

Aegla laevis talcahuano (Aeglidae), 132f
Alto Cauca Highland district, 174
Alto Patía district, 174
Amorim, D. S., 31–32, 31f
Andalucía district, 175
Andean-Patagonian, 43
Andean region, 43–45
  biogeographical analysis, 52f
  biotic relationship, 47–52
  case study, 58–61
  dispersal-vicariance analysis, 59f
  endemic and characteristic taxa, 45–46
  geological evolution, 53–58
  Listroderina cladogram, 60f
  niche conservatism, 46–47
  regionalization, 14–27, 52–53
  three perspectives of, 55f
Angol district, 76–77
Anomophthalmus insolitus (Curculionidae), 153f
Apiaceae (*Azorella filamentosa*), 99f
Araucaria araucana (Araucariaceae), 78f
Arequipa district, 184
Argentinean, 43
Artigas, J. N., 23
Asteraceae
  Chaetanthera serrata, 73f
  Chuquiraga avellanedae, 148
  Robinsonia, 113f, 116–118, 117f, 118f
  Triptilion achilleae, 73f
Atacama province, 195–196
  biotic relationships, 199
  cenocrons, 202
  Desventuradas Archipelago district, 199
  endemic and characteristic taxa, 196–197
  Interior Desert district, 199–200
  Northern Andean district, 200
  Northern Coast district, 201
  Northern Precordilleran district, 201
  Tamarugal district, 202
  vegetation, 197, 198f
Austral High Andean district, 153–154
Austral kingdom, 29–30
  Amorim and Tozzoni approach, 31–32, 31f
  area cladograms, 39f
  biotic relationship, 30–34
  case study, 37–41
  endemic and characteristic taxa, 30
  geological evolution, 35–37, 36f
  Giribet and Edgecombe approach, 33–34
  Moreira-Muñoz approach, 32–33
  Morrone approach, 32, 32f
  regionalization, 35
  Ribeiro and Eterovic approach, 34
Austral tracks, 47, 48f, 50f, 145
Awa district, 175
Aysén Cordillera district, 84
Azorella filamentosa (Apiaceae), 99f

B

Biogeographical regionalization, 8–9, 11–14
Biotic hybridization, 7
Bolivian district, 193

C

Cabrera, A. L., 15, 20, 143
Cajamarca district, 193–194
Callao district, 184
Cañón Chicamocha district, 175
Cañón del Cauca district, 175
Cardonales district, 185
Cascellius (Carabidae), 66f
Case studies
  areas of endemism in Patagonian steppes based on insect taxa, 157–160
  cladistic biogeographical analysis of Central Chile, 122–123
  cladistic biogeographical analysis of Subantarctic subregion, 68, 70f, 71, 71f, 72f
  diversification of plant genus *Nolana*, 186–189
  event-based biogeographical analysis of Andean region, 58–61
  event-based biogeographical analysis of Austral areas, 37–41, 38f
  evolutionary biogeography of páramo flora, 180–182
  integrative biogeographic analysis of fly genus *Palpibracus*, 92–96, 93f, 94f, 95f
  integrative biogeographic analysis weevils of Falkland Islands, 106–111
  modes of speciation of plant genus *Robinsonia* in Juan Fernández Islands, 116–118
  post-glacial recolonization of marsupial *Dromiciops gliroides* in Valdivian Forest, 86–88
  quaternary biogeography of grass *Munroa argentina*, 166–168
Catatumbo Mountains Forest district, 175
Cauca and Valle Western Cordillera Andean Forest district, 175
Cauca Pacific Slope Subandean Forest district, 176
Central Andean Cordillera district, 128
Central Chilean subregion, 119–120
  area cladograms, 124f
  biotic relationship, 121
  case study, 122–123
  Coquimban province, 123–130
  endemic and characteristic taxa, 120–121
  regionalization, 122, 123f
  Santiagan province, 130–136
Central Patagonian subprovince, 147
Chaetanthera serrata (Asteraceae), 73f
Chilean/Austral, 43
Chillán Cordillera district, 77
Chubut district, 148
Chuquiraga avellanedae (Asteraceae), 148
Cladistic biogeographical analysis, 5–7, 31, 31f
Coastal Desert district, 185
Coleoptera (*Nannomacer germaini*), 73, 74f
Colombian páramos distribution, 181f
Comechingones province, 213
  biotic relationships, 215
  endemic and characteristic taxa, 213–214
  vegetation, 214–215, 214f
Conepatus rex inca (Mephitidae), 183f
Coquimban province, 123–125
  biotic relationship, 127–128
  cenocrons, 130
  Central Andean Cordillera district, 128
  Desert district, 129
  endemic and characteristic taxa, 125–126
  Intermediate Desert district, 130
  vegetation, 126–127, 126f, 127f
Cox, C. B. C., 14
Crinodendron hookerianum (Elaeocarpaceae), 81f
Curculionidae
  Anomophthalmus insolitus, 153f
  Falklandius antarcticus, 68, 71f
  Germainiellus, 66f
  Lanteriella microphthalma, 103f
  Listroderes robustior, 197f
  Listroderes robustus, 125f
  Megalometis spinifer, 108f
  Trachodema tuberculosa, 51f
Cuyan district, 194
Cuyan High Andean province, 202–203
  biotic relationships, 204
  cenocrons, 204–205
  endemic and characteristic taxa, 203
  representative plant species, 205f
  vegetation, 204
Cyanoliseus patagonus byroni (Psittacidae), 132f

D

Desert province, 182–183
  biotic relationships, 184
  case study, 186–189
  endemic and characteristic taxa, 183

regionalization, 184–185
vegetation, 184
Desventuradas Archipelago district, 199
Dicksonia berteriana (Dicksoniaceae), 113f
Dispersal-vicariance (DIVA) analysis, 186–187, 187f
Domínguez, M. C., 159
Dromiciops gliroides (Microbiotheriidae), 87f, 88f

E

Eastern Andean district, 176
Eastern Cordillera Páramos district, 176
Edgecombe, G. D., 33–34
Elaeocarpaceae (*Crinodendron hookerianum*), 81f
Engler, A., 13
Epilobium conjugens (Onagraceae), 91f
Epilobium denticulatum (Onagraceae), 46f
Epilobium pedicelare (Onagraceae), 192, 192f
Eterovic, A., 34
Evolutionary biogeography, 1–2
biogeographical regionalization, 8–9
biotas identification, 2–5
cenocrons, 7–8
flowchart, 3f
geobiotic scenario, 8
nomenclatural conventions, 9
parsimony analysis steps, 4f
regionalization, 7
testing biotas, 5–7

F

Falkland Islands province, 100–101
biotic evolution of, 110f
biotic relationship, 104, 105f
case study, 106–111
cenocrons, 106
Curculionidae, 108f
district, 104–105
endemic and characteristic taxa, 102–103
South Georgia Island district, 105
vegetation, 103–104
Falklandius antarcticus (Curculionidae), 68, 71f
Farallones de Cali district, 176
Frontino district, 176–177
Fuegian
district, 144
subprovince, 150

G

Germainiellus (Curculionidae), 66f
Gigantodax brophyi (Simuliidae), 98f
Gigantodax wrighti (Simuliidae), 46f
Giribet, G., 33–34
Gondwanic cenocron, 165–166

H

Henicopidae (*Paralamyctes*), 33, 33f
Holarctic cenocron, 166
Holarctic plant genera, 180
Huancaspata district, 194

I

International Code of Area Nomenclature (ICAN), 9
Intraspecific phylogeography analysis, 7–8

J

Juan Fernández province, 111–112
biotic relationship, 114
case study, 116–118
cenocrons, 111–116
endemic and characteristic taxa, 112–114
panoramic view, 112, 112f
plant lineages representation, 115f
vegetation, 114
Jujuyan district, 194

K

Kuschel, G., 13, 15, 19

L

Lamas, G., 184, 192
Lanteriella microphthalma (Curculionidae), 103f
Late Cretaceous–Early Paleocene, 53, 56f
Listroderes robustior (Curculionidae), 197f
Listroderes robustus (Curculionidae), 125f
Llanquihue district, 84

M

Magellanic Forest province, 88–90
biotic relationship, 92
case study, 92–96
endemic and characteristic taxa, 90

Magellanic Forest province (*Continued*)
  Nothofagus forest in, 91f
  vegetation, 90–92
Magellanic Moorland province, 96–97
  biotic relationship, 100
  endemic and characteristic taxa, 98
  vegetation, 98–100, 99f
Maps
  in Andean region, 46f, 54f
  Argentina, 17f, 21f
  Argentinean Patagonia, 26f
  in Atacama province, 197f
  Austral kingdom, 35f
  in Central Chilean subregion, 93f, 121f, 122f
  Chile, 18f, 24f
  in Coquimban province, 125f
  in Cuyan High Andean province, 203, 204f
  in Desert province, 183f
  Falkland Islands, 102f
  Juan Fernández Islands, 117f
  in Magellanic Forest province, 91f
  in Magellanic Moorland province, 98f
  in Maule province, 73f
  in Monte province, 208f, 210f
  Neotropical/Wallace regions, 12f
  in Páramo province, 171f
  Patagonian province/subregion, 139f, 141f
  in Puna province, 192f
  in Santiagan province, 132f
  South America, 20f, 22f, 25f, 165f
  Southern Chile, 16f
  in Subantarctic subregion, 66f, 69f, 70f, 93f
  in Valdivian Forest province, 81f
  weevil species, 102f
Maule province, 72–73
  biotic relationship, 76
  endemic and characteristic taxa, 73–74
  regionalization, 76–79
  vegetation, 74–76, 75f
Megalometis spinifer (Curculionidae), 108f
Mephitidae (*Conepatus rex inca*), 183f
Meridional Subandean Patagonia district, 154
Microbiotheriidae (*Dromiciops gliroides*), 87f, 88f
Middle Miocene–Late Miocene, 55–56, 57f
Misodendrum angulatum (Misodendraceae), 81f
Mollendo district, 185
Monte province, 205–207
  biotic relationships, 209–210
  cenocrons, 213
  endemic and characteristic taxa, 207–208
  Eremean district, 210–211
  Northern district, 211
  plant species characteristic, 209f
  Prepuna district, 211–212
  Southern district, 212
  vegetation, 208–209
Moreira-Muñoz, A., 14, 32–33
Morrone, J. J., 32, 32f, 141
Müller, P., 23
Munroa argentina (Poaceae), 166–168, 167f
Muscidae (*Palpibracus*), 92–96

N

Nannomacer germaini (Coleoptera), 74, 74f
Navarro, G., 25, 27
Nolana (Solanaceae), 186–189
Nomenclatural conventions, 9
Nothocascellius hyadesii (Carabidae), 98f
Nothofagus pumilio (Nothofagaceae), 83f, 91f

O

Onagraceae
  Epilobium conjugens, 91f
  Epilobium denticulatum, 46f
  Epilobium pedicelare, 192, 192f

P

Palpibracus (Muscidae), 92–96
Pangeic cenocron, 165
Paralamyctes (Henicopidae), 33, 33f
Paramillo del Sinú district, 177
Páramo province, 169–170, 171f
  biotic relationships, 173–174
  case study, 180–182
  endemic and characteristic taxa, 170
  regionalization, 174–179
  vegetation, 170–173, 172f
Páramos Huila-Tolima district, 177
Parsimony analysis, 2–5, 4f, 6f
Pasco district, 194–195
Patagonian province, 139–141
  areas of endemism in, 158f
  biotic relationships, 143
  case study, 157–160
  cenocrons, 155–157

Central district, 143–144
Coleoptera of, 144f
endemic and characteristic taxa, 141–142
phylogeographic analyses in, 156f
plains and plateaus characteristic, 140f
plant taxa, 146f
regionalization, 143–155
shrub steppes of, 142f
vegetation, 142–143
Western subprovince, 154–155
Patagonian steppes, 151
Patagonian subregion, 137–138
biotic relationships, 138
endemic and characteristic taxa, 138
province. *See* Patagonian province
regionalization, 138–139
Payunia
district, 143
Northern district, 151
Southern district, 152
subprovince, 150–151
Pehuén district, 77–79
Perijá district, 177–178
Platnickia elegans (Zodariidae), 121f
Poaceae (*Munroa argentina*), 166–168, 167f
Porculla district, 185
Prepuna district, 211–212
Psittacidae (*Cyanoliseus patagonus byroni*), 132f
Puna province, 189–191
biotic relationships, 192
Central district, 194
endemic and characteristic taxa, 191–192
regionalization, 192–195
vegetation, 192, 193f

Q

Quindío Páramo district, 178

R

Ribeiro, G. C., 34
Ringuelet, R. A., 15–17, 23
Rivas-Martínez, S., 23, 25, 27
Robinsonia (Asteraceae), 113f, 116–118, 117f, 118f
Roig, F. A., 27

S

San Agustín district, 178
San Jorge Gulf district, 148–149
San Juan Cloud Forest district, 178
San Lucas Mountains district, 178
Santa Cruz district, 149
Santiagan province, 130–131
biotic relationship, 132
cenocrons, 136
Central Coastal Cordillera district, 133–134
Central Valley district, 134–135
endemic and characteristic taxa, 131–132
Southern Andean Cordillera district, 135–136
vegetation, 132, 133f
Sclater-Wallace system, 11
Septentrional Subandean Patagonia district, 154
Septentrional track, 145
Shimbe district, 195
Sierra Nevada district, 178–179
Simuliidae
Gigantodax brophyi, 98f
Gigantodax wrighti, 46f
Solanaceae (*Nolana*), 186–189
South American transition zone, 14, 161–163
Atacama province, 195–202
biotic relationships, 164
case study, 166–168
cenocrons, 165–166
Comechingones province, 213–215
Cuyan High Andean province, 202–205
Desert province, 182–189
endemic and characteristic taxa, 163–164
Monte province, 205–213
Páramo province, 169–182
Puna province, 189–195
regionalization, 165
Subandean subprovince, 152–153
Subantarctic subregion, 63–65
biotic relationship, 66–68, 67f, 69f
case study, 68, 70f, 71, 71f, 72f
endemic and characteristic taxa, 65–66
Falkland Islands province, 100–111
Juan Fernández province, 111–118
Magellanic Forest province, 88–96
Magellanic Moorland province, 96–100
Maule province, 72–79
regionalization, 68, 70f
Valdivian Forest province, 79–88
Surco district, 185
Synanthropic cenocron, 166

T

Tachira district, 179
Takhtajan, A., 23, 25
Tamarugal district, 202
Temperate Gondwanaland, 14
Temuco district, 79
Tolima district, 179
Tovar, O., 23
Tozzoni, H. S., 31–32, 31f
Trachodema tuberculosa (Curculionidae), 51f
Triptilion achilleae (Asteraceae), 73, 73f
Triptilion globosum (Asteraceae), 125f
Triptilion spinosum (Asteraceae), 121f

U

Udvardy, M. D. F., 23
Urtubey, E., 164, 164f

V

Valdivian Cordillera district, 86
Valdivian district, 85
Valdivian Forest province, 79–80
  Aysén Cordillera district, 84
  biotic relationship, 83
  case study, 86–88
  cenocrons, 86
  endemic and characteristic taxa, 81
  Llanquihue district, 84
  South Chiloé district, 85
  vegetation, 82–83, 82f, 83f
Verbena tridens (Verbenaceae), 149

W

Western Cordillera Northern Andean Forests district, 179

X

Xeric vegetation, 126, 126f

Y

Yepes, J., 15

Z

Zodariidae (*Platnickia elegans*), 121f